Building Services Engineering Spreadsheets

Building Services Engineering Spreadsheets

David V. Chadderton

MSc, CP Eng, C Eng, MIE Aust, MCIBSE, MAIRAH

E & FN SPON
An Imprint of Thomson Professional

London · Weinheim · New York · Tokyo · Melbourne · Madras

Published by E & FN Spon, an imprint of Thomson Professional
2–6 Boundary Row, London SE1 8HN, UK

Thomson Science and Professional, 2–6 Boundary Row, London SE1 8HN, UK

Thomson Science and Professional, Pappelallee 3, 69469 Weinheim, Germany

Thomson Science and Professional, 115 Fifth Avenue, New York, NY 10003, USA

Thomson Science and Professional, ITP-Japan, Kyowa Building, 3F, 2-2-1 Hirakawacho, Chiyoda-ku, Tokyo 102, Japan

Thomson Science and Professional, 102 Dodds Street, South Melbourne, Victoria 3205, Australia

Thomson Science and Professional, R. Seshadri, 32 Second Main Road, CIT East, Madras 600 035, India

First edition

© 1997 David V. Chadderton

Typeset in India by Pure Tech India Ltd, Pondicherry
Printed in Great Britain by The Alden Press, Oxford

ISBN 0 419 22620 6

A catalogue record for this book is available from the British Library

Library of Congress Catalog Card Number: 97–66255

∞ Printed on permanent acid-free text paper, manufactured in accordance with ANSI/NISO Z39.48-1992 and ANSI/NISO Z39.48-1984 (Permanence of Paper).

Contents

Preface

Spreadsheets are very powerful software programs. Almost every personal computer either has a spreadsheet loaded on its hard disk, or could take one. This book and disk bridge the gap between having access to the means of handling large amounts of data, and benefiting from the facility. This is an easily understandable entry into spreadsheet use for engineers and other building industry professionals.

Acknowledgements

My thanks are extended towards those who made helpful comments on parts of this work. They are: Malcolm Hudson of Buckle & Partners, Consulting Engineers, Southampton; David Ferris, Southampton Technical College; and Eric Curd, John Moores University, Liverpool. I should like to thank those unnamed engineers and past students who were part of the stimulation for the creation of this book. Figures 6.4, 6.5, 6.6, 6.7 and 6.8 are reproduced with permission of Woods of Colchester Limited. E & FN Spon are particularly thanked for publishing my efforts. My wife has been very understanding and uncomplaining during the long hours of wrestling with the keyboard. Thank you, Maureen.

The use of SI units

Système International units are used throughout, and Table A shows the basic and derived units with their symbols and common equalities.

Table A   SI units

Quantity	Unit	Symbol	Equality
mass	kilogram	kg	
	tonne	t	$1.0\,t = 1000\,kg$
length	metre	m	
area	square metre	m^2	
volume	litre	l	
	cubic metre	m^3	$1.0\,m^3 = 10^3\,l$
time	second	s	
	hour	h	$1.0\,h = 3600\,s$
energy	joule	J	$1.0\,J = 1.0\,N\,m$
force	newton	N	$1.0\,kg = 9.807\,N$
power	watt	W	$1.0\,W = 1.0\,J/s$
			$1.0\,W = 1.0\,N\,m/s$
			$1.0\,W = 1.0\,V\,A$
pressure	pascal	Pa	$1.0\,Pa = 1.0\,N/m^2$
	newton per square metre	N/m^2	$1.0\,bar = 100\,000\,N/m^2$
	bar	bar	$1.0\,bar = 1000\,mbar$
			$1.0\,bar = 100\,kPa$
frequency	hertz	Hz	$1.0\,Hz = 1.0\,cycle/s$
temperature	degree Celsius	°C	
	Kelvin	K	$K = °C + 273$
luminous flux	lumen	lm	
illuminance	lux	lx	$1.0\,lx = 1.0\,lm/m^2$
Electrical units:			
resistance, R	ohm	Ω	
potential, V	volt	V	
current, I	ampere	A	$I = V/R$

Table B   Multiples and submultiples

Quantity	Name	Symbol
10^{12}	tera	T
10^{9}	giga	G
10^{6}	mega	M
10^{3}	kilo	k
10^{-1}	deci	d
10^{-3}	milli	m
10^{-6}	micro	μ
10^{-12}	pico	p

Table C   Physical constants

gravitational acceleration	$g = 9.807\,\text{m/s}^2$
specific heat capacity of air	SHC = 1.012 kJ/kg K
specific heat capacity of water	SHC = 4.186 kJ/kg K
density of air at 20°C, 1013.25 mbar	1.205 kg/m^3
density of water at 4 °C	1000 kg/m^3
exponential	e = 2.718

Symbols

Symbol	Description	Units
A	area	m^2
α	surface absorption coefficient	
$\bar{\alpha}$	mean surface absorption coefficient	
b	barometric pressure	bar
$^{\circ}$	angle	degree
d	diameter	m or mm
d.b.	dry-bulb temperature	$^{\circ}C$
dp	pressure drop	Pa
dp_{1-2}	pressure change from node 1 to node 2	Pa
EL	equivalent length	m
F	surface factor	
FSP	fan static pressure	Pa
FTP	fan total pressure	Pa
FVP	fan velocity pressure	Pa
F_1, F_2	heat loss factors	
F_u	thermal transmittance factor	
F_v	ventilation factor	
F_y	admittance factor	
f	decrement factor	
g	gravitational acceleration	m/s^2
H	height	m
h	time	hours
h	specific enthalpy	kJ/kg
h_{fg}	latent heat of vapourization	kJ/kg
h_{so}	outside surface heat transfer coefficient	$W/m^2\,K$
h_{si}	inside surface heat transfer coefficient	$W/m^2\,K$
I	electrical current	A
I	solar irradiance	W/m^2
J	quantity of energy	J
k	thermal conductivity	W/m K
kg	mass	kilogram
kJ	quantity of energy	kilojoule
kVA	kilovolt-ampere	kVA
kW	power	kilowatt
kWh	energy consumed	kilowatt hour
L	length	m
LH	latent heat of evaporation	kW

Symbol	*Description*	*Units*
l	length	m
l	volume, litre	l
log	logarithm to base 10	
LTHW	low-temperature hot water	
M	mass flow rate	kg/s
	quantity of energy, megajoule	MJ
MTHW	medium-temperature hot water	
MW	power, megawatt	MW
m	length	metre
N	air change rate	per hour
N	force, newton	N
N	number of occupants	
p	pressure	pascal, Pa
p_a	atmospheric pressure	Pa
π	ratio of circumference to diameter of a circle	
PS	percentage saturation	%
p_s	saturated vapour pressure	Pa
p_s	static pressure	Pa
p_t	total pressure	Pa
p_v	vapour pressure	Pa
p_v	velocity pressure	Pa
p_{sl}	saturated vapour pressure at sling wet bulb air temperature, t_{sl}	Pa
Q	volume flow rate	m³/s or l/s
Q_u	steady state fabric heat loss	W
$\tilde{Q}_u$	cyclic variation in fabric heat loss	W
Q_v	ventilation heat exchange	W
R	resistance, electrical	Ω
R	thermal resistance	m² K/W
R	specific gas constant	kJ/kg K
R	reflectivity of glass	
R	room sound absorption constant	m²
RH	relative humidity	%
ρ	density	kg/m³
R_a	air space thermal resistance	m² K/W
R_{si}	internal surface thermal resistance	m² K/W
R_{so}	external surface thermal resistance	m² K/W
SH	sensible heat transfer	kW
SHC	specific heat capacity	kJ/kg K
$\Sigma ( )$	summation	
SQRT (C7)	square root of contents of cell C7	
SR	regain of static pressure	Pa
s	time	seconds
T	absolute temperature	kelvin, K
T	total irradiance	W/m²
T	transmissivity of glass	
t	temperature	°C

Symbol	Description	Units
t	dry bulb air temperature	°C
t_a	air temperature	°C
t_{ai}	inside air temperature	°C
t_{ao}	outside air temperature	°C
t_c	resultant temperature at centre of room	°C
t_e	environmental temperature	°C
t_{ei}	inside environmental temperature	°C
$\tilde{t}_{eo}$	swing in environmental temperature	°C
t_f	flow temperature	°C
t_g	glass temperature	°C
t_m	mean temperature	°C
t_r	mean radiant temperature	°C
t_r	return temperature	°C
t_r	room temperature	°C
t_s	supply air temperature	°C
t_s	surface temperature	°C
t_{sl}	sling wet bulb air temperature	°C
U	thermal transmittance	W/^{2} K
V	volume	m^3 or l
V	electrical potential	V
V	vertical surface	
v	velocity	m/s
v	specific volume	m^3/kg
w.b.	wet bulb temperature	°C
Y	admittance	W/m^2 K
10^7	10 raised to the power of 7, 10^7	

Introduction

How often do you use a spreadsheet program? The answer to this question is probably: not very often. You may not even be very familiar with the use of a spreadsheet. Engineers understand what a spreadsheet is for, and realize that it is a powerful software tool, but they rarely used it, if at all. Paper, pen and a scientific calculator have always been good enough for engineering problems anyway. When a large task is to be performed, such as calculating the heat gains to a 100-room commercial building, dedicated software costing thousands of pounds or dollars is used.

Who needs a spreadsheet? Only accountants use them! Why is this the view of engineers and engineering students?

What use are spreadsheets to engineers? Who would use them? What can be learned from their use? They look complicated. It must take a long time to learn how to use them. An alternative view may be that spreadsheets are too simple, and that they can only do addition problems like bank statements and company accounts. They are no good at real engineering mathematics, are they? Can you think of any more reasons why spreadsheets are not as common as calculators?

The reasons for such doubts include the fact that there are few or no application spreadsheets for building services engineering design calculation problems. If someone produced useful spreadsheet applications and made them available at a reasonable price, would you use them? The answer to this question must be a resounding yes. Spreadsheets are very powerful software programs that cost megabucks to develop and bring to the marketplace. Almost every personal computer either has a spreadsheet loaded on its hard disk, or could take one. Spreadsheets can cost from a few pounds for public domain programs or share-ware disks up to £500 for the market leaders. They all perform the same tasks – some more colourfully than others. Very involved mathematical problems including trigonometry, exponentials and logarithms are easily handled. Data that the user enters can be automatically copied to other parts of the sheet and processed. The spreadsheet handles text and numbers with equal equanimity. Discrete parts of the sheet can be printed to create graphical output and reports. Laptop and palmtop computers have spreadsheets in read-only memory and on disk. Placing such a small computer onto architects' drawings or any workspace allows for entry of data directly through the keyboard. This avoids the need for the manual entry of data onto forms for processing on a later occasion. There is no need for data to be held on paper before or after processing, other than for presentation and reporting purposes. The spreadsheet's response to new data is immediate. There is no need for expensive digitizing equipment. There is no need for the designer to be tied to any one computer, drawing-board or workstation. The data entry, filing and processing can be entirely portable.

Once a problem has a spreadsheet for it, solving complex and lengthy calculations takes little time and effort. Altering one number will cause the entire worksheet to recalculate and prepare new graphs or charts. The spreadsheet is an enormous electronic sheet of 'paper'. Its size is almost too large to comprehend. It is certainly too extensive for most people's calculation requirements. Spreadsheets are excessive in size relative to the computer random access memory that is normally available. Spreadsheet applications that are written on one software product can usually be read by another spreadsheet program. Spreadsheets can be integrated with word processing and database software. Data files that are produced as lists of numbers by word processing, database, laboratory experimental equipment or building energy management software can be processed automatically by a spreadsheet to analyse the figures. Are you getting the idea?

The potential use of a spreadsheet is almost certainly limited only by the imagination of the user. Good application software does take a lot of preparation. The programs need to be tested on sample data to check the validity of the answers. They need to be made user friendly. A wide range of applications are desirable. You would use them if they were available, right? Well, here are some to keep you happily tapping keyboards. There are no longer any excuses. Spreadsheets are here to stay. Their standards have been well established and have maintained consistency. Effort that is invested in using these applications will be most rewarding in the long term. The knowledge and techniques learned will be transferable to other problems.

Each chapter in this book has a variety of questions. Some questions are descriptive, and a few require manual analysis prior to entering data onto the worksheet. Descriptive questions can be answered by reference to the relevant parts of this book. Knowledge can be enhanced by the use of further references, discussions with colleagues or tutors, and from practical experience. All the answers have been checked by the author but not necessarily by the reviewers. The answers should be correct, because they have been calculated on the spreadsheets. If there seem to be errors, either incorrect input data was used, by the author or by the reader, or a genuine mistake has been made. A source of apparent error could arise from the reader's use of an edited worksheet. A copy of the original spreadsheet should be used to check solutions before claiming an inexactitude. Edited worksheets are the responsibility of the user and not the author. Readers are thanked in advance if they are able to inform us of incorrect items.

The preparation work that has been put into these applications will save hard-pressed lecturers from such a task. All the methods and calculations are fully explained. This collection should allow a depth of study that is not often, if ever, possible in class situations, where the timetable controls all activities. This is an advanced learning resource. It is suitable for various modes of study. It is sufficiently complete to be used as a stand-alone means of study. It can be used for distance learning where the student has occasional contact with the providing organization. It is ideal for open learning centres where students have open access to library, tutor and computer facilities. It is particularly suitable for an introduction by the lecturer followed by self-study by any number of students. It is superb for in-company training, where recruits have the use of computers and reference books but there is little time available for tutoring.

SPREADSHEETS IN THIS BOOK

The spreadsheet applications in this book have been produced with As-Easy-As software. The program retails at £45.00, exclusive of delivery and VAT in the EU.

As-Easy-As is published by, and can be purchased from: Atlantic Coast plc, The Shareware Village, Colyton, Devon, EX13 6HA, UK. Telephone: +44 (0)1297 552222. Facsimile: +44 (0)1297 553366.

Why have a spreadsheet?

1 Computer and spreadsheet use

INTRODUCTION

The users of this book will range from those who are new to using a computer, through those who have not used a spreadsheet previously, and up to those who are experienced in computer spreadsheets. This chapter reminds those who need to be reminded of, and instructs those who do not know, the basic rules for using a computer, for handling files and for using a spreadsheet. All users are strongly advised to be rigorous in their saving of files under unique name and numbering systems, and to organize their files efficiently. There is nothing worse for an engineering computer user than wasting anything from 15 minutes to a whole day of work through the basic oversight of not saving a working file of the master version.

The spreadsheet applications have been produced with As-Easy-As software running under DOS. This product is compatible with the other spreadsheet products that are commonly available. There should be no difficulty with retrieving these files from disk with any of the popular products, whether they be DOS based or Windows based. The author and the reviewers have run these files from other software products, including Microsoft^R Windows^{TM} 95 versions.

LEARNING OBJECTIVES

Study of this chapter will enable the user to:

1. know how to handle, store and file the original application files;
2. know how to transfer, copy and store files;
3. understand and use directories and file conventions;
4. create working job files and disks;
5. become proficient in the use of a spreadsheet;
6. know how to edit spreadsheet application files;
7. use spreadsheets to solve a variety of building services engineering design work;
8. understand the difference between numeric and alphabetic entries in spreadsheet cells;
9. use the graphical facilities of a spreadsheet;
10. print ranges of cells from a spreadsheet.

Key terms and concepts

COMPUTER USE

This section explains how to use the disk provided with the book. The very first thing that you must do is make a duplicate copy of the disk. This will ensure that the original files that you have purchased will not be damaged, changed or deleted. Once you have made the copy, seal the original disk in an envelope, write the name of the disk, the date and the fact that this is the master version on the envelope. Store the master disk in a safe place, where other people cannot find it but you can. A locked disk-storage box can be used for all your master disks.

Experienced computer users may not need to spend much time reading these notes. New users must read them. Years of painful experience have been distilled into a few brief recommendations. Ignore them at your peril!

Have with you four things: this book, the master 3.5 in disk and one or, better still, two blank 3.5 in double-sided double-density or high-density floppy disks and a disk label or two. Lay the master disk on top of the pages of this book, and not in the disk drive slot of the computer. The computer needs to be an IBM model or one that is fully compatible with IBM personal computers. It has two or more disk drives. Drive C is the internal hard disk that contains the disk operating system, or DOS, and all the frequently used software. Drive A is the 3.5 in floppy disk slot in the front panel, or in the side of a portable machine. Drive B might be an additional floppy disk drive and drive D another internal hard drive.

Switch the computer and display screen on. This is fairly obvious, but people do forget. They then call for help because the hardware is not working! The disk operating system will start up or 'boot' the computer into use. Some information will be briefly displayed on the screen, and will pass too quickly to be read unless you press the pause key, followed by the enter key to restart the scrolling text. The screen now has the default drive prompt, C:\>, on display. Now it is your turn to make the computer do something. Try these;

1. Directory

C:\>dir

You type the letters DIR in lower case or capitals at the C:\> prompt. The screen display changes to something like that shown in Table 1.1.

Table 1.1 Directory of a disk drive

C:\>DIR				
Volume in drive C is HARD DISK				
Volume Serial Number is 382D-1BF0				
Directory of C:\				
CONFIG	SYS	118	28-06-92	3:58p
DOS	\<DIR>		28-06-90	3:58p
SYS	\<DIR>		28-06-90	3:58p
MOUSE	COM	15108	06-01-90	6.33a
AUTOEXEC	BAT	84	06-01-92	6.33a
LOTUS	\<DIR>		03-11-93	2.30p
WINDOWS	\<DIR>		29-11-90	11:20p
WORKS	\<DIR>		06-12-90	9:42a
	65 File(s)		15562368 bytes free	

The first column is the name of a file. A file is an area of a magnetic disk that stores information. This is just like a file in a drawer of a filing cabinet. A directory, or index, is needed to know the names of all the files. You invoked the DIRECTORY command of the disk operating system, DOS, to display the contents of the hard disk drive.
2. The directory also shows the space available for new files. In this case, the free space amounts to 15 562 368 bytes. This is over 15.5 megabytes (15.5 MB). Some software requires 5 MB before it can be loaded. The master disk contains less than 1 MB of files.
3. Carefully insert the master disk into drive A. It only fits in one way. Look for the arrow on the disk case.
4. Type A:DISKCOPY and press the ENTER key.
5. The screen displays the instruction to insert the SOURCE disk. This is the one containing the master files to be copied. This is already in drive A, so press the ENTER key. The master disk is now being read into the random access memory (RAM) chips.

6. When the RAM is nearly full, you are requested to take out the master disk and replace it with the TARGET disk. This is the blank unformatted disk that you have ready. Remove the master and insert the blank disk. Press ENTER.

7. The blank disk is now being formatted and will then have the copied files transferred onto it.

8. The computer may not be able to transfer all the files in one go. You will be asked to insert the master disk again. When this occurs, remove the TARGET disk and put back the SOURCE or master disk. Press ENTER.

9. The second batch of master files is now being read into RAM. When this is completed, you are asked to insert the TARGET disk again. Remove the SOURCE disk and insert the TARGET disk. Press ENTER.

10. The DISKCOPY process should be completed in these two processes.

11. You will then be asked if you wish to DISKCOPY another disk. If there were the recommended two blank disks, then yes you do want to DISKCOPY another one. Type Y for yes, then ENTER. From now on, the ENTER key instruction will not be stated. You have the idea by now, and do not need reminding.

12. There is no need to use the master disk again, so lock it away as recommended.

13. Now DISKCOPY the first copy disk onto the new blank unformatted disk exactly as just done from the master: that is, go back to step 5. The new SOURCE disk is already in drive A.

14. You now have two disks containing the original program files, plus the master, which is safely stored out of sight. Label one disk Building Services Engineering Spreadsheets, original files, disk 1.

15. Label the second copy Building Services Engineering Spreadsheets, working files, disk 2.

16. Disk 1 will be kept in the original form, but disk 2 will be used for modifying the files for your own use. Further copies of the files and disks can be created for the various jobs that you undertake.

17. When you have finished the DISKCOPY process for drive A and have labelled the floppy disks, respond N for no to the DISKCOPY another? question. The C:\> prompt returns.

18. If you have your own personal computer, then the files can be copied onto the hard disk drive C. You must create a directory for the worksheets so that they can be easily identified in the root directory. This will be explained now.

19. If you are using someone else's computer, or a college computer that anyone can use, ignore the next copy instruction (number 20). Your spreadsheet files will exist only on your own floppy disks, and this will maintain their security for your use only.

20. Make a subdirectory of As-Easy-As for the worksheets on drive C: by entering MD ASEASY. Type MD, a space, and then the name of the new subdirectory to be created: ASEASY.

    Then change to that directory with:

    C:\ASEASY>

    (If another spreadsheet is being used, change to that directory, such as CD 123 for Lotus 1-2-3.) Make a subdirectory within the ASEASY directory for all your building services engineering working files:

C:\ASEASY>MD BSEWORK

This makes the subdirectory called BSEWORK, which is within the ASEASY directory on drive C. After pressing ENTER the display returns to C:\ASEASY>.

21. Type DIR to check that the directory BSEWORK appears on the list. BSE-WORK is now a subdirectory of the root directory on drive C. Its proper address is C:\ASEASY\BSEWORK.

22. The hard disk directory symbol C:\ASEASY> is on the screen. Change to the subdirectory BSEWORK by typing CD BSEWORK. The screen display changes thus:

C:\ASEASY>CD BSEWORK
C:\ASEASY\BSEWORK>

23. Put disk 1 into your disk storage box and leave it there. Insert disk 2, containing working copies of the original files, into drive A.

24. You are now going to COPY all the files from disk 2 in drive A into the subdirectory BSEWORK on drive C. This will allow faster access to the spreadsheets. Type the following but do not press ENTER:

C:\ASEASY\BSEWORK>COPY A:*.* C:\ASEASY\BSEWORK

Check that you have typed this correctly before pressing the ENTER key. Notice the space between Y and A: and the space between * and C:. These must not be missed out. The instruction is to COPY all the files that are found from drive A to the subdirectory C:\ASEASY\BSEWORK. The wildcard symbol *.* after A: signifies that all files are to be transferred. The first asterisk stands for any filename. The dot and second asterisk mean any file extension type. These may be word processing files (.WPS), batch processing files (.BAT) or, in this case, spreadsheet files that have .WKS after the filename. The COPY instruction is always from somewhere to somewhere else. Advanced users will know short cuts here.

25. When you are sure that the instruction is correct, press ENTER. This will cause all the files to be copied onto drive C. The files that are being copied will have their names displayed on the screen while copying takes place.

26. When the copying has finished, leave disk 2 in drive A. This will be used to hold a duplicate of the files on drive C. Never rely on only one copy of a file.

27. When finishing a computer session, SAVE the latest version of the file on drive A and remove the disk before exiting from the software. Never switch a computer off while a floppy disk remains in drive A, just in case the magnetic disk becomes corrupted.

28. This may be a convenient time to leave the keyboard. The screen display is:

C:\ASEASY\BSEWORK>

Change the current directory from BSEWORK back to the root directory with:

C:\ASEASY\BSEWORK>CD \

The screen returns to:

C:\>

The computer may now be switched off if necessary.

28. Data on the storage space remaining on a disk can be checked with the check disk command, CHKDSK. From the root directory, type:

C:\>CHKDSK

The display then shows the volume name of the disk, its capacity in bytes, and the bytes available for further use. Typically, the volume name is HARD DISK, total disk space is 32 598 016 bytes, or 32.6 MB and there are 7 587 840 bytes available (7.6 MB). The random access memory (RAM) available is shown as 655 360 bytes, and there may be 561 328 bytes free for program use. All these figures are samples. Your own computer configuration and space depend on what is there.

29. Now transfer to drive A with:

C:\>A:

Then look in the directory of drive A. A sample is shown in Table 1.2, but this is not what will be found on the real disk.

Table 1.2   Sample directory of a 3.5 in floppy disk

A:\>Dir				
Directory of A:\				
Volume in drive A is SPREAD1				
Volume Serial Number is 382D-1BF0				
Directory of A:\				
ASSGT1	WPS	1312	20-07-92	8:37p
COMB1	WP1	6432	29-07-92	10:40a
UVALUE	WKS	4575	29-07-92	10:41a
	3 File(s)		1444864 Bytes free	

Check that the files on the disk are the ones that you expected. Do not leave your own files on the hard drive C of a computer that is not yours and expect the files to be there on your next visit. Anyone can alter and remove them.

30. Now check disk A with:

A:\CHKDSK

This is to make sure that there is enough storage space for further files. Otherwise, a disk full error is displayed on screen when you try to SAVE another file onto it. You either have to delete some unwanted files to make room, or use another disk. This is usually accompanied by an embarrassed state of panic!

31. Return to drive C with:

    A:\>C:
    C:\>

SPREADSHEET USE

The spreadsheet to be used will be on drive C, or the network drive name in an office or university, so change to its directory and implement it. If this is As-Easy-As in the ASEASY subdirectory, then type:

C:\>CD ASEASY
C:\ASEASY>ASEASY

The program is now activated. If this is a 123 product in the 123 subdirectory, then type:

C:\>CD 123
C:\123>123

The program is now activated.

The files on your disk are in As-Easy-As format. As-Easy-As is compatible with the major software products. A typical filename is COMB1.WKS. COMB1 is the file-name and .WKS is the extension used for worksheets. Several other spreadsheet programs can read these files. Refer to the list of suitable software further on in this chapter.

Familiarize yourself with the spreadsheet software that you are using. Learn the main commands and where to find them. Write your own lists of commonly needed commands, and carry them with you at all times. Write down the sequence of keystrokes, user identity codes and passwords that you will use on the computers available. Keep a copy of the keyboard guide with the disk or computer. This will show the special use of each command key: for example, the / key is used to activate the menus at the top of the screen. There is little worse than sitting at a blank screen, with an urgent task to perform, and finding that you have forgotten how to switch on the computer or the software. Very disconcerting.

It is not the intention of this book to teach the use of spreadsheets. The spreadsheet software manual and on-screen tutorials will do this for you. A brief introduction on how to use a spreadsheet is given, and this should allow you to make a quick start.

Beware of the power of a spreadsheet. Whenever you enter data or change the old data and then save the edited version, the old sheet is lost. If your new changes are mistaken, you have a problem. Always keep a backup copy or two of the old version before you commence work. If you make a mistake, then the original is intact. Files take up very little disk space, so do not be afraid to have multiple copies. Maintain identical copies of all your files on two 3.5 in disks. Disks can become corrupted by physical damage and electromagnetic interference. Occasionally, a computer will refuse to read a disk, and will give a read error in drive A message. All that can be done for such a disk is to format it again and lose all its contents. Having only one disk with your valuable data is not enough. A hard disk and two floppies is much safer, although not foolproof.

Never use an original spreadsheet file for work or for making changes. Always copy the original file into a new filename. Make several copies of the original files, with each copy having a different name. You can make changes to the copies and use them to store specific data and solutions. Keep a record of all your filenames and their contents. If you are really well organized, you will use a database to record filenames and their contents for quick retrieval. However, a simple word processed document containing filenames and contents will do. As a last resort, you can use a pencil and paper (this is low tech but reliable).

Names of files should represent their content. Limit file names to eight characters. There is no need to list the file extension such as .WKS for worksheet files and .WPS for word processor documents, as this is usually obvious to the user. A filename has to have a letter as the first character, otherwise the computer reads it as a number and finds an error. Numbers and letters can be used in any combination after the first character. Use easily remembered combinations that explain the content. A shortened word followed by the series number is understood by other computer users. Do not use special characters and mathematical symbols in filenames, such as !, *, /, ;, +, =, and others. Stay with letters and numbers. These are alphanumeric filenames. Some computers will accept more than eight alphanumeric characters in a filename. There is little value in writing sentences as filenames, and they may not transfer to another computer or software, so the effort of typing them can be wasted. Never put a blank space in a file name with the space bar as this is a fatal error. Suitable filenames are shown in Table 1.3.

Table 1.3   Typical names of computer files

Original file	Working copy	Job-specific files	
UVALUE.WKS	UVALUE1.WKS	UVALUE2	Daily use master
		UVALUE10	Common values
		UVALUE21	Problem 21
		UVAL6815	Job number 68, file 15

Files take up very little hard and floppy disk space. Make as many job-specific files as necessary. Only delete files that are definitely redundant or are very unlikely to be needed next month, next year or after the examinations. If the computer gives you disk full error signals, then you have forgotten to occasionally type DIR and CHKDSK after the disk drive prompt C:\>, A:\> and B:\> to check the space available on the hard drive and floppy disk. DIR is the MS-DOS operating system directory command, and CHKDSK checks the space used. A list of the files and directories that are held in the root directory of drive C, A or B is displayed.

Look in the directory of the files on the 3.5 in disk number 2 that you have slotted into drive A. It has been given the volume label SPREAD1 after the FORMAT command. The column on the left is the list of the filenames that are present. The second column is the type of file, such as WPS for word processor, WKS for worksheet, SYS for system, EXE for executable file, ALT for an alternative file that could be used, BAT for a batch processing file, COM for a command file, OLD for a redundant file or <DIR> for a directory. The third column gives the space taken by each file, in bytes. The fourth column is the date when the file was last changed. Disk A may, for example, have three files that take up 1312 + 6432 + 4575 bytes of space, 12 319 bytes. The empty disk had 12 319 + 1 444 864 bytes of usable space, 1 457 183 bytes. The files take up only (12 319/ 1 457 183) × 100% = 0.8% of the available space. This complete book is stored on one 3.5 in disk, with space to spare. Drive C normally has a capacity of at least 10 MB. A disk space of 30 MB is common, but 105 MB or more are available. The computer in Table 1.1 has a hard disk capacity of 30 MB, and there are 15.562 368 MB of free space available for additional files. A file may be 10 kB or more, 10 000 bytes, so if you find that a floppy disk has less than 100 kB or a hard drive has less than 1 MB, it is time to clear some space.

Data that are entered into a spreadsheet are read by the software as either numeric or alphabetic data. The spreadsheet program reads an entry that commences with a number as numeric data and that which begins with a letter as alphabetic data. The computer will not correct your errors, your lack of understanding of this concept, or your typing mistakes. You may think that the data format of 2 August 1997 is an appropriate alphanumeric entry into a cell. The computer reads it as a number, and finds that it has errors. There are spaces and letters mixed up in the number. The screen display will be ##########. This is not what you meant. You have to inform the computer that this is not a number, by forcing it to be text. Commence the input to the cell with the text symbol '. Whatever is typed after the ' is alphanumeric text, or words. Another problem occurs if the data format is to be 21/08/97, meaning 21 August 1997. A cell containing 21/08/97 is interpreted as a formula, and the computer may divide the numbers to display 0.028 rather than the intended date. Some software may recognize 21/08/97 or 21 August 1997 as a date, but you need to be sure what your system will do. Typing '21/08/97 is correctly interpreted as alphanumeric characters.

AS-EASY-AS USE

1. To activate As-Easy-As from drive C, return to the root directory with CD \. It may be necessary to find the name of the software in the root directory of the hard disk. Type C:\>DIR/P to list the directory one page at a time. Look for the subdirectory that is named ASEASY, or whatever you have renamed it.
2. Change to that directory with C:\>CD ASEASY.
3. List the files in the ASEASY directory with

   C:\ASEASY>DIR/P

4. Spreadsheet software includes on-screen help menus or, in some cases, a tutorial package called View. Check that it exists in the directory. The file VIEW.EXE should be in the list. It is accessed from the main directory with C:\name>VIEW. New users should work through the tutorial.
5. Find the executable file for As-Easy-As. This should be ASEASY.EXE.
6. Execute the spreadsheet with

   C:\ASEASY>ASEASY

7. The licence display appears briefly. It can be held with the PAUSE key. Press ENTER when you are ready to continue. The blank spreadsheet screen then appears.

8. The cursor is in the top-left hand corner of the sheet. This is in column A and row 1. This is the first cell of the worksheet, and is referenced as A1. The current cell location of the cursor is shown in the top-left corner of the screen.

9. Use the right arrow cursor movement key to move the cursor to the right. Notice that the cell reference at the top left corner changes with the cursor position.

10. Use the down arrow key to move to the lower rows. Cell H20 is probably the bottom right corner cell on the screen.

11. When the cursor is moved beyond H20, the cell column and row numbers scroll along.

12. The number of columns width and rows deep of the spreadsheet that you use are more likely to be limited by your imagination than by the physical limits of the software. The available random access memory (RAM) also limits the size of worksheet that can be manipulated. Do not bother to try and find out what the limits of the electronic sheet of paper are. It is not worth the effort, as it is unbelievably huge. A sheet of A4 paper will normally display up to 34 rows and 8 columns, or 272 cells. As-Easy-As can address 8192 rows and 256 columns. This would be $8192 \times 256 = 2\,097\,152$ cells or 7710 sheets of A4 paper. If each cell were to be eight characters wide and each character needs two bytes, the RAM requirement would be $(8192 \times 256 \times 8 \times 2)/10^6$, or 34 MB. Many personal computers have between 655 kB and 4 MB of RAM, so the potential spreadsheet size is excessive. It is unlikely that a worksheet will be larger than 10 000 cells: that is, 1000 rows and 10 columns, producing a RAM requirement of $(1000 \times 10 \times 8 \times 2)/10^3$, or 160 kB.

13. To move the cursor back to cell A1, either use the arrow keys or press the HOME key.

14. Notice the READY indicator at the top right corner of the screen. When this changes to a red flashing WAIT signal, do not press any key.

15. The current date and time are shown at the bottom left of the screen.

16. The NUM LOCK indicator is at the bottom right of the screen. Test it by pressing the NUM LOCK key.

17. While the cursor is flashing and the READY indicator is displayed, data can be entered into any cell.

18. Move to cell D5. Type the word psychrometric, and press ENTER.

19. The cell contents are shown at the top of the screen as D5: 'psychrometric. The ' indicates that alphabetic characters have been recognized, and that all the contents of that cell are a word and not a number. New users may make the mistake of entering alphabetic characters or a number preceded by an ' and expect the computer to think of the cell content as a number. The user can see a number on the screen but the computer will not, and an error message is displayed when a calculation is attempted. Beware the alphabetic 'number'!

20. Words are left-justified in the cell: they commence at the left end of the cell.

21. The word psychrometric appears in cell D5, but the last four letters have spilled out of the width of the column and are taking up space in the next column. This is poor practice, and will lead to the contents of cell E5 over-writing the excess contents of D5.

22. Move to E5. Type abc and press ENTER. Move between D5 and E5 and notice how some letters are hidden.

23. Go to E5, press the space bar and ENTER, not simultaneously. Cell E5 now contains only an empty space character, but the cell appears to be clear.

24. Got to D5. Psychrometric is still present on the top of the screen but it does not all appear on the sheet. The blank space character in E5 is overwriting the excess letters. The erroneous content of E5 will need to be erased in order to recover the word psychrometric. This will be explained later.

25. Go to cell A5. Type 36.249 and ENTER. This is recognized as a number and is shown in full.

26. Put these numbers into cells:

cell A6 type 4000	display 4000
cell A7 type 4000000	display 4000000
cell A8 type 4000000000	display 4.0E+09
cell A9 type 0.005	display 0.005
cell A10 type 0	display 0

The numbers are right-justified: that is, they are close to the right-hand end of the cell. This can be changed if necessary. The number 4 000 000 000 has been displayed in its exponent form, as it would not fit the width of the cell. Such numbers are expressed as

$$4\,000\,000\,000 = 4 \times 10^9 = 4.0E + 09$$

The E + 09 means the number 10 raised to a positive power of 9. Negative powers are also used.

27. Experiment with different cell contents.

28. Go to cell D5. Any cell can be edited without the need to type the whole cell. This is particularly useful when a formula has been entered into a cell. Press the F2 key. This is the EDIT CELL function key. The cell contents are repeated on the next line down from the top of the screen. The flashing cursor has been transferred from the cell area of the worksheet to the right end of the line in the editor. Use the left or right arrow keys to move along the line in the desired direction. The backspace DELETE key is used to delete leftwards. Delete the letters to the left of the first letter r in psychrometric. This leaves the new (erroneous) word rometric. Press ENTER. The newly edited word is displayed in D5. This is a useful function, which will save time and allow temporary or permanent changes to be made to cell contents.

29. The menu commands are activated by pressing the / key. Two lines of menus appear. The upper line holds the directory titles to the lower line of commands.

30. The WORKSHEET command heading is highlighted. Use the right and left arrow keys to select other command headings.

31. To return control to the worksheet cells, press the ESCAPE (ESC) key. This removes the menus. The flashing cursor returns to the worksheet cell where it was last. ESC always retreats to the previous level of menu, previous status or a lower level of command. Test these keystrokes by moving around the menus in the forward and backward directions.

32. Select FILE on the upper menu row. The lower row shows RETRIEVE a file, SAVE a file and other options.

33. To leave the worksheet select QUIT. There is no need to SAVE the experimental cell contents so press ENTER. Now the choice is NO to return to the worksheet or YES to end the session and leave As-Easy-As. Select YES with the arrow key and then ENTER.

34. The current directory symbol C:\ASEASY> now appears on the cleared screen, and the exit from As-Easy-As has been completed.

35. Type ASEASY to restart the spreadsheet.

36. To retrieve a file, press / to bring up the menu, right arrow to File, press ENTER and Retrieve is highlighted. ENTER again and the filename to be retrieved is requested. The file will either be C:\ASEASY\BSEWORK\AXIAL.WKS or A:\BSEWORK\AXIAL.WKS, where A: or C: is the drive location, and BSEWORK is the directory containing the file AXIAL.WKS. If directories exist, they will be highlighted on the second line of the screen. Select the directory and press ENTER. The files within that directory are now listed. Select the correct file with the arrow keys and press ENTER. Should the filename not be in the menu, type it and it will be added to the path. The worksheet appears on the screen very quickly.

37. Use the arrow keys to move around the worksheet to investigate its contents. Press the HOME function key to jump to cell A1.

38. Move to the cell with the current date. Press the F2 function key. The cell contents are repeated on the second line of the screen. Use the left arrow and backspace delete keys to remove unwanted characters. Type the current date. When it is correct, press ENTER and the new date is displayed in the edited cell.

39. Move the cursor to the cells that contain the user's name, job reference, date and filename. Either edit these cells or overwrite them with new information.

40. Go to the other cells where new input data is required and type the information.

41. As soon as new numbers are entered, the whole of the worksheet is automatically recalculated. Try changing an input number and watch the resulting effect. It does not take much time.

42. Some input data can be copied from other parts of the worksheet rather than being typed. Move the cursor to the cell that contains the desired number and press /, right arrow to COPY and ENTER. The cell or range of cells that are to be copied from are requested. Move the cursor up or down and watch how the range of cells is selected. Reverse the cursor movement to remove the selection and return to the original target cell. Press ENTER to confirm that the cell selection has been made. Now move the cursor to the cell where the copied data is to be located and press ENTER. The cursor returns to the original cell, and its contents are copied into the destination cell. All the features of the cell were copied: that is, the displayed number or text, the formula in the cell, and the format of the data.

43. To change the format of a cell or range of cells, press /, right arrow, RANGE, ENTER, FORMAT, ENTER. Choose the format to be used: for example, FIXED, ENTER, 3 decimal places, ENTER, confirm the range of cells to be formatted, ENTER and view the displayed result. Generally, minimize the use of displayed decimal places to zero to three as necessary for ease of reading. The real number remains in use for calculations and only the display is

affected. Only use three decimal places when the third place means something real, such as one millimetre.

44. Save your work frequently. As soon as new data has been entered, SAVE the worksheet. Use a unique filename. Repeat the SAVE instruction often and overwrite the old version. Select /, right arrow, FILE, ENTER, right arrow, SAVE, and ENTER. The path and filename are requested. Decide whether to file the worksheet on drive A or C, or both, and the directory to be used. Press the ESCAPE key once to make changes to the displayed path and file name. Press ESCAPE a second time to go back to the main spreadsheet directory. A third ESC use will enable a new path to be typed. The fourth press of ESC removes the file command. Remember that filing your work is vital. Computers do fail, networks go off line, fuses melt, micro circuit breakers operate when wiring becomes overloaded, electrical power cables are brought down during gales, mechanical diggers break underground cables, and someone else will switch the PC power socket off (Oh, I am sorry, I thought that was the compact disk plug. I do hope that I have not done any damage!). The PC user emits a repressed scream.

45. Each worksheet has the date of its use and the time of day displayed automatically. This is to ensure that the user is aware of when modifications and new data have been employed. The @TODAY function brings the serial number of the current date and time of day into that cell. Try it in two blank cells, for example C7 and C8. @TODAY shows 34421.73535. This corresponds to 05:38:55 p.m. on Monday 28 March 1994. Format cell C7 to show the date by pressing the sequence of keys /, RANGE, FORMAT, DATE, Number1 for (DD-MM-YY), accept the C7..C7 range with the Enter key. The current date is displayed. Select cell C8 and press /, RANGE, FORMAT, DATE, TIME, Number 1 for (HH:MM:SS AM/PM), accept the cell range C8..C8 with the enter key. The current time is displayed.

46. Graphs can be viewed and changed with /, GRAPH, ENTER, right arrow, NAME, ENTER, USE, ENTER. The names of the charts that are in the worksheet are displayed along the third row at the top of the screen. Use the right arrow to highlight the desired chart and press ENTER. The chart appears. It is removed by pressing either the ENTER or the ESC key. Quit the graph menu to return to the worksheet.

47. The graph type is changed by moving the cursor to highlight TYPE in the GRAPH menu and pressing ENTER. Select the way in which the data is to be represented and ENTER. Line graphs have been used in this book.

48. The data range that is charted is specified by selecting the letter X, for the horizontal chart axis, and A to F for the vertical *y*-axes. Up to six *y*-axis data ranges can be used. Each *y* range produces a separate line. When COLOUR is selected in the OPTION menu, each line and its label is shown in a different colour. To change the range of data that is used, select the range letter, press ENTER, and the cells are displayed. Either type the range of cells to be used as A36..A63, or whatever is required, or move the cursor to the first cell of the desired range, typing '.' (without the quotation marks), and then move the cursor to the end of the range. Press ENTER, and the selected range is highlighted. Check that it is correct, and view the graph.

49. Notice the SAVE instruction in the Graph menu. Do not use this to save the graph settings, as this is done automatically. When the graph is finished and in

a printable form, use the SAVE command. This saves the picture for printing. A graph is easily printed from As-Easy-As from the drop-down menus. To print it in an early 123 product, a separate program is used. This is PGRAPH, and it is has its own directory.

50. The OPTIONS menu is for producing labels and titles, and for creating different markers for each data point. LEGEND is for typing the description of the data range at the bottom of the graph. FORMAT is used to create lines and marker symbols with FORMAT, ENTER, GRAPH, ENTER, BOTH, ENTER, QUIT. TITLE facilitates the use of two lines of text at the top of the graph. GRID, ENTER, BOTH, ENTER draws both x and y gridlines. SCALE, ENTER, SKIP, ENTER, 5, ENTER is used to alter the space between the numbers that are displayed along the x-axis. Choose a suitable skip value so that the numbers are clearly spaced along the axis. COLOUR, ENTER draws the graph in multicolours. DATA LABELS is used, if required, to locate labels on each line. The position for each label is selected, and some experimentation may be necessary to achieve a satisfactory result. QUIT and then VIEW the graph. Making changes is quick and easy.

51. Only one graph can be current at one time. When a new graph has been created, it is saved and named, and it becomes the current graph. The NAME command in the GRAPH menu is selected. The CREATE command puts the current graph settings into the name that is typed by the user. Once charts are created with different names, they can be viewed with the USE command. When changes are made to the current graph, the CREATE command is used to save the new settings.

52. Save the latest version of the worksheet, with a new file name if necessary. Save the charts as .PIC files for printing.

53. Quit the worksheet.

54. The presence of the graph printing program PGRAPH.EXE should be confirmed from the main directory with C:\123>DIR/p. To run it, enter C:\123>PGRAPH. Use IMAGE-SELECT to choose the picture to be printed from the list of files that were saved. The settings, paper size, fonts and hardware set up can be easily changed if necessary. The default settings are probably suitable.

SUITABLE SPREADSHEETS

The disk provided with this book contains worksheet application files that were produced in As-Easy-As. The application files can be read by Lotus 1-2-3 release 2.01. This is a version that is likely to be available to most users of both personal computers and networks. Later versions of Lotus software, including Windows applications, are compatible with these files. Application files are usually upgradable to later and Windows versions. Lotus 1-2-3 refers to the UK distributor of Lotus products, Lotus Development (UK) Limited, 825 Yeovil Road, Slough, Berkshire SL1 4JA.

There are several spreadsheet programs that will read the files on the disk with this book. They have been seen to run on the following software:

1. Microsoft Works version 2. This is an integrated software package, which has an excellent word processor, spreadsheet, database and communications

ability. The author has used Works version 2 to write this book, and others. The close linking of the word processor with the spreadsheet, and the mouse operation, make the package a pleasure to use. On-screen tutorials are available at all times. The graphics reproduction facility is very good.

2. As-Easy-As version 5.01. This is the recommended shareware spreadsheet.
3. Microsoft Excel. The author uses Microsoft Office Windows® software, and has run application files easily including Windows® 95.
4. Reviewers have run some application files on Quattro Pro without any difficulty.

MATHEMATICAL SOFTWARE

Software that is dedicated to solving mathematical problems is needed to convert coordinate data from graphs such as fan performances, into polynomial curve-fit equations. The user of this book does not have to perform this work, as it has been done by the author in the preparation of some of the fan curve data in Chapter 6.

PUBLIC DOMAIN SOFTWARE

Public domain (PD) software is available for only the cost of servicing the customer. There is no charge for the software itself. The total cost of acquiring PD software can be less than £10. Some software can be accessed through the Internet. A supplier of public domain and shareware software is Eigen PD Software, PO Box 722, Swindon SN2 6YB, UK; telephone +44 (0) 793 611270.

SHAREWARE

Shareware is software for trial use for a limited period of time. If the user wishes to make use of it, a registration fee is payable. In return, the instruction manual and the most recent version of the program are supplied. Shareware can be tried for less than £10. To purchase the shareware, an additional £10 to £100 can be payable for the latest version, the instruction manual and product support. Shareware and public domain software can provide very good value for money.

Some academics scoff.

2 Thermal transmittance

INTRODUCTION

Finding the thermal transmittance of a range of different types of structure is one of the earliest tasks for the design engineer. The user of this worksheet file, UVALUE.WK1, types the material common or trade name and its thickness directly onto the worksheet screen, perhaps with a laptop or palmtop computer, which is resting on the plans of the building. Thermal conductivity data for most applications is held on the worksheet for viewing and copying. Further material data is easily added by the user. The worksheet has been arranged for the calculation of thermal transmittances for various typical constructions. These are configured into groups for walls, floors, ceilings, roofs and windows. Each group has between one and four ready-made data areas, which require only the new material entries. A construction can have up to eight layers of material, depending upon how many are relevant for that type.

The output data of thermal resistance, transmittance and conductance are displayed on the screen. The thermal transmittance is checked against an allowable range of values. If it falls outside the expected span, a warning is given. The warning limits can easily be changed by the user. When a paper copy is required of those parts of the worksheet that are relevant to the job or assignment that is being undertaken, the user specifies the print areas in the normal manner for the spreadsheet that is being used. Each section of the work and the formulae used is explained. There are many worked examples.

LEARNING OBJECTIVES

Study and use of this chapter will enable the user to:

1. know where the data for thermal transmittance is found;
2. become familiar with commonly used values;
3. understand how thermal transmittance is calculated;
4. calculate the thermal transmittance of the elements of building constructions accurately and reliably;
5. maintain a personal or office-wide file of thermal transmittance data;
6. have a convenient and portable method of storing and using thermal transmittance data and calculations;
7. calculate the weight of a structural element per unit surface area;
8. know the practical values for thermal transmittance;

9. know the structural weight of different elements of the construction;

10. be able to provide the structural weight information to the structural design engineer;

11. understand the equations that are used for each type of constructional element;

12. calculate the thermal transmittance for flat surface elements such as walls, glazing, internal floors and ceilings;

13. calculate the thermal transmittance of pitched roofs;

14. calculate the thermal transmittance of solid floors in contact with the earth;

15. understand the use of hyperbolic tangent and arctangent;

16. understand how the thermal transmittance of suspended timber and concrete floors is found for the suspended and solid floor components;

17. calculate the thermal transmittance for suspended floors;

18. know what conductance is and how to find it;

19. calculate the average thermal transmittance of single- double- and triple-glazed windows and doors which have wood, polyvinyl chloride or metal frames.

Key terms and concepts

air space thermal resistance	20	frame	38
arctangent	30	frame proportion	38
cavity	21	glazed	38
ceiling	26	high emissivity	35
conductance	22	hyperbolic arctangent	30
convective resistance	21	hyperbolic tangent	30
COPY command	19	internal wall	24
density	22	low emissivity	21
door	38	material	23
double-glazed	38	mouse pointer	19
earth resistance	29	overall weight	22
earth thermal transmittance	35	pitched roof	27
edge insulation	40	radians	32
element	20	radiation resistance	34
emissivity	21	resistance	20
exponential	30	roof	27
exposure	21	roof pitch angle	27
external wall	22	shape factor	29
file	19	sloping layers	29
flat layers	39	solid floor	29
flat roof	26	source cell	19
floor	25	surface film thermal resistance	20
foundations	34	suspended floor	34

THERMAL TRANSMITTANCE CALCULATION

This worksheet is intended to provide a well-organized method of storing the data on construction elements, material thermal conductivity, surface film and air space thermal resistances that the user requires. Files can be created for each design project that is carried out and then referred to at a later date when other applications arise. Data can easily be added to the worksheet when new materials become available. There is really no limit to the user's addition of material names, trade names and thermal conductivities, unless the computer's random access memory or disk space become filled, which is rather unlikely. Sufficient data is provided on the worksheet for most applications and design projects. All the stored data can be checked with the references and amended should this be necessary. The data for materials commences in cell H2 and that for surfaces and air spaces starts in cell L2. Each element area of the worksheet is provided with the data that is used in the worked examples through this chapter. This means that the user will not need to make many changes for some assignments. The original numbers can be left in place; only the new information has to be entered. When a mouse pointer is available, the user moves the cursor to the source data cell, selects COPY from the menu commands, moves the cursor to the target cell, and enters the replacement data with the mouse button. The user should start by viewing the extent of the worksheet to become familiar with where to find the element types and the reference data. The cursor arrow keys, PAGE UP key, PAGE DOWN key and mouse pointer are used to move around the worksheet.

All dedicated software requires the user to input the material components. Such input from the engineer is either by means of handwritten data sheets prior to keyboard entry or by direct typing onto a visual display screen, which may be formatted to appear as a spreadsheet layout. Some programs have code letters for each material, and the user needs to become familiar with unique combinations of letters that other software suppliers do not use. Programs are normally associated with a personal computer that is fixed in position and possibly a large digitizing board for the architectural drawings. An equipment system like this is immovable, and not always accessible by engineers who only wish to use it for half an hour or so. While a powerful computer system has advantages, its software costs from the hundreds and into the thousands of pounds sterling. There is absolutely no way in which a user can ever know how the answers are produced by a large system. The programming is confidential to the supplier, who will only discuss input and output data. The alternative to a computer system is the paper, pencil and calculator technique, which also requires the data source book to be open.

The advantage of this book's method of thermal transmittance calculation is that every part of the procedure is fully understandable, it is an open learning resource, and it can be customized. Another very powerful reason for making use of this worksheet is its portability. Most computers have a spreadsheet program on their hard disk. This will either be Lotus 1-2-3 or a spreadsheet program that can read worksheet files that were created by Lotus 1-2-3. This means that the user's floppy disk of application files is all that is needed to operate the calculation in almost any office, road vehicle, home or building site location.

The worksheet is set up to analyse:

1. four external walls;
2. two internal walls;
3. two internal floors;
4. two ceilings;
5. two flat roofs;
6. one external floor;
7. two pitched roofs;
8. solid ground floor that is in contact with the earth;
9. suspended floor;
10. single-, double- and triple-glazed framed window or door.

An element may have up to eight layers of material, depending upon what is realistic for that application. For example, an exterior cavity wall can consist of plaster, concrete blockwork, rigid glass-fibre slab, a ventilated cavity, external brickwork and a layer of cement render. This has five layers of material, an air space and two surface film thermal resistances. It is unlikely that this number of layers of construction materials will be exceeded, but it could occur, and a space for a sixth layer is provided. All the thermal conductivity and density data for the materials listed has been taken from the *Chartered Institution of Building Services Engineers Guide A* (CIBSE, 1986a). The user can update the information and add data from trade or research organizations as it becomes available. Extra explanatory notes can be typed into blank cells of the worksheet as required. Be careful not to overwrite cells that contain information or formulae, or the original contents may be lost.

The thermal transmittance (U value) of a structure is found from the thickness and thermal conductivity of each solid material, the thermal resistance of any air space within the structure, and the thermal resistance of the layer of air that is attached to the indoor and outdoor surfaces. The terms used are:

1. material thickness, L mm;
2. material thermal conductivity, k W/m K;
3. air space thermal resistance, R_a m^2 K/W;
4. internal surface thermal resistance, R_{si} m^2 K/W;
5. external surface thermal resistance, R_{so} m^2 K/W.

Figure 2.1 depicts these items.

The overall thermal resistance R of a structure that comprises structural elements that represent a series of resistance values is found by adding the individual resistances:

$$R = R_{so} + R_1 + R_a + R_2 + R_3 + R_{si}$$

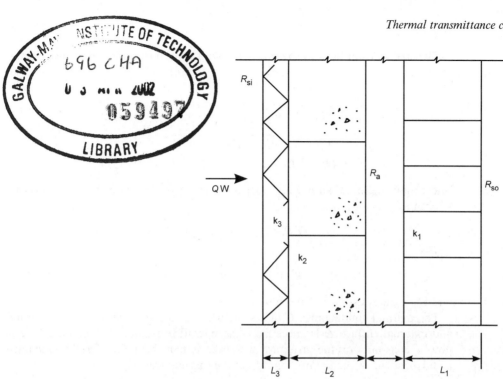

Fig. 2.1   Data for the calculation of thermal transmittance.

where R_1, R_2 and R_3 are the resistances of the solid elements, which are found from

$$R_1 = L \text{ mm} \times \frac{1 \text{ m}}{10^3 \text{ mm}} \times \frac{1 \text{ mK}}{k \text{ W}}$$

$$= \frac{L}{10^3 \times k} \frac{\text{m}^2\text{K}}{\text{W}}$$

R_2 and R_3 are found by the same method. R_{so}, R_a and R_{si} are read from data tables, because they are found from experimental testing and not by calculation. The internal surface resistance depends upon the emissivity of the indoor surface finish, its orientation, the room air velocity over the surface, and the presence of radiant heat exchanges within the room. Similar heat exchange variables also affect the external surface, except that it is the outdoor environment, with solar radiation, wind and rain, that is involved. The degree of exposure of the external surface to the prevailing climate is classified into three categories:

1. sheltered, for up to the third floor in city centres;
2. normal, for most suburban and rural buildings, and the fourth to eighth floors of buildings in city centres;
3. severe, for buildings on coastal or hill sites, floors above the fifth in suburban or rural districts, and floors above the ninth in city centres.

The air space thermal resistance results from the convective and radiative heat exchange that takes place across the cavity. Typical values are $0.06 \text{ m}^2 \text{ K/W}$ for R_{so}, $0.18 \text{ m}^2 \text{ K/W}$ for R_a and $0.12 \text{ m}^2 \text{ K/W}$ for R_{si}.

Thermal transmittance, $U \, \text{W/m}^2 \, \text{K}$ is found from the reciprocal of the thermal resistance:

$$U = \frac{1}{R} \frac{W}{\text{m}^2 \, \text{K}}$$

$$= \frac{1}{R_{so} + R_1 + R_a + R_2 + R_3 + R_{si}} \frac{W}{\text{m}^2 \, \text{K}}$$

$$= (R_{so} + R_1 + R_a + R_2 + R_3 + R_{si})^{-1} \, \text{W/m}^2 \, \text{K}$$

There may be more than one air space and three layers of material, so this may be written as

$$U = (\Sigma R + R_{so} + R_a + R_{si})^{-1} \, \text{W/m}^2 \, \text{K}$$

where

$$\Sigma R = R_1 + R_2 + R_3 + \ldots + R_n \, \text{m}^2 \, \text{K/W}$$

and R_n is the resistance of the final layer n.

The thermal conductance C of a structure is found by subtracting the internal and external surface resistances from the overall resistance. The conductance is now a constant for the structure wherever it may be built. The conductance is independent of the factors that affect the thermal transmittance:

$$C = \frac{1}{R - R_{so} - R_{si}} \frac{W}{\text{m}^2 \, \text{K}}$$

Structural design engineers need to know the weight W of the structure. This is expressed in kilograms per square metre of the internal surface area from the density, $\rho \, \text{kg/m}^3$, and thickness of each material.

$$W = L \, \text{mm} \times \frac{1 \, \text{m}}{10^3 \, \text{mm}} \times \rho \frac{\text{kg}}{\text{m}^3}$$

$$= \frac{L\rho}{10^3} \frac{\text{kg}}{\text{m}^2}$$

Most structures have several layers. The overall weight is

$$W = (L_1\rho_1 + L_2\rho_2 + \ldots + L_n\rho_n) \times \frac{1}{10^3} \frac{\text{kg}}{\text{m}^2}$$

where the number of the final layer of materials is n. The calculations for each type of structure are shown below.

External walls

Figure 2.1 depicted a typical external wall of three layers of material. Thermal insulation and surface finishes may increase the number of materials to six.

EXAMPLE 2.1

Calculate the thermal transmittance, conductance and structural weight for an insulated cavity brick and block wall from the data in Table 2.1.

Table 2.1   Data for Example 2.1

Layer	L (mm)	k (W/m K)	ρ (kg/m³)	R (m² K/W)	
Outside surface				R_{so}	0.06
Cement render	15	0.5	1300		
Brickwork	105	0.84	1700		
Air space				R_a	0.18
Glass-fibre quilt	60	0.04	12		
Light concrete block	120	0.19	600		
Light plaster	15	0.16	600		
Inside surface				R_{si}	0.12

The thermal resistance of each layer of material is

$$\text{cement render, } R_1 = \frac{L_1}{10^3 \times k_1} \frac{\text{m}^2\,\text{K}}{\text{W}}$$

$$= \frac{15}{10^3 \times 0.5} \frac{\text{m}^2\,\text{K}}{\text{W}}$$

$$= 0.03\,\text{m}^2\,\text{K/W}$$

$$\text{brickwork, } R_2 = \frac{105}{10^3 \times 0.84} \frac{\text{m}^2\,\text{K}}{\text{W}}$$

$$= 0.125\,\text{m}^2\,\text{K/W}$$

$$\text{glass-fibre quilt, } R_3 = \frac{60}{10^3 \times 0.04} \frac{\text{m}^2\,\text{K}}{\text{W}}$$

$$= 1.5\,\text{m}^2\,\text{K/W}$$

$$\text{light concrete block, } R_4 = \frac{120}{10^3 \times 0.19} \frac{\text{m}^2\,\text{K}}{\text{W}}$$

$$= 0.632\,\text{m}^2\,\text{K/W}$$

$$\text{light plaster, } R_5 = \frac{15}{10^3 \times 0.16} \frac{\text{m}^2\,\text{K}}{\text{W}}$$

$$= 0.094\,\text{m}^2\,\text{K/W}$$

The thermal transmittance is

$$U = \frac{1}{R} \frac{\text{W}}{\text{m}^2\,\text{K}}$$

$$= (R_{so} + R_1 + R_2 + R_a + R_3 + R_4 + R_5 + R_{si})^{-1}\,\text{W/m}^2\,\text{K}$$

$$= (0.06 + 0.03 + 0.125 + 0.18 + 1.5 + 0.632 + 0.094 + 0.12)^{-1}\,\text{W/m}^2\,\text{K}$$

$$= (2.741)^{-1}\,\text{W/m}^2\,\text{K}$$

$$= 0.365\,\text{W/m}^2\,\text{K}$$

The conductance of the wall is

$$C = \frac{1}{R - R_{so} - R_{si}} \frac{\text{W}}{\text{m}^2\,\text{K}}$$

$$= \frac{1}{2.741 - 0.06 - 0.12} \frac{W}{m^2 \, K}$$

$$= 0.391 \, W/m^2 \, K$$

The overall weight of the wall structure for $n = 5$ is

$$W = (L_1\rho_1 + \ldots + L_5\rho_5) \times \frac{1}{10^3} \frac{kg}{m^2}$$

$$= (15 \times 1300 + 105 \times 1700 + 60 \times 12 + 120 \times 600 + 15 \times 600) \times \frac{1}{10^3} \frac{kg}{m^2}$$

$$= 279\,720 \times \frac{1}{10^3} \frac{kg}{m^2}$$

$$= 279.72 \, kg/m^2$$

Internal walls

An internal wall separates two rooms within a building.

EXAMPLE 2.2

Calculate the thermal transmittance, conductance and structural weight for an internal wall between two rooms of a house from the data provided: $R_{si} = 0.12 \, m^2$ K/W; 15 mm light plaster, $k = 0.16 \, W/m\,K$, $W = 600 \, kg/m^3$; 100 mm lightweight concrete block, $k = 0.19$, $W = 600 \, kg/m^3$; 15 mm light plaster, $k = 0.16 \, W/m\,K$, $W = 600 \, kg/m^3$.

The solution data is organized into a layout similar to that used for the worksheets, and is shown in Table 2.2.

Table 2.2   Solution data for Example 2.2

Layer material	l (mm)	k (W/m K)	R (m² K/W)	ρ (kg/m³)	W (kg/m²)
R_{si} internal surface			0.12		
Lightweight plaster	15	0.16	0.094	600	9
Light concrete block	100	0.19	0.526	600	60
Lightweight plaster	15	0.16	0.094	600	9
R_{si} inside surface			0.12		
			$\Sigma R = 0.954$		$\Sigma W = 78$

An example of the thermal resistance calculation is

$$\text{lightweight plaster, } R_1 = \frac{L_1}{10^3 \times K_1} \frac{m^2 \, K}{W}$$

$$= \frac{15}{10^3 \times 0.16} \frac{m^2 \, K}{W}$$

$$= 0.094 \, m^2 \, K/W$$

The thermal transmittance is

$$U = \frac{1}{\Sigma R} \frac{W}{m^2\,K}$$
$$= (0.12 + 0.094 + 0.526 + 0.094 + 0.12)^{-1}\,W/m^2\,K$$
$$= (0.954)^{-1}\,W/m^2\,K$$
$$= 1.048\,W/m^2\,K$$

The conductance of the wall is

$$C = \frac{1}{R - R_{so} - R_{si}} \frac{W}{m^2\,K}$$
$$= \frac{1}{0.954 - 0.12 - 0.12} \frac{W}{m^2\,K}$$
$$= 1.4\,W/m^2\,K$$

The weight of a layer of plaster is

$$W = L\rho \times \frac{1}{10^3} \frac{kg}{m^2}$$
$$= 15 \times 600 \times \frac{1}{10^3} \frac{kg}{m^2}$$
$$= 9\,kg/m^2$$

The overall weight of the internal wall, $\Sigma W = 78\,kg/m^2$.

Internal floors

An internal floor separates two rooms within a building.

EXAMPLE 2.3

Calculate the thermal transmittance, conductance and structural weight for an intermediate cast concrete floor in an office building, where heat flow is downwards to the office below, from the data provided: $R_{si} = 0.14\,m^2\,K/W$; 5 mm thermoplastic tile, $k = 0.5\,W/m\,K$, $W = 1050\,kg/m^3$; 50 mm cement screed, $k = 0.41\,W/m\,K$, $W = 1200\,kg/m^3$; 150 mm cast concrete, $k = 1.13\,W/m\,K$, $W = 2000\,kg/m^3$; $R_{si} = 0.14\,m^2\,K/W$.

The solution data is shown in Table 2.3.

Table 2.3 Solution data for Example 2.3

Layer material	l (mm)	k (W/m K)	R (m² K/W)	ρ (kg/m³)	W (kg/m²)
R_{si} internal surface			0.14		
Thermoplastic tile	5	0.5	0.01	1050	5
Cement screed	50	0.41	0.122	1200	60
Cast concrete	150	1.13	0.133	2000	300
R_{si} inside surface			0.14		
			$\Sigma R = 0.545$		$\Sigma W = 365$

$$U = (0.545)^{-1}\,W/m^2\,K$$
$$= 1.84\,W/m^2\,K$$

$$C = (0.545 - 0.14 - 0.14)^{-1}\,\text{W/m}^2\,\text{K}$$
$$= 3.77\,\text{W/m}^2\,\text{K}$$
$$W = 365\,\text{kg/m}^2$$

Ceilings

An internal ceiling separates two rooms within a building.

EXAMPLE 2.4

Calculate the thermal transmittance, conductance and structural weight for an internal ceiling in a multi-storey building where the heat flow is upwards, from the data provided: $R_{si} = 0.1\,\text{m}^2\,\text{K/W}$; 5 mm thermoplastic tile, $k = 0.5\,\text{W/m K}$, $W = 1050\,\text{kg/m}^3$; 50 mm cement screed, $k = 0.41\,\text{W/m K}$, $W = 1200\,\text{kg/m}^3$; 150 mm cast concrete, $k = 1.13\,\text{W/m K}$, $W = 2000\,\text{kg/m}^3$; $R_{si} = 0.1\,\text{m}^2\,\text{K/W}$.

The solution data is shown in Table 2.4.

Table 2.4   Solution data for Example 2.4

Layer material	l (mm)	k (W/m K)	R (m² K/W)	ρ (kg/m³)	W (kg/m²)
R_{si} internal surface			0.1		
Thermoplastic tile	5	0.5	0.01	1050	5
Cement screed	50	0.41	0.122	1200	60
Cast concrete	150	1.13	0.133	2000	300
R_{si} inside surface			0.1		
			$\Sigma R = 0.465$		$\Sigma W = 365$

$$U = (0.465)^{-1}\,\text{W/m}^2\,\text{K}$$
$$= 2.15\,\text{W/m}^2\,\text{K}$$
$$C = (0.465 - 0.1 - 0.1)^{-1}\,\text{W/m}^2\,\text{K}$$
$$= 3.77\,\text{W/m}^2\,\text{K}$$
$$W = 365\,\text{kg/m}^2$$

Flat roofs

A flat roof is calculated in the same manner as for any flat surface.

EXAMPLE 2.5

Calculate the thermal transmittance, conductance and structural weight for an external flat roof over a restaurant from the data provided: $R_{so} = 0.04\,\text{m}^2\,\text{K/W}$; 19 mm asphalt, $k = 0.5\,\text{W/m K}$, $W = 1700$, kg/m³; 75 mm cement screed, $k = 0.41\,\text{W/m K}$, $W = 1200\,\text{kg/m}^3$; 150 mm cast concrete, $k = 1.13\,\text{W/m K}$, $W = 2000\,\text{kg/m}^3$; 13 mm dense plaster $k = 0.5\,\text{W/m K}$, $W = 1300\,\text{kg/m}^3$; $R_{si} = 0.1\,\text{m}^2\,\text{K/W}$.

The solution data is shown in Table 2.5.

Table 2.5 Solution data for Example 2.5

Layer material	l (mm)	k (W/m K)	R (m^2 K/W)	ρ (kg/m^3)	W (kg/m^2)
R_{so} external surface			0.04		
Asphalt	19	0.5	0.038	1700	32
Cement screed	75	0.41	0.183	1200	90
Cast concrete	150	1.13	0.133	2000	300
Dense plaster	13	0.5	0.026	1300	17
R_{si} inside surface			0.1		
			$\Sigma R = 0.52$		$\Sigma W = 439$

$$U = (0.52)^{-1} \, \text{W/m}^2 \, \text{K}$$
$$= 1.92 \, \text{W/m}^2 \, \text{K}$$
$$C = (0.52 - 0.04 - 0.1)^{-1} \, \text{W/m}^2 \, \text{K}$$
$$= 2.63 \, \text{W/m}^2 \, \text{K}$$
$$W = 439 \, \text{kg/m}^2$$

Pitched roofs

The total thermal resistance of a pitched roof has three components; the sloping external surface and layers, an air space, and a horizontal ceiling. The flat ceiling and air space resistances are added, but the sloping surface is first multiplied by the cosine of the roof slope angle from the horizontal. Figure 2.2 shows the arrangement.

The thermal resistance of the sloping surface is reduced by the cosine of the roof pitch angle. The hypotenuse, or sloping side of the triangle, has a greater length, larger surface area and increased heat loss compared with that of the horizontal

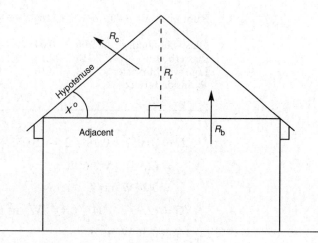

Fig. 2.2 Data for a pitched roof.

ceiling. The horizontal component of the hypotenuse thermal resistance is added to the horizontal ceiling thermal resistance.

$$\text{cosine } X° = \frac{\text{adjacent side}}{\text{hypotenuse}}$$

$$\text{adjacent} = \text{hypotenuse} \times \cos X°$$

The lengths of the sides of the triangle are not needed, because the same effect is obtained by using the thermal resistance of the two surfaces;

$$\text{modified } R_c = R_c \cos X° \text{ m}^2\text{ K/W}$$

The equation for the thermal resistance of a pitched roof is

$$R = R_c \cos X° + R_r + R_b \text{ m}^2\text{ K/W}$$

EXAMPLE 2.6

Calculate the thermal transmittance, conductance and structural weight for a tiled roof that has a pitch of 40° from the horizontal. The roof is on a two-storey house. Use only the data provided. $R_{so} = 0.04 \text{ m}^2\text{ K/W}$; 20 mm concrete tiles, $k = 0.84$ W/m K, $W = 1900 \text{ kg/m}^3$; 5 mm roofing felt, $k = 0.5$ W/m K, $W = 1700 \text{ kg/m}^3$; roof space $R_r = 0.18 \text{ m}^2\text{ K/W}$; 100 mm glass fibre quilt, $k = 0.04$ W/m K, $W = 12$ kg/m^3; 10 mm plasterboard, $k = 0.16$ W/m K, $W = 950 \text{ kg/m}^3$; 8 mm lightweight plaster, $k = 0.16$ W/m K, $W = 600 \text{ kg/m}^3$; $R_{si} = 0.1 \text{ m}^2\text{ K/W}$.

The solution data is shown in Table 2.6.

Table 2.6   Solution data for Example 2.6

Layer material	*l* (mm)	*k* (W/m K)	*R* (m² K/W)	ρ (kg/m³)	*W* (kg/m²)
R_{so} external surface			0.04		
Concrete tile	20	0.84	0.024	1900	38
Roofing felt	5	0.5	0.01	1700	9
			$\Sigma R_c = 0.074$		
Roof pitch $X° = 40°$		$R_c \cos X° = 0.057$			
R_r roof space			0.18		
Glass-fibre quilt	100	0.04	2.5	12	1
Plasterboard	10	0.16	0.063	950	10
Lightweight plaster	8	0.16	0.05	600	5
R_{si} inside surface			0.1		
			$\Sigma R = 2.95$		$\Sigma W = 63$

$$U = (0.057 + 0.18 + 2.5 + 0.063 + 0.05 + 0.1)^{-1} \text{ W/m}^2\text{ K}$$

$$= (2.95)^{-1} \text{ W/m}^2\text{ K}$$

$$= 0.34 \text{ W/m}^2\text{ K}$$

$$C = (2.95 - 0.04 - 0.1)^{-1} \text{ W/m}^2\text{ K}$$

$$= 0.36 \text{ W/m}^2\text{ K}$$

$$W = 63 \text{ kg/m}^2$$

Solid ground floors

The thermal transmittance $U \, \mathrm{W/m^2 \, K}$ for solid ground floors that are in contact with the earth and have four exposed edges is found from (CIBSE, 1986a)

$$U = \frac{2 \times 1.4}{0.5B \times \pi} \times \exp\left(\frac{0.5B}{L}\right) \times \mathrm{arctanh}\left(\frac{0.5B}{0.5B + 0.5W}\right)$$

where B = breadth, lesser dimension, of the floor (m); L = length, greater dimension, of the floor (m); W = width of the surrounding wall, usually 0.3 m; 1.4 = average thermal conductivity of earth in contact with the floor (W/m K); and the third term on the right-hand side is a dimensionless shape factor, S:

$$S = \frac{0.5B}{0.5B + 0.5W}$$

The shape factor S is easily calculated from the building data. All floors have four exposed edges, even when the area is very large. The thermal transmittance of perimeter zones is higher than that of inner parts of the building, but the effect of this is not easy to quantify.

EXAMPLE 2.7

Calculate the values of the shape factor for solid ground floors that are in contact with earth, for floors that range in plan size from 500 m × 100 m down to 4 m × 4 m. The floors are bounded by a 300 mm thick exterior wall.

$$S = \frac{0.5B}{0.5B + 0.5W}$$

For a very long floor that is 100 m wide, for example an aircraft hanger:

$$S = \frac{0.5 \times 100 \, \mathrm{m}}{0.5 \times 100 \, \mathrm{m} + 0.5 \times 0.3 \, \mathrm{m}}$$
$$= \frac{50 \, \mathrm{m}}{50 \, \mathrm{m} + 0.15 \, \mathrm{m}}$$
$$= 0.997 \text{ dimensionless}$$

For a very small floor of 4 m x 4 m wide, such as a single room that has four exposed edges to the floor:

$$S = \frac{0.5 \times 4 \, \mathrm{m}}{0.5 \times 4 \, \mathrm{m} + 0.5 \times 0.3 \, \mathrm{m}}$$
$$= \frac{2 \, \mathrm{m}}{2 \, \mathrm{m} + 0.15 \, \mathrm{m}}$$
$$= 0.93$$

The dimensionless shape factor S will almost always be between 0.93 and 0.997, and it will frequently be around 0.99.

The arctanh S function means the value of the trigonometric angle, in radians, which corresponds to the numerical value of the shape factor.

Consistency in the analysis of the mathematics will be maintained by using x in the place of the shape factor S until real data can be substituted:

$$\text{arctanh}\, S = \text{arctanh}\, x$$

$$\tanh S = \tanh x$$

Tanh is the hyperbolic tangent function, and arctanh is the inverse hyperbolic tangent. Some calculators and spreadsheets may be able to operate arctanh functions. These will appear as $\tanh^{-1} x$ on a key or in the spreadsheet functions list. Hyperbolic tangent, $\tanh x$, is not the same as tangent, $\tan x$.

$$\tanh^{-1} x = \text{arctanh}\, x$$

$$= X \text{ radians}$$

X radians is the angle that corresponds to the hyperbolic tangent value that is shown as x. As many users will not have access to an easy solution to this function, it will be simplified and applied to any method of calculation and spreadsheet program.

A hyperbolic tangent is defined as (Bird and May, 1991)

$$\tanh x = \frac{e^x - e^{-x}}{e^x + e^{-x}}$$

The exponential e is 2.718 28, and it can be viewed on a scientific calculator by entering 1 in the display and then pressing the e^x key; e^{-x} is equal to the reciprocal of e^x:

$$e^{-x} = \frac{1}{e^x}$$

To find the value of X radians:

$$X = \text{arctanh}\left(\frac{0.5B}{0.5B + 0.5W}\right)$$

$$= \text{arctanh}\, S$$

The exponential ratio for $\tanh x$ is used, with S as the variable:

$$\tanh S = \frac{e^S - e^{-S}}{e^S + e^{-S}}$$

The value of S is known, as it is the dimensionless shape factor from the building data. This allows the hyperbolic tangent value to be calculated.

EXAMPLE 2.8

Calculate the hyperbolic tangent for a $4\,\text{m} \times 4\,\text{m}$ floor when the dimensionless shape factor is 0.93.

$S = 0.93$, so

$$\tanh S = \frac{e^S - e^{-S}}{e^S + e^{-S}}$$

$$= \frac{e^{0.93} - e^{-0.93}}{e^{0.93} + e^{-0.93}}$$

$$= \frac{2.5345 - 0.3946}{2.5345 + 0.3946}$$

$$= \frac{2.1399}{2.9291}$$

$$= 0.7306$$

These calculations have been rounded to the first four decimal places.

Arctanh S is the angle X radians that corresponds to the known hyperbolic tangent value, 0.7306 in Example 2.8. It is found by further use of the definition of the hyperbolic tangent:

$$\tanh x = \frac{e^{-x} - e^{-x}}{e^{x} + e^{-x}}$$

where $\tanh x$ is known but the value of x is unknown. The unknown x appears too many times for a simple solution, so multiply both sides of the equation by the same thing, so that the balance and value of the equation are not altered. It is convenient to use e^{-x} in this manner:

$$\tanh x \times \frac{e^{-x}}{e^{-x}} = \frac{e^{x} - e^{-x}}{e^{x} + e^{-x}} \times \frac{e^{-x}}{e^{-x}}$$

Notice that both sides of the equation are multiplied by the same fraction, e^{-x}/e^{-x}, which is equal to unity. Multiply out the right-hand side. Remember that when numbers of the same base, e in this case, are multiplied, their powers are added:

$$\tanh x \times 1 = \frac{(e^{x} - e^{-x}) \times e^{-x}}{(e^{x} + e^{-x}) \times e^{-x}}$$

$$\tanh x = \frac{e^{x} \times e^{-x} - e^{-x} \times e^{-x}}{e^{x} \times e^{-x} + e^{-x} \times e^{-x}}$$

$$= \frac{e^{0} - e^{-2x}}{e^{0} + e^{-2x}}$$

and as $e^{0} = 1$,

$$\tanh = \frac{1 - e^{-2x}}{1 + e^{-2x}}$$

This can be simplified further by substituting any symbol (Z will be used for convenience) for e^{-2x}.

$$\tanh(x) = \frac{1 - Z}{1 + Z}$$

where $Z = e^{-2x}$

The value of Z needs to be found, and then x. Multiply the denominator across to the other side of the equality, simplify the equation and find the value of Z:

$$(1 + Z) \times \tanh x = 1 - Z$$

Remove the brackets:

$$\tanh x + Z \times \tanh x = 1 - Z$$

Collect the Z terms together:

$$Z + Z \times \tanh x = 1 - \tanh x$$

Take Z outside a bracket:

$$Z \times (1 + \tanh x) = 1 - \tanh x$$

Divide both sides by $(1 + \tanh x)$:

$$Z = \frac{1 - \tanh x}{1 + \tanh x}$$

but $Z = e^{-2x}$, so

$$e^{-2x} = \frac{1 - \tanh x}{1 + \tanh x}$$

This is an equation that is in the standard form of a logarithm, $e^x = N$, and this can be written as

$$\log_e N = x$$

where

$$N = \frac{1 - \tanh x}{1 + \tanh x}$$

and the power, or logarithm, is $-2x$, not just x.

A logarithm to base e is a natural logarithm, $\ln x$, so

$$\ln \left(\frac{1 - \tanh x}{1 + \tanh x} \right) = -2x$$

Take the -2 to the other side:

$$x = -\tfrac{1}{2} \times \ln \left(\frac{1 - \tanh x}{1 + \tanh x} \right)$$

In this application, the value of $\tanh x$ is the known shape factor S, and the angle x is that which satisfies the specific value of S. This means that angle x is the inverse hyperbolic function of $\tanh x$, or arctanh x. For practical use, the final equation will be written as

$$X = -\tfrac{1}{2} \times \ln \left(\frac{1 - S}{1 + S} \right) \text{ radians}$$

EXAMPLE 2.9

Calculate the thermal transmittance of a solid ground floor that is 10 m long and 6 m wide, when the surrounding walls have a thickness of 300 mm and the floor has four exposed edges.

$$U = -\frac{2 \times 1.4}{0.5B \times \pi} \times \exp\left(\frac{0.5B}{L} \right) \times \text{arctanh} \left(\frac{0.5B}{0.5B + 0.5W} \right)$$

The shape factor is

$$S = \frac{0.5B}{0.5B + 0.5W}$$

$$= \frac{0.5 \times 6}{0.5 \times 6 + 0.5 \times 0.3}$$

$$= 0.952$$

$$\text{arctanh}\left(\frac{0.5B}{0.5B + 0.5W}\right) = X \text{ radians}$$

so

$$\text{arctanh } S = X \text{ radians}$$

and

$$X = -\tfrac{1}{2} \times \ln\left(\frac{1 - S}{1 + S}\right) \text{ radians}$$

$$= -\tfrac{1}{2} \times \ln\left(\frac{1 - 0.952}{1 + 0.952}\right) \text{ radians}$$

$$= -\tfrac{1}{2} \times \ln(0.024\,59) \text{ radians}$$

$$= -0.5 \times (-3.7054) \text{ radians}$$

$$= 1.853 \text{ radians}$$

$$U = \frac{2 \times 1.4}{0.5B \times \pi} \times \exp\left(\frac{0.5B}{L}\right) \times \text{arctanh}\left(\frac{0.5B}{0.5B + 0.5W}\right)$$

$$= \frac{2 \times 1.4}{0.5 \times 6 \times \pi} \times \exp\left(\frac{0.5 \times 6}{10}\right) \times \text{arctanh } S$$

$$= 0.297 \times \exp(0.3) \times X$$

$$= 0.297 \times e^{0.3} \times 1.853 \text{ W/m}^2 \text{ K}$$

$$= 0.743 \text{ W/m}^2 \text{ K}$$

$$= 0.74 \text{ W/m}^2 \text{ K}$$

to the normal two significant decimal places. This agrees with the thermal transmittance published in CIBSE (1986a).

EXAMPLE 2.10

Calculate the thermal transmittance of a solid ground floor that is 40 m long and 20 m wide, when the surrounding walls have a thickness of 300 mm and the floor has four exposed edges.

The explanations are not given for this example.

$$S = \frac{0.5B}{0.5B + 0.5W}$$

$$= \frac{0.5 \times 10}{0.5 \times 10 + 0.5 \times 0.3}$$

$$= 0.971$$

$$X = -\tfrac{1}{2} \times \ln\left(\frac{1 - S}{1 + S}\right) \text{ radians}$$

$$= -\tfrac{1}{2} \times \ln\left(\frac{1 - 0.971}{1 + 0.971}\right) \text{ radians}$$

$$= 2.11 \text{ radians}$$

$$U = \frac{2 \times 1.4}{0.5 \times 10 \times \pi} \times \exp\left(\frac{0.5 \times 10}{20}\right) \times 2.11$$

$$= 0.48\,\text{W/m}^2\,\text{K}$$

Suspended ground floors

Suspended ground floors may be constructed from timber or a reinforced concrete slab. A ventilated air space separates the floor from the earth, gravel or concrete foundations. The thermal resistance of the solid earth path to the outdoor air is found in the same manner as for a solid concrete floor. The thermal resistance of the suspended flooring materials is found as it would be for any other plane slab construction. Heat is emitted from the underside of the warmed suspended floor by radiation and convection across the ventilated underfloor cavity. Figure 2.3 shows the data used.

The variables employed are:

R_c = equivalent thermal resistance due to convection (m^2 K/W)
R_e = thermal resistance of the earth path (m^2 K/W)
R_g = thermal resistance of the suspended floor (m^2 K/W)
R_r = equivalent thermal resistance due to radiation (m^2 K/W)
R_{si} = thermal resistance of the floor surface (m^2 K/W)
R_{t1} = transform thermal resistance (m^2 K/W)
R_{t2} = transform thermal resistance (m^2 K/W)
R_v = equivalent thermal resistance due to ventilation (m^2 K/W)

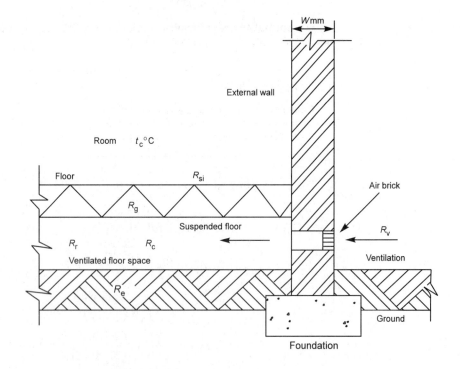

Fig. 2.3 Data for a suspended ground floor.

Some of these have fixed values for most applications:

$R_{si} = 0.14\,\mathrm{m^2\,K/W}$ for a high-emissivity floor with heat flow in the downward direction

$R_r = 0.2\,\mathrm{m^2\,K/W}$

$R_c = 0.67\,\mathrm{m^2\,K/W}$

$$R_{t1} = \frac{R_r R_c}{R_r + 2R_c}\ \mathrm{m^2\,K/W}$$

$$= \frac{0.2 \times 0.67}{0.2 + 2 \times 0.67}\ \mathrm{m^2\,K/W}$$

$$= 0.09\,\mathrm{m^2\,K/W}$$

$$R_{t2} = \frac{R_c^2}{R_r + 2R_c}\ \mathrm{m^2\,K/W}$$

$$= \frac{(0.67)^2}{0.2 + 2 \times 0.67}\ \mathrm{m^2\,K/W}$$

$$= 0.29\,\mathrm{m^2\,K/W}$$

$R_v = 0.63B\,\mathrm{m^2\,K/W}$

$B = $ breadth of floor(m)

The thermal transmittance of the solid floor included the surface film resistance R_{si} although this was not explicitly stated. The thermal resistance of the earth path is found by subtracting the surface film component of the overall resistance:

$$R_e = \frac{1}{U} - R_{si}\ \mathrm{m^2\,K/W}$$

The total thermal resistance of the suspended floor is found from

$$R = R_{si} + R_g + R_{t1} + \left(\frac{1}{R_{t1} + R_e} + \frac{1}{R_{t2} + R_v}\right)^{-1}\ \mathrm{m^2\,K/W}$$

EXAMPLE 2.11

A house has floor plan dimensions of $10\,\mathrm{m} \times 10\,\mathrm{km}$. The floor consists of 12 mm thick timber floorboards that are supported on wooden joists, a ventilated floor cavity, and an over-site concrete slab. The thermal conductivity of the floor board is $0.14\,\mathrm{W/m\,K}$. The thermal transmittance of the solid floor, which is in contact with the earth, is $0.62\,\mathrm{W/m^2\,K}$. Calculate the overall thermal transmittance of the bare board floor.

Take the normal values:

$R_{si} = 0.14\,\mathrm{m^2\,K/W}$

$R_r = 0.2\,\mathrm{m^2\,K/W}$

$R_c = 0.67\,\mathrm{m^2\,K/W}$

$R_{t1} = 0.09\,\mathrm{m^2\,K/W}$

$R_{t2} = 0.29\,\mathrm{m^2\,K/W}$

$R_v = 0.63 \times B\,\mathrm{m^2\,K/W}$

$$B = 10 \, \text{m}$$
$$R_v = 0.63 \times 10 \, \text{m}^2 \, \text{K/W}$$
$$= 6.3 \, \text{m}^2 \, \text{K/W}$$

The solid floor thermal transmittance is $0.62 \, \text{W/m}^2 \, \text{K}$, and its resistance is

$$R_e = \frac{1}{U} - R_{si} \, \text{m}^2 \, \text{K/W}$$

$$= \frac{1}{0.62} - 0.14 \, \text{m}^2 \, \text{K/W}$$

$$= 1.473 \, \text{m}^2 \, \text{K/W}$$

The resistance R_g is due only to the timber floorboards;

$$R_g = \frac{0.012 \, \text{m}}{0.14} \times \frac{\text{m K}}{\text{W}}$$

$$= 0.086 \, \text{m}^2 \, \text{K/W}$$

The total thermal resistance of the suspended floor is

$$R = R_{si} + R_g + R_{t1} + \left(\frac{1}{R_{t1} + R_e} + \frac{1}{R_{t2} + R_v} \right)^{-1} \text{m}^2 \, \text{K/W}$$

$$= 0.14 + 0.086 + 0.09 + \left(\frac{1}{0.09 + 1.473} + \frac{1}{0.29 + 6.3} \right)^{-1} \text{m}^2 \, \text{K/W}$$

$$= 0.316 + (0.64 + 0.152)^{-1} \, \text{m}^2 \, \text{K/W}$$

$$= 1.579 \, \text{m}^2 \, \text{K/W}$$

$$U = \frac{1}{R} \, \text{W/m}^2 \, \text{K}$$

$$= \frac{1}{1.579} \, \text{W/m}^2 \, \text{K}$$

$$= 0.63 \, \text{W/m}^2 \, \text{K}$$

EXAMPLE 2.12

The $10 \, \text{m} \times 10 \, \text{m}$ bare board suspended timber floor in the house in Example 2.11 is to be insulated with 100 mm thick glass-fibre quilt, which will be laid onto nylon netting between the joists. Calculate the thermal transmittance of the insulated floor. The thermal conductivity of glass-fibre quilt is $0.04 \, \text{W/m K}$.

The thermal resistance of the timber floor and insulation is

$$R_g = \frac{0.012 \, \text{m}}{0.14 \, \text{W/m K}} + \frac{0.1 \, \text{m}}{0.04 \, \text{W/m K}}$$

$$= 2.59 \, \text{m}^2 \, \text{K/W}$$

All the other values remain unaltered. The new resistance is found from

$$\text{new } R = \text{old } R - \text{old } R_g + \text{new } R_g$$

$$R = 1.579 - 0.086 + 2.59 \, \text{m}^2 \, \text{K/W}$$

$$= 4.083 \, \text{m}^2 \, \text{K/W}$$

$$U = \frac{1}{4.083} \, \text{W/m}^2 \, \text{K}$$

$$= 0.25 \, \text{W/m}^2 \, \text{K}$$

EXAMPLE 2.13

A community hall has floor plan dimensions of 60 m × 40 m and a wall thickness of 300 mm. The floor consists of 15 mm thick timber floorboards, which are supported on wooden joists, 75 mm thick expanded polystyrene slab, a ventilated floor cavity, and an over-site concrete slab. The thermal conductivity of the floorboards is 0.14 W/m K, and that of the polystyrene is 0.035 W/m K. Calculate the overall thermal transmittance of the floor.

For the solid floor component:

$$S = \frac{0.5B}{0.5B + 0.5W}$$

$$= \frac{0.5 \times 40}{0.5 \times 40 + 0.5 \times 0.3}$$

$$= 0.993$$

$$X = -\tfrac{1}{2} \times \ln\left(\frac{1 - S}{1 + S}\right) \text{ radians}$$

$$= -\tfrac{1}{2} \times \ln\left(\frac{1 - 0.993}{1 + 0.993}\right) \text{ radians}$$

$$= 2.83 \text{ radians}$$

$$U = \frac{2 \times 1.4}{0.5 \times 40 \times \pi} \times \exp\left(\frac{0.5 \times 40}{60}\right) \times 2.83$$

$$= 0.176 \, \text{W/m}^2 \, \text{K}$$

Use the normal values to find the overall thermal transmittance:

$R_{si} = 0.14 \, \text{m}^2 \, \text{K/W}$
$R_r = 0.2 \, \text{m}^2 \, \text{K/W}$
$R_c = 0.67 \, \text{m}^2 \, \text{K/W}$
$R_{t1} = 0.09 \, \text{m}^2 \, \text{K/W}$
$R_{t2} = 0.29 \, \text{m}^2 \, \text{K/W}$
$R_v = 0.63B \, \text{m}^2 \, \text{K/W}$
$B = 40 \, \text{m}$
$R_v = 0.63 \times 40 \, \text{m}^2 \, \text{K/W}$
$\quad = 25.2 \, \text{m}^2 \, \text{K/W}$

The solid floor thermal transmittance is $0.176 \, \text{W/m}^2 \, \text{K}$, and its resistance is

$$R_e = \frac{1}{U} - R_{si} \, \text{m}^2 \, \text{K/W}$$

$$= \frac{1}{0.176} - 0.14 \, \text{m}^2 \, \text{K/W}$$

$$= 5.54 \, \text{m}^2 \, \text{K/W}$$

The resistance R_g of the timber floor and polystyrene is

$$R_g = \frac{0.015}{0.14} + \frac{0.075}{0.035} \text{ m}^2 \text{ K/W}$$
$$= 2.25 \text{ m}^2 \text{ K/W}$$

The total thermal resistance of the suspended floor is

$$R = R_{si} + R_g + R_{t1} + \left(\frac{1}{R_{t1} + R_e} + \frac{1}{R_{t2} + R_v}\right)^{-1} \text{ m}^2 \text{ K/W}$$

$$= 0.14 + 2.25 + 0.09 + \left(\frac{1}{0.09 + 5.54} + \frac{1}{0.29 + 25.2}\right)^{-1} \text{ m}^2 \text{ K/W}$$

$$= 7.09 \text{ m}^2 \text{ K/W}$$

$$U = \frac{1}{R} \frac{\text{W}}{\text{m}^2 \text{ K}}$$

$$U = \frac{1}{7.09} \frac{\text{W}}{\text{m}^2 \text{ K}}$$

$$= 0.14 \text{ W/m}^2 \text{ K}$$

Single-, double- and triple-glazed framed window or door

Glazed windows and doors are complex openings through an external structural wall. They comprise frame and glazing materials that may be present in several layers and have air spaces. The designer usually needs to know the mean thermal transmittance, conductance and weight of the composite assembly. These are found by the proportional area method. For the U value:

$$U = [F\% \times U_f + (100 - F\%) \times U_g] \times \frac{1}{100} \frac{\text{W}}{\text{m}^2\text{K}}$$

where $F\%$ = percentage of wall opening occupied by the frame; U_f = thermal transmittance of the frame (W/m^2 K); and U_g = thermal transmittance of the glazing (W/m^2 K).

EXAMPLE 2.14

Calculate the thermal transmittance, conductance and structural weight of a single-glazed window in a beechwood frame. The frame occupies 10% of the wall opening. Use the following data: $R_{so} = 0.06 \text{ m}^2 \text{ K/W}$; 32 mm beech frame, $k = 0.165 \text{ W/m K}$, $W = 600 \text{ kg/m}^3$; 4 mm plane glass, $k = 1.05 \text{ W/m K}$, $W = 2500 \text{ kg/m}^3$; $R_{si} = 0.12 \text{ m}^2 \text{ K/W}$.

The solution data is shown in Table 2.7.

$$U_f = (0.374)^{-1} \text{ W/m}^2 \text{ K}$$
$$= 2.67 \text{ W/m}^2 \text{ K}$$
$$U_g = (0.184)^{-1} \text{ W/m}^2 \text{ K}$$
$$= 5.44 \text{ W/m}^2 \text{ K}$$

Table 2.7 Solution data for Example 2.14

Layer material	*l* (mm)	*k* (W/m K)	*R* (m² K/W)	*ρ* (kg/m³)	*W* (kg/m²)
Window frame:					
R_{so} external surface			0.06		
Beech frame	32	0.165	0.194	700	22
R_{si} inside surface			0.12		
		frame $\Sigma R = 0.374$			$\Sigma W = 22$
Glass:					
R_{so} external surface			0.06		
Plane glass	4	1.05	0.0038	2500	10
S_{si} inside surface			0.12		
		glass $\Sigma R = 0.184$			$\Sigma W = 10$

$$U = [F\% \times U_f + (100 - F\%) \times U_g] \times \frac{1}{100} \frac{W}{m^2\,K}$$

$$= (10 \times 2.67 + (100 - 10) \times 5.44) \times \frac{1}{100} \frac{W}{m^2\,K}$$

$$= 5.16\,W/m^2\,K$$

$$C_f = (0.374 - 0.06 - 0.12)^{-1}\,W/m^2\,K$$

$$= 5.15\,W/m^2\,K$$

$$C_g = (0.184 - 0.06 - 0.12)^{-1}\,W/m^2\,K$$

$$= 262.5\,W/m^2\,K$$

$$W = [F\% \times W_f + (100 - F\%) \times W_g] \times \frac{1}{100} \frac{kg}{m^2}$$

$$= (100 \times 22 + (100 - 10) \times 10) \times \frac{1}{100} \frac{kg}{m^2}$$

$$= 11.2\,kg/m^2$$

DATA REQUIREMENT

The user will enter new data for:

1. Your name in cell B4.
2. Name of the job or assignment in cell B5.
3. Reference number or job details in cell B6.
4. Name of the file in cell B9.

Flat surfaces

5. Data entry for external wall 1: external surface thermal resistance in cell E33.
6. Description of the first external layer of material for wall 1 in cell B34.
7. Thickness of material layer 1 in millimetres in cell C34.
8. Thermal conductivity of material layer 1 in cell D34.
9. Density of material 1 in cell F34.

10. Repeat the entries for the second and third layers of material in the inward direction from the external surface.
11. Where layers of material do not exist, leave the material name cell in column B blank. Ensure that there is zero in the thickness cell in column C and that the thermal conductivity has a value of 1.0 W/m K in column D, otherwise an error is shown in the resistance column E as ERR. This would be due to the division of thickness by zero thermal conductivity.
12. Thermal resistance of the air space in cell E37; zero if it does not exist.
13. Repeat the entries for material layers 4, 5 and 6 as for layer 1.
14. Internal wall surface thermal resistance in cell E41.
15. Repeat the corresponding entries for external walls 2, 3 and 4, internal wall types 1 and 2, internal floor types 1 and 2, internal ceiling types 1 and 2, and external flat roof types 1 and 2, in the cell range A55 to F327. Remember that internal walls, floors and ceilings have an R_{si} value on each side of the structure. The correct terms and symbols are on the worksheet.
16. Repeat the previous data entries for external floor 1 in the cell range A341 to F353. An external floor is one that has the underside exposed to the outdoor air. It is not normally a ground floor, because this has a ventilated air space that is warmer than the outdoor air. An external floor is usually a first or higher floor with a pedestrian concourse, highway or car park beneath it.

Pitched roof

17. Roof pitch angle from the horizontal in degrees in cell D369.
18. External surface resistance of the roof in cell E373.
19. Description of the first exterior layer of sloping material in cell B374.
20. Thickness of sloping material 1 in cell C374.
21. Thermal conductivity of sloping material 1 in cell D374.
22. Density of sloping material 1 in cell F374.
23. Repetition of data entry for sloping materials 2, 3 and 4.
24. Thermal resistance of the roof space between the sloping and flat layers in cell E381.
25. Repeat data entry for horizontal materials 5, 6, 7 and 8.
26. Thermal resistance of the internal surface in cell E387.
27. Repeat the data entry for pitched roof 2 in the cell range A404 to F424.

Solid ground floor

28. Length of floor in cell D448.
29. Breadth of the floor in cell D447.
30. Reduction in thermal transmittance for edge insulation in cell D450.
31. Thickness of the wall that surrounds the solid floor in cell D453.

Suspended ground floor

32. Length of the floor in cell D470.
33. Breadth of the floor in cell D471.
34. Reduction in thermal transmittance for edge insulation in cell D472.
35. Thickness of the wall that surrounds the floor in cell D474.

36. Description of the first layer of material commencing with that which is nearest to the outdoor air in cell B491.
37. Thickness of material layer 1 in cell C491.
38. Thermal conductivity of material 1 in cell D491.
39. Density of material 1 in cell F491.
40. Repeat data entry for material layers 2–5 in cells B492 to F495. The final layer is the internal surface finish, such as carpet.
41. Thermal resistance of the internal surface in cell E496.

Single-glazed window or door

42. The percentage of the opening in the wall that comprises the frame of the window or door in cell D525. A frame is that part of the construction that is not glazing. Windows are likely to have around 10% of the wall opening taken up by the frame. A door may be all of one type of material, such as wood, glass, metal or a plastic, or it may have glazed panels within a frame.
43. Thermal resistance of the external surface in cell E529.
44. Description of the exterior material of the window frame in cell B530.
45. Thickness of the exterior material of the window frame in cell C530.
46. Thermal conductivity of material layer 1 in cell D530.
47. Density of material layer 1 in cell F530.
48. Description of any joining or thermal break material, layer 2, between the interior and exterior parts of the frame in cell B531.
49. Thickness of layer 2 in cell C531.
50. Thermal conductivity of layer 2 in cell D531.
51. Density of layer 2 in cell F531.
52. Repetition of the data for the interior materials of the frame in cells B532 to F532.
53. Internal surface thermal resistance for the frame in cell E533.
54. External surface thermal resistance of the glass in cell E539.
55. Description of the glass in cell B540.
56. Thickness of the glass in cell C540.
57. Thermal conductivity of the glass in cell D540.
58. Density of the glass in cell F540.
59. Internal surface thermal resistance in cell E541.
60. Repetition of the data for double- and triple-glazed windows and doors in the cell range A562 to F650.

OUTPUT DATA

The output data is as follows.

1. The current date is given in cell B7.
2. The time of opening this worksheet or when it was last used to calculate with new data is given in cell B8.

External wall type 1

3. The thermal resistance of the layers of material is given in cell range E34 to E40.

4. The total thermal resistance of the structure is given in cell D43.
5. The thermal transmittance of the structure is given in cell D45.
6. The conductance between the internal and external surfaces of the structure is given in cell D47.
7. The weight of the structure in kilograms per square metre of surface area is given in cell D49.
8. The check that compares the thermal transmittance with the lowest expected value is given in cell C51.
9. The check that compares the thermal transmittance with the highest expected value is given in cell C52. The output data for external wall types 2, 3 and 4 is given in the cell range E60 to C130. All the further output data is arranged in the same format as for an external wall except where the differences are stated.

Internal wall types 1 and 2

The output data is given in the cell range E138 to C182.

Internal floor types 1 and 2

The output data is given in the cell range E190 to C234.

Internal ceiling types 1 and 2

The output data is given in the cell range E242 to C286.

External flat roof types 1 and 2

The output data given in the cell range E294 to C338.

External floor

The output data is given in the cell range E346 to C364.

Pitched roof type 1

10. The resistance of the sloping materials is given in the cell range E374 to E377.
11. The total thermal resistance of the sloping layers of material and external surface is given in cell E379.
12. The resistance of the horizontal layers of material is given in the cell range E383 to E386.
13. The total thermal resistance of the horizontal layers of material and the internal surface is given in cell E389.
14. The resistance, transmittance, conductance, weight and validity check are shown in the cell range D399 to C401. The calculation of the weight of the roof disregards the slope angle.
15. The output data for pitched roof type 2 is given in the cell range E411 to C438.

Solid ground floor

16. The shape factor is shown in cell D452.
17. The thermal transmittance of the floor is given in cell D456.
18. The lower and upper validity checks on the thermal transmittance are shown in cells C458 and C459.

Suspended floor

19. The shape factor is given in cell D473.
20. The thermal transmittance of the earth path is given in cell D475.
21. The thermal resistance of the earth path is given in D476.
22. The thermal resistance of the flooring materials is given in the cell range E491 to E495.
23. The thermal resistance of the suspended floor slab materials and internal surface resistance is shown in cell D499.
24. The thermal resistance of the whole floor is given in cell D501.
25. The thermal transmittance of the whole floor is given in cell D503.
26. The conductance of the floor is shown in cell D505.
27. The structural weight for $1\,\text{m}^2$ of floor surface area is given in cell D507.
28. The lower and upper validity checks on the thermal transmittance are shown in cells C509 and C510.

Single-glazed windows

29. The thermal resistance of the frame materials is given in the cell range E530 to E532.
30. The thermal resistance of the glazing is shown in cell E540.
31. The thermal transmittance of the frame is shown in cell D546.
32. The thermal transmittance of the glazing is shown in cell D548.
33. The average value of the thermal transmittance for the whole window opening in the structural wall is shown in cell D550.
34. The conductance of the glazing is shown in cell D552.
35. The conductance of the frame is shown in cell D554.
36. The structural weight per square metre of window surface area for the complete glazed window is shown in cell D556.
37. The lower and upper validity checks on the window thermal transmittance are shown in cells C558 and C559.

Double- and triple-glazed windows and doors

The output data is shown in the cell range E579 to C668.

FORMULAE

The worksheet is formatted to display three decimal places for resistance and transmittance so that the user can decide the significance of the third place. Thermal transmittance should be expressed only to the first two decimal places after its initial calculation. At least one case of each type of different formula is explained: for example, the calculation of the thermal resistance of a material. Such calculations are repeated throughout the worksheet, and they only have different line numbers. The block of output data for each complete structural element is similarly repetitious. There is a worked example of each formula through application to a practical case. These examples are on the original copy of the file UVALUE.WK1.

Cell B7

@TODAY

The @TODAY date and time function produces the serial number of the current date and time. This cell has been given the date format so that the date of use is displayed automatically.

Cell B8

@TODAY

This cell has the time of day format. Whenever the worksheet is opened or new data is entered for calculation, the current time is updated.

External wall

Cell E34

(F3) +C34/1000/D34

The thermal resistance of the first layer of material is calculated. The cell range E34 to E649 has identical formulae for the appropriate line number. These cells are formatted to three decimal places.

$$R = L \text{ mm} \times \frac{1 \text{ m}}{1000 \text{ mm}} \times \frac{1}{k} \frac{\text{m K}}{\text{W}}$$

$$= \frac{L}{1000k} \frac{\text{m}^2 \text{ K}}{\text{W}}$$

$$E34 = \frac{C34}{1000 \times D34} \frac{\text{m}^2 \text{ K}}{\text{W}}$$

Cell D43

(F3) @SUM (E33..E41)

The thermal resistance of external wall 1 is found by summation of the cells that contain the resistance of up to six layers of material, the surface and air space resistances:

$$\Sigma R = R_{so} + R_1 + \ldots + R_a + \ldots + R_6 + R_{si} \text{ m}^2 \text{ K/W}$$

$$D43 = E33 + E34 + \ldots + E41 \text{ m}^2 \text{ K/W}$$

$$= @SUM(E33..E41) \text{ m}^2 \text{ K/W}$$

Cell D45

(F3) 1/D43

The thermal transmittance is found from the reciprocal of the overall resistance.

$$U = \frac{1}{R} \frac{\text{W}}{\text{m}^2 \text{K}}$$

$$D45 = \frac{1}{D43} \frac{\text{W}}{\text{m}^2 \text{K}}$$

Cell D47

(F3) 1/(D43−E41−E33)

The thermal conductance of the complete structural element is found by subtracting the internal and external surface resistances from the overall resistance. The conductance is now a constant for the structure wherever it may be built. The internal surface resistance depends upon the emissivity of the indoor surface finish, its orientation, the room air velocity over the surface, and the presence of radiant heat exchanges within the room. Similar heat exchange variables also affect the external surface, except that it is the outdoor environment, with solar radiation, wind and rain, that is involved. The conductance can be seen to be independent of such considerations.

$$C = \frac{1}{R - R_{\text{so}} - R_{\text{si}}} \frac{\text{W}}{\text{m}^2 \text{K}}$$

$$D47 = \frac{1}{D43 - E33 - E41} \frac{\text{W}}{\text{m}^2 \text{K}}$$

Cell D49

(F1) (C34*F34+C35*F35+C38*F38+C39*F39+C40*F40)/1000

The weight of the complete structure per unit surface area is calculated. This will be used by the structural engineer to find the forces upon the supporting floor slabs, beams, columns and foundations for their design. The cell is formatted to only one decimal place. For a single material:

$$W = L\,\text{mm} \times \rho \frac{\text{kg}}{\text{m}^3} \times \frac{1\,\text{m}}{10^3\,\text{mm}}$$

$$= \frac{L\rho}{10^3} \frac{\text{kg}}{\text{m}^2} \text{ of wall surface area}$$

$$= (L_1\rho_1 + \ldots + L_6\rho_6) \times \frac{1}{10^3} \frac{\text{kg}}{\text{m}^2}$$

$$D49 = (C34 \times F34 + \ldots + C40 \times F40) \times \frac{1}{10^3} \frac{\text{kg}}{\text{m}^2}$$

Cell C51

@IF (D45<=0.1,"TOO LOW","above 0.1 W/m2 K, OK.")

The thermal transmittance is verified to be within possible limits by comparison with adjustable lower and upper values. The lower possible limit for an external wall has been set to $0.1\,\text{W/m}^2\,\text{K}$. The user may choose to raise or reduce this, depending upon current design criteria. If the Building Regulations or codes of practice specify an upper limit of, say, $0.4\,\text{W/m}^2\,\text{K}$ for the thermal transmittance of a wall, then it is unlikely that the designer would wish to utilize a structure that is insulated to better than $0.3\,\text{W/m}^2\,\text{K}$. This could be set for the lower validity check. To edit the cell and change the lower limit:

1. Move the cursor to cell C51.
2. Press the F2 function key to bring the formula into the edit command line.
3. Move the cursor leftwards to the space between the 0.1 and the letter W.
4. Press the delete key to remove the number 1.
5. Type the number desired, such as 3.
6. Move the cursor leftwards again and replace the number 1 that is just prior to "TOO LOW".
7. Replace the 1 with the same number that was just used, 3 in this case.
8. When the formula is correct, press the enter key and the edited formula is used.
9. Save the edited version of the worksheet under a new file name.

The formula compares the thermal transmittance against a low limit with an @IF command. If the thermal transmittance is equal to or less than $0.1\,\text{W/m}^2\,\text{K}$, it is too low, and a declaration is produced on the screen. If the thermal transmittance is above $0.1\,\text{W/m}^2\,\text{K}$, it satisfies this test and is deemed to be acceptable. The @IF command causes one of two outputs depending upon whether the test is passed or failed. Immediately after making the test $U \leqslant 0.1\,\text{W/m}^2\,\text{K}$, a 'Yes, true' outcome will cause the next statement "TOO LOW" to be printed on the screen in cell C51. A 'No, false' outcome will cause the next statement "above 0.1 W/m2 K, OK" to be printed on the screen in cell C51.

Cell C52

@IF (D45>=2, "TOO HIGH", "below 2 W/m2 K, OK.")

The upper possible limit for an external wall has been set to $2\,\text{W/m}^2\,\text{K}$. The user may change this depending upon current design criteria. If the Building Regulations or codes of practice state an upper limit of, say, $0.4\,\text{W/m}^2\,\text{K}$ for the thermal transmittance of a wall, then it is unlikely that the designer would wish to utilize a structure that is worse than $0.4\,\text{W/m}^2\,\text{K}$. This could be set for the upper validity check. To edit the cell and change the upper limit, repeat the steps as for the lower limit, putting 0.4 in place of the 2, and save the new file that has been created.

The formula compares the thermal transmittance with a high limit with an @IF command. If the thermal transmittance is equal to or greater than $2\,\text{W/m}^2\,\text{K}$, it is too high, and a declaration is produced on the screen. If the thermal transmittance is below $2\,\text{W/m}^2\,\text{K}$, it satisfies this test and is deemed to be acceptable. The @IF command causes one of two outputs depending upon whether the test is passed or failed. Immediately after making the test $U \geqslant 2\,\text{W/m}^2\,\text{K}$, a 'Yes, true' outcome will cause the next statement "TOO HIGH" to be printed on the screen in cell C52. A 'No, false' outcome will cause the next statement "below 2 W/m2 K, OK" to be printed on the screen in cell C52.

Walls, floors and ceilings

Cell range E60 to C364

The formulae in these cells repeat those for external wall type 1 and are applied to walls, floors and ceilings.

Pitched roof

Cell E374

(F3) +C374/1000/D374

The thermal resistance of the first external layer of material on a pitched roof is calculated:

$$R = L\,\text{mm} \times \frac{1\,\text{m}}{10^3\,\text{mm}} \times \frac{1}{k}\,\frac{\text{m K}}{\text{W}}$$

$$= \frac{L}{10^3 k}\,\frac{\text{m}^2\,\text{K}}{\text{W}}$$

$$\text{E374} = \frac{\text{C374}}{10^3 \times \text{D374}}\,\frac{\text{m}^2\,\text{K}}{\text{W}}$$

Cell E379

(F3) @SUM(E373..E377)

The resistances of the sloping layers of material are summed:

$$R_\text{c} = R_\text{so} + R_1 + R_2 + R_3 + R_4\,\text{m}^2\,\text{K/W}$$

$$\text{E379} = \text{E373} + \text{E374} + \text{E375} + \text{E376} + \text{E377}\,\text{m}^2\,\text{K/W}$$

Cell E383

(F3) +C383/1000/D383

The thermal resistance of the first uppermost layer of material on a flat ceiling is calculated (this may be floorboard):

$$R = L\,\text{mm} \times \frac{1\,\text{m}}{10^3\,\text{mm}} \times \frac{1}{k}\,\frac{\text{m K}}{\text{W}}$$

$$= \frac{L}{10^3 k}\,\frac{\text{m}^2\,\text{K}}{\text{W}}$$

$$\text{E383} = \frac{\text{C383}}{10^3 \times \text{D383}}\,\frac{\text{m}^2\,\text{K}}{\text{W}}$$

Cell E388

(F3) @SUM(E383..E387)

The resistances of the horizontal layers of material are summed:

$$R_\text{b} = R_5 + R_6 + R_7 + R_8 + R_\text{si}\,\text{m}^2\,\text{K/W}$$

$$\text{E388} = \text{E383} + \text{E384} + \text{E385} + \text{E396} + \text{E387}\,\text{m}^2\,\text{K/W}$$

Cell D392

(F3) +E379*@COS (D369*@PI/180)+E381+E389

The total thermal resistance of a pitched roof has three components: the sloping external surface and layers, an air space, and a horizontal ceiling. The flat ceiling and air space resistances are added, but the sloping surface is first multiplied by the cosine of the roof slope angle from the horizontal, $X°$. Figure 2.2 showed the arrangement. The thermal resistance of the sloping surface is reduced by the cosine of the roof pitch angle. The hypotenuse, or sloping side of the triangle, has a greater length, larger surface area and increased heat loss than the horizontal ceiling. Only the horizontal component of the slope thermal resistance can be added to the horizontal ceiling thermal resistance.

$$\cos X° = \frac{\text{adjacent}}{\text{hypotenuse}}$$

$$\text{adjacent} = \text{hypotenuse} \times \cos X°$$

The lengths of the sides of the triangle are not needed, because the same effect is obtained by using the thermal resistance of the two surfaces:

$$\text{modified } R_c = R_c \times \cos X° \, \text{m}^2 \, \text{K/W}$$

The roof pitch angle, $X°$, is entered into cell D369. The worksheet will evaluate the angle in radians. There are 2π radians in $360°$, so there are π radians in $180°$. In radians:

$$\text{modified } R_c = R_c \cos\left(X° \times \frac{\pi}{180°}\right) \frac{\text{m}^2 \, \text{K}}{\text{W}}$$

The whole equation is

$$R = R_c \cos\left(X° \times \frac{\pi}{180°}\right) + R_a + R_b \frac{\text{m}^2 \text{K}}{\text{W}}$$

$$\text{D392} = \text{E379} \times \cos(\text{D369} \times \pi/180) + \text{E381} + \text{E389} \, \text{m}^2 \, \text{K/W}$$

Cells D394 to C401

These formulae are the same as those that have been described for a wall.

Cells E411 to C438

These formulae are the same as those that have been described for pitched roof type 1.

Solid floor

Cell D452

(F3) (0.5*D449)/(0.5*D449+0.5*D453)

The shape factor of a solid floor that is in direct contact with the earth is calculated from

$$S = \frac{0.5B}{0.5B + 0.5W}$$

where B = breadth of floor (m); W = thickness of the wall that surrounds the floor (m).

$$D452 = \frac{0.5 \times D449}{0.5 \times D449 + 0.5 \times D453}$$

The shape factor is dimensionless, and it will virtually always be between 0.91 and 0.997; see Example 2.15.

Cell D456

(F3) (2*1.4*(@EXP (0.5*D449/D448))/(0.5*D449*@PI))*(-0.5*@LN ((1-D452)/(1+D452)))*(100−D450)/100

The thermal transmittance of the solid floor is calculated from

$$U = \frac{2 \times 1.4}{0.5B \times \pi} \times \exp\left(\frac{0.5B}{L}\right) \times \operatorname{arctanh}\left(\frac{0.5B}{0.5B + 0.5W}\right)$$

where B = breadth, lesser dimension, of the floor (m); L = length, greater dimension, of the floor (m); W = width of the surrounding wall, usually 0.3 m; 1.4 = average thermal conductivity of earth in contact with the floor (W/m K); and

$$\operatorname{arctanh}\left(\frac{0.5B}{0.5B \times 0.5W}\right) = -\tfrac{1}{2} \times \ln\left(\frac{1-S}{1+S}\right)$$

The explanation and use of this formula was given in the earlier section on solid floors. The thermal transmittance is evaluated from

$$U = \frac{2 \times 1.4}{0.5 \times B \times \pi} \times \exp\left(\frac{0.5B}{L}\right) \times -\tfrac{1}{2} \times \ln\left(\frac{1-S}{1+S}\right)$$

$$D456 = \frac{2 \times 1.4}{0.5 \times D449 \times \pi} \times \exp\left(\frac{0.5 \times D449}{D448}\right) \times -\tfrac{1}{2} \times \ln\left(\frac{1-D452}{1+D452}\right)$$
$$\times \frac{100 - D450}{100} \; \frac{W}{m^2\,K}$$

Cells C458 and C459

The lower and upper limits of the thermal transmittance for a solid floor are tested as are those for a wall.

Suspended floor

Cell D473

(F3) (0.5*D471)/(0.5*D471+0.5*D474)

A suspended floor has three principal components: a timber or concrete floor, a ventilated air space, and the ground, which includes the foundations and any layer of concrete or gravel. The ground layers behave as a solid floor, and this part is calculated to be the earth thermal tranmittance component of the overall U value. The shape factor of this part of the floor is calculated in the same manner as for a solid ground floor, as previously described for cell D450.

Cell D475

(F3) (2*1.4*(@EXP(0.5*D471/D470))/(0.5*D471*@PI))* (−0.5*@LN ((1−D473)/(1+D473)))*(100−D472)/100

The earth thermal transmittance is calculated in the same manner as for a solid ground floor, as previously described for cell D456.

Cell D474

(F3) 1/D475−E485

The earth thermal resistance is found by subtracting the value of the indoor surface resistance from the total resistance of the solid floor.
Earth resistance:

$$R_e = \frac{1}{U_e} - R_{si} \frac{m^2\,K}{W}$$

$$D474 = \frac{1}{D475} - E485 \frac{m^2\,K}{W}$$

Cell E491

(F3) +C491/1000/D491

The thermal resistance of the layer of material that is nearest to the ground is calculated from

$$R = L\,mm \times \frac{1\,m}{10^3\,mm} \times \frac{1}{k} \frac{m\,K}{W}$$

$$= \frac{L}{10^3 k} \frac{m^2\,K}{W}$$

$$E491 = \frac{C491}{10^3 \times D491} \frac{m^2\,K}{W}$$

Cells E492 to E495

These calculate the thermal resistance of subsequent layers in the suspended floor and culminate with the final floor surface finish material.

Cell D499

(F3) @SUM(E491..E496)

The thermal resistance of the suspended floor slab is found from

$$R_g = R_1 + \ldots + R_5 + R_{si}\, m^2\,K/W$$

$$D499 = E491 + \ldots + E495 + E496\, m^2\,K/W$$

Cell D501

(F3)+E485+D499+E483+1/((1/(E483+D476))+(1/(E484+E482)))

The thermal resistance of the whole floor is found from a combination of the suspended floor resistance, the effective resistance of the heat transfers across the ventilated air space, and the earth resistance:

$$R = R_{si} + R_g + R_{t1} + \left(\frac{1}{R_{t1} + R_e} + \frac{1}{R_{t2} + R_v} \right)^{-1} \text{m}^2 \, \text{K/W}$$

$$\text{D501} = \text{E485} + \text{D499} + \text{E483} + \left(\frac{1}{\text{E483} + \text{D476}} + \frac{1}{\text{E484} + \text{E482}} \right)^{-1} \text{m}^2 \, \text{K/W}$$

Cell D503

(F3) 1/D501

The thermal transmittance of the whole floor is

$$U = \frac{1}{R} \, \frac{\text{W}}{\text{m}^2 \, \text{K}}$$

$$\text{D503} = \frac{1}{\text{D501}} \, \frac{\text{W}}{\text{m}^2 \, \text{K}}$$

Cell D505

(F3) 1/(D501−E485)

The conductance of the whole floor is found by subtracting the internal surface resistance from the overall thermal resistance and then taking the reciprocal. The internal surface resistance is assumed to be high emissivity, because a floor is not likely to be a polished reflector.

$$C = \frac{1}{R - R_{si}} \, \frac{\text{W}}{\text{m}^2 \, \text{K}}$$

$$\text{D505} = \frac{1}{\text{D501} - \text{E485}} \, \frac{\text{W}}{\text{m}^2 \, \text{K}}$$

Cell D507

(F1) (C491*F491+C492*F492+C493*F493+C494*F494+C495*F495)/1000

The weight of the suspended parts of the floor is calculated for 1 m² of floor surface area. The weight of any concrete or gravel that is lying directly on the earth is not included.

$$W = (L_1 \rho_1 + \ldots + L_5 \rho_5) \times \frac{1}{10^3} \, \text{kg/m}^2$$

$$\text{D506} = (\text{C491} \times \text{F491} + \ldots + \text{C495} \times \text{F495})/1000 \, \text{kg/m}^2$$

Cells C509 and C510

The lower and upper checks on the validity of the thermal transmittance of the suspended floor are made as for the other structures.

Single glazing

Cell E530

(F3) + C530/1000/D530

The thermal resistance of the outermost layer of material in the frame of a single glazed window or door is calculated:

$$R = L\,\text{mm} \times \frac{1\,\text{m}}{10^3\,\text{mm}} \times \frac{1}{k}\frac{\text{m k}}{\text{W}}$$

$$= \frac{L}{10^3 k}\frac{\text{m}^2\,\text{K}}{\text{W}}$$

$$\text{E530} = \frac{\text{C530}}{10^3 \times \text{D530}}\frac{\text{m}^2\,\text{K}}{\text{W}}$$

Cells E531, E532 and E540

The thermal resistances of the second and third layers of material of the frame and the single layer of glass are calculated as in cell E530.

Cell D546

(F3) 1/(@SUM(E529..E533)

The thermal transmittance of the frame is found from

$$U = \frac{1}{R_{\text{so}} + R_1 + R_2 + R_3 + R_{\text{si}}}\frac{\text{W}}{\text{m}^2\,\text{K}}$$

$$\text{D546} = \frac{1}{\text{E529} + \ldots + \text{E533}}\frac{\text{W}}{\text{m}^2\,\text{K}}$$

Cell D548

(F3) 1/@SUM(E539..E541)

The thermal transmittance of the glazing is found from

$$U = \frac{1}{R_{\text{so}} + R_1 + R_{\text{si}}}\frac{\text{W}}{\text{m}^2\,\text{K}}$$

$$\text{D548} = \frac{1}{\text{E539} + \text{E540} + \text{E541}}\frac{\text{W}}{\text{m}^2\,\text{K}}$$

Cell D550

(F3) ((D525*D546)+(100−D525)*D548)/100

The thermal transmittance of the window and frame is found by the proportional area method. This produces the average U value for the opening through the structural wall.

$$U = [F\% \times U_{\text{f}} + (100 - F\%) \times U_{\text{g}}] \times \frac{1}{100}\frac{\text{W}}{\text{m}^2\,\text{K}}$$

where $F\%$ = percentage of wall opening occupied by the frame; U_f = thermal transmittance of the frame (W/m^{2}K); and U_g = thermal transmittance of the glazing (W/m^{2}K).

$$D550 = (D525 \times D546 + (100 - D525) \times D548) \times \frac{1}{100}\ \frac{W}{m^2\,K}$$

Cell D552

(F3) 1/(1/D548−E539−E541)

The conductance of the glazing is

$$C = \frac{1}{R_g - R_{so} - R_{si}}\ \frac{W}{m^2\,K}$$

$$= \frac{1}{\dfrac{1}{U_g} - R_{so} - R_{si}}\ \frac{W}{m^2\,K}$$

$$D554 = \frac{1}{\dfrac{1}{D548} - E539 - E541}\ \frac{W}{m^2\,K}$$

Cell D554

(F3) 1/(1/D546−E529−E533)

The conductance of the frame is

$$C = \frac{1}{R_f - R_{so} - R_{si}}\ \frac{W}{m^2\,K}$$

$$= \frac{1}{\dfrac{1}{U_f} - R_{so} - R_{si}}\ \frac{W}{m^2\,K}$$

$$D554 = \frac{1}{\dfrac{1}{D546} - E529 - E533}\ \frac{W}{m^2\,K}$$

Cell D556:

(F1)(D525*(C530*F530+C531*F531+C532*F532)+(100−D525)*(C540*F540))/10^5

The weight of the three layers of frame material and the single glazing is

$$W = [F\% \times (L_1\rho_1 + \cdots + L_3\rho_3) + (100 - F\%) \times L_4\rho_4]\frac{1\,kg}{10^3\,m^2} \times \frac{1}{100}$$

$$D556 = [D525 \times (C530 \times F530 + C531 \times F531 + C532 \times F532 + (100 - D525)$$
$$\times C540 \times F540]\frac{1\,kg}{10^5\,m^2}$$

Cells C558 and C559

The lower and upper limits for the thermal transmittance of the single-glazed window are tested as they are for walls.

Double and triple glazing

Cells E579 to C668

The calculations for double- and triple-glazed windows and doors are in the same format as that for single glazing.

WORKED EXAMPLES

EXAMPLE 2.15

Calculate the shape factor of solid floors that have breadths of 3 m, 10 m, 50 m and 100 m when the building has a perimeter wall 300 mm thick.

For a floor whose smaller dimension, $B = 3$ m:

$$W = 300 \, \text{mm} \times \frac{1 \, \text{m}}{10^3 \, \text{mm}}$$

$$= 0.3 \, \text{m}$$

$$S = \frac{0.5B}{0.5B + 0.5W}$$

$$= \frac{0.5 \times 3}{0.5 \times 3 + 0.5 \times 0.3}$$

$$= 0.91$$

For $B = 10$ m:

$$S = \frac{0.5 \times 10}{0.5 \times 10 + 0.5 \times 0.3}$$

$$= 0.971$$

For $B = 50$ m:

$$S = \frac{0.5 \times 50}{0.5 \times 50 + 0.5 \times 0.3}$$

$$= 0.994$$

For $B = 100$ m:

$$S = \frac{0.5 \times 100}{0.5 \times 100 + 0.5 \times 0.3}$$

$$= 0.997$$

EXAMPLE 2.16

Calculate the thermal transmittance, conductance and structural weight for a new design of housing wall from the data in Table 2.8.

Table 2.8 Data for Example 2.16

Layer material	l (mm)	k (W/m K)	R (m^2 K/W)	ρ (kg/m^3)	W (kg/m^2)
R_{so} external surface			0.06		
Formed aluminium sheet	2	160	0	2800	6
R_a air space			0.18		
Light concrete block	200	0.19	1.053	600	120
Polyurethane board	50	0.025	2	30	2
Pine timber board	12	0.14	0.086	660	8
R_{si} inside surface			0.12		
			$\Sigma R = 3.499$		$\Sigma W = 136$

$$U = \frac{1}{\Sigma R} \frac{W}{m^2 K}$$

$$= (3.499)^{-1} \, W/m^2 \, K$$

$$= 0.286 \, W/m^2 \, K$$

$$C = \frac{1}{3.499 - 0.06 - 0.12} \frac{W}{m^2 \, K}$$

$$= 0.3 \, W/m^2 \, K$$

$$W = 136 \, kg/m^2$$

EXAMPLE 2.17

Calculate the thermal transmittance, conductance and structural weight for a tile-hung wall from the data in Table 2.9.

Table 2.9 Data for Example 2.17

Layer material	l (mm)	k (W/m K)	R (m^2 K/W)	ρ (kg/m^3)	W (kg/m^2)
R_{so} external surface			0.06		
Tile hanging	25	0.84	0.03	1900	48
Roofing felt	8	0.5	0.016	1700	14
R_a air space			0.12		
Medium concrete block	200	0.51	0.392	1400	280
Mineral fibre slab	100	0.035	2.857	30	3
Plasterboard	15	0.16	0.094	950	14
R_{si} inside surface			0.12		
			$\Sigma R = 3.689$		$\Sigma W = 359$

$$U = (3.689)^{-1} \, W/m^2 \, K$$

$$= 0.27 \, W/m^2 \, K$$

$$C = \frac{1}{3.689 - 0.06 - 0.12} \frac{W}{m^2 \, K}$$

$$= 0.285 \, W/m^2 \, K$$

$$W = 359 \, kg/m^2$$

EXAMPLE 2.18

Calculate the thermal transmittance, conductance and structural weight for a cavity wall from the data in Table 2.10.

Table 2.10   Data for Example 2.18

Layer material	l (mm)	k (W/m K)	R (m² K/W)	ρ (kg/m³)	W (kg/m²)
R_{so} external surface			0.06		
Medium concrete block	100	0.51	0.196	1400	140
Phenolic foam	25	0.04	0.625	30	1
R_a air space			0.18		
Light concrete block	100	0.38	0.263	1200	120
Light plaster	13	0.16	0.081	600	8
R_{si} inside surface			0.12		
			$\Sigma R = 1.525$		$\Sigma W = 269$

$$U = (1.525)^{-1} \, \text{W/m}^2 \, \text{K}$$
$$= 0.656 \, \text{W/m}^2 \, \text{K}$$
$$C = \frac{1}{1.525 - 0.06 - 0.12} \, \frac{\text{W}}{\text{m}^2 \, \text{K}}$$
$$= 0.743 \, \text{W/m}^2 \, \text{K}$$
$$W = 269 \, \text{kg/m}^2$$

EXAMPLE 2.19

Calculate the thermal transmittance, conductance and structural weight for a plaster-board cavity partition wall that separates two rooms on the first floor of a house. Use the data in Table 2.11.

Table 2.11   Data for Example 2.19

Layer material	l (mm)	k (W/m K)	R (m² K/W)	ρ (kg/m³)	W (kg/m²)
R_{so} internal surface			0.12		
Light plaster	15	0.16	0.094	600	9
Plasterboard	20	0.16	0.125	950	19
R_a air space			0.18		
Plasterboard	20	0.16	0.125	950	19
Light plaster	15	0.16	0.094	600	9
R_{si} inside surface			0.12		
			$\Sigma R = 0.858$		$\Sigma W = 56$

$$U = (0.858)^{-1} \, \text{W/m}^2 \, \text{K}$$
$$= 1.166 \, \text{W/m}^2 \, \text{K}$$

$$C = \frac{1}{0.858 - 0.12 - 0.12} \frac{W}{m^2 \, K}$$
$$= 1.618 \, W/m^2 \, K$$
$$W = 56 \, kg/m^2$$

EXAMPLE 2.20

Calculate the thermal transmittance, conductance and structural weight for an intermediate cast concrete floor from the data in Table 2.12. Heat flows from the top-floor boardroom downwards into an office.

Table 2.12 Data for Example 2.20

Layer material	l (mm)	k (W/m K)	R (m^2 K/W)	ρ (kg/m^3)	W (kg/m^2)
R_{si} internal surface			0.14		
Wool carpet	10	0.055	0.182	160	2
Wool underlay	10	0.045	0.222	160	2
Cement screed	50	0.41	0.122	1200	60
Cast concrete	150	1.13	0.133	2000	300
R_a air space			0.22		
Fibreboard ceiling panel	10	0.06	0.167	300	3
R_{si} inside surface			0.14		
			$\Sigma R = 1.326$		$\Sigma W = 367$

$$U = (1.326)^{-1} \, W/m^2 \, K$$
$$= 0.754 \, W/m^2 \, K$$
$$C = \frac{1}{1.326 - 0.14 - 0.14} \frac{W}{m^2 \, k}$$
$$= 0.956 \, W/m^2 \, K$$
$$W = 367 \, kg/m^2$$

EXAMPLE 2.21

Calculate the thermal transmittance, conductance and structural weight for an intermediate cast concrete ceiling over an office from the data in Table 2.13. Heat flow is upwards.

Table 2.13 Data for Example 2.21

Layer material	l (mm)	k (W/m K)	R (m^2 K/W)	ρ (kg/m^3)	W (kg/m^2)
R_{si} internal surface			0.1		
Wool carpet	10	0.055	0.182	160	2
Cellular rubber underlay	10	0.065	0.154	270	3

Table 2.13 (*contd.*)

Layer material	l (mm)	k (W/m K)	R (m² K/W)	ρ (kg/m³)	W (kg/m²)
Cement screed	50	0.41	0.122	1200	60
Cast concrete	150	1.13	0.133	2000	300
R_a air space			0.17		
Fibreboard ceiling panel	10	0.06	0.167	300	3
R_{si} inside surface			0.1		
			$\Sigma R = 1.128$		$\Sigma W = 368$

$$U = (1.128)^{-1}\,\text{W/m}^2\,\text{K}$$
$$= 0.887\,\text{W/m}^2\,\text{K}$$
$$C = \frac{1}{1.128 - 0.1 - 0.1}\,\frac{\text{W}}{\text{m}^2\,\text{K}}$$
$$= 1.078\,\text{W/m}^2\,\text{K}$$
$$W = 368\,\text{kg/m}^2$$

EXAMPLE 2.22

Calculate the thermal transmittance, conductance and structural weight for an insulated external flat roof on a top-floor residence from the data in Table 2.14.

Table 2.14 Data for Example 2.22

Layer material	l (mm)	k (W/m K)	R (m² K/W)	ρ (kg/m³)	W (kg/m²)
R_{so} external surface			0.04		
Asphalt	19	0.5	0.038	1700	32
Concrete decking	75	1.13	0.066	2000	150
R_a air space			0.17		
Glass-fibre quilt	100	0.04	2.5	12	1
Plasterboard	10	0.16	0.063	950	10
Light plaster	13	0.16	0.081	600	8
R_{si} inside surface			0.1		
			$\Sigma R = 3.058$		$\Sigma W = 201$

$$U = (3.058)^{-1}\,\text{W/m}^2\,\text{K}$$
$$= 0.327\,\text{W/m}^2\,\text{K}$$
$$C = \frac{1}{3.058 - 0.04 - 0.1}\,\frac{\text{W}}{\text{m}^2\,\text{K}}$$
$$= 0.343\,\text{W/m}^2\,\text{K}$$
$$W = 201\,\text{kg/m}^2$$

EXAMPLE 2.23

Calculate the thermal transmittance, conductance and structural weight for an external floor that is over an outdoor car park, from the data in Table 2.15.

Table 2.15 Data for Example 2.23

Layer material	l (mm)	k (W/m K)	R (m² K/W)	ρ (kg/m³)	W (kg/m²)
R_{so} external surface			0.07		
Cast concrete	150	1.13	0.133	2000	300
Cement screed	25	0.41	0.061	1200	30
Thermoplastic tile	10	0.5	0.02	1050	11
R_{si} inside surface			0.14		
			$\Sigma R = 0.424$		$\Sigma W = 341$

$$U = (0.424)^{-1}\,\text{W/m}^2\,\text{K}$$
$$= 2.358\,\text{W/m}^2\,\text{K}$$
$$C = \frac{1}{0.424 - 0.07 - 0.14}\,\frac{\text{W}}{\text{m}^2\,\text{K}}$$
$$= 4.673\,\text{W/m}^2\,\text{K}$$
$$W = 341\,\text{kg/m}^2$$

EXAMPLE 2.24

Calculate the thermal transmittance, conductance and structural weight of a well-insulated tiled roof that has a pitch of 80° from the horizontal. The roof is on a two-storey house. Use only the data in Table 2.16.

Table 2.16 Data for Example 2.24

Layer material	l (mm)	k (W/m K)	R (m² K/W)	ρ (kg/m³)	W (kg/m²)
R_{so} external surface			0.04		
Concrete tile	20	0.84	0.024	1900	38
Roofing felt	5	0.5	0.01	1700	8.5
Glass-fibre slab	100	0.035	2.857	25	2.5
			$\Sigma R_c = 2.931$		
Roof pitch $X = 80°$		$(R_c \times \cos X°) = 0.509$			
R_r roof space			0.18		
Timber boards	15	0.14	0.107	650	9.8
Glass-fibre quilt	100	0.04	2.5	12	1.2
Plasterboard	12	0.16	0.075	950	11.4
Light plaster	6	0.16	0.038	600	3.6
R_{si} inside surface			0.1		
			$\Sigma R_b = 2.82$		$\Sigma W = 75$

$$U = (R_c \cos X° + R_r + R_b)^{-1}\,\text{W/m}^2\,\text{K}$$
$$= (0.509 + 0.18 + 2.82)^{-1}\,\text{W/m}^2\,\text{K}$$
$$= (3.509)^{-1}\,\text{W/m}^2\,\text{K}$$
$$= 0.285\,\text{W/m}^2\,\text{K}$$

$$C = \frac{1}{3.509 - 0.04 - 0.1} \frac{W}{m^2 K}$$
$$= 0.297 \, W/m^2 \, K$$

$W = 75 \, kg/m^2$. The roof pitch has been ignored.

EXAMPLE 2.25

Calculate the thermal transmittance, conductance and structural weight of a double-glazed window in a beechwood frame. The frame occupies 10% of the wall opening. There is an air space between the two parts of the frame. Use only the data in Table 2.17.

Table 2.17   Data for Example 2.25

Layer material	l (mm)	k (W/m K)	R (m² K/W)	ρ (kg/m³)	W (kg/m²)
Window frame:					
R_{so} external surface			0.06		
Outer beech frame	25	0.165	0.152	700	17.5
R_a air space			0.18		
Inner beech frame	25	0.165	0.152	700	17.5
R_{si} inside surface			0.12		
		Frame $\Sigma R = 0.664$		$\Sigma W = 35$	
Glass:					
R_{so} external surface			0.06		
Plane glass	6	1.05	0.0057	2500	15
R_a air space			0.18		
Plane glass	6	1.05	0.0057	2500	15
R_{si} inside surface			0.12		
		Glass $\Sigma R = 0.371$		$\Sigma W = 30$	

$$\text{frame } U = (0.664)^{-1} \, W/m^2 \, K$$
$$= 1.506 \, W/m^2 \, K$$

$$\text{frame } C = \frac{1}{0.664 - 0.06 - 0.12} \frac{W}{m^2 K}$$
$$= 2.006 \, W/m^2 \, K$$

$$\text{glass } U = (0.371)^{-1} \, W/m^2 \, K$$
$$= 2.695 \, W/m^2 \, K$$

$$\text{glass } C = \frac{1}{0.371 - 0.06 - 0.12} \frac{W}{m^2 K}$$
$$= 5.236 \, W/m^2 \, K$$

$$\text{window } U = [F\% \times U_f + (100 - F\%)U_g] \times \frac{1}{100} \frac{W}{m^2 K}$$
$$= [10 \times 1.506 + (100 - 10) \times 2.695] \times \frac{1}{100} \frac{W}{m^2 K}$$
$$= 2.576 \, W/m^2 \, K$$

$$\text{window } W = [F\% \times W_\mathrm{f} + (100 - F\%) \times W_\mathrm{g}] \times \frac{1}{100} \frac{\mathrm{kg}}{\mathrm{m}^2}$$

$$= [10 \times 35 + (100 - 10) \times 30] \times \frac{1}{100} \frac{\mathrm{kg}}{\mathrm{m}^2}$$

$$= 30.5 \,\mathrm{kg/m}^2$$

EXAMPLE 2.26

Calculate the thermal transmittance, conductance and structural weight of a triple-glazed window in a beechwood frame. The frame occupies 10% of the wall opening. There are air spaces between the parts of the frame. Use only the data in Table 2.18.

Table 2.18 Data for Example 2.26

Layer material	l (mm)	k (W/m K)	R (m² K/W)	ρ (kg/m³)	W (kg/m²)
Window frame:					
R_so external surface			0.06		
Outer beech frame	25	0.165	0.152	700	17.5
R_a air space			0.18		
Middle beech frame	25	0.165	0.152	700	17.5
R_a air space			0.18		
Inner beech frame	25	0.165	0.152	700	17.5
R_si inside surface			0.12		
		Frame $\Sigma R = 0.996$		$\Sigma W = 52.5$	
Glass:					
R_so external surface			0.06		
Plane glass	6	1.05	0.0057	2500	15
R_a air space			0.18		
Plane glass	6	1.05	0.0057	2500	15
R_a air space			0.18		
Plane glass	6	1.05	0.0057	2500	15
R_si inside surface			0.12		
		Glass $\Sigma R = 0.557$		$\Sigma W = 45$	

$$\text{frame } U = (0.996)^{-1} \,\mathrm{W/m}^2\,\mathrm{K}$$

$$= 1.004 \,\mathrm{W/m}^2\,\mathrm{K}$$

$$\text{frame } C = \frac{1}{0.996 - 0.06 - 0.12} \frac{\mathrm{W}}{\mathrm{m}^2\,\mathrm{K}}$$

$$= 1.225 \,\mathrm{W/m}^2\,\mathrm{K}$$

$$\text{glass } U = (0.557)^{-1} \,\mathrm{W/m}^2\,\mathrm{K}$$

$$= 1.795 \,\mathrm{W/m}^2\,\mathrm{K}$$

$$\text{glass } C = \frac{1}{0.557 - 0.06 - 0.12} \frac{\mathrm{W}}{\mathrm{m}^2\,\mathrm{K}}$$

$$= 2.653 \,\mathrm{W/m}^2\,\mathrm{K}$$

$$\text{window } U = [10 \times 1.004 + (100 - 10) \times 1.795] \times \frac{1}{100} \frac{W}{m^2 K}$$

$$= 1.716 \, W/m^2 \, K$$

$$\text{window } W = [10 \times 52.5 + (100 \times 10) \times 45] \times \frac{1}{100} \frac{kg}{m^2}$$

$$= 45.75 \, kg/m^2$$

Questions

1. List the factors that are included in the assessment of the thermal transmittance of a structure.
2. Explain how the external climate affects thermal transmittance.
3. Explain how the different modes of heat transfer are included in the calculation of thermal transmittance.
4. Write a single equation that can be used to calculate the thermal transmittance of a cavity brick and block wall that is plastered on the internal surface and has a layer of thermal insulation within the cavity. State the meaning of each term in the equation.
5. Explain how the severity of exposure to the weather is taken into account in the assessment of thermal transmittance.
6. Discuss how heat transfer takes place across an air space within any type of structure and how it is influenced by the constructional design of the structure.
7. State the reasons for using a conductance figure during the thermal design of a structure.
8. State why the weight of a structure may be calculated along with its thermal transmittance.
9. Calculate the thermal transmittance, conductance and structural weight for an internal wall between two rooms of a house from: $R_{si} = 0.12 \, m^2 \, K/W$; 12 mm dense plaster, $k = 0.5 \, W/m \, K$, $W = 1300 \, kg/m^3$; 100 mm medium-weight concrete block, $k = 0.51 \, W/m \, K$, $W = 1400 \, kg/m^3$; 12 mm dense plaster, $k = 0.5 \, W/m \, K$, $W = 1300 \, kg/m^3$.
10. Calculate the thermal transmittance, conductance and structural weight for an intermediate cast concrete floor in a hotel, where heat flow is downwards to the room below, from: $R_{si} = 0.14 \, m^2 \, K/W$; 5 mm thermoplastic tile, $k = 0.5 \, W/m \, K$, $W = 1050 \, kg/m^3$; 30 mm cement screed, $k = 0.41 \, W/m \, K$, $W = 1200 \, kg/m^3$; 200 mm cast concrete, $k = 1.4 \, W/m \, K$, $W = 2100 \, kg/m^3$; ceiling void, $R_a = 0.22 \, m^2 \, K/W$; 12 mm fibreboard ceiling tiles, $k = 0.06 \, W/m \, K$, $W = 300 \, kg/m^3$; $R_{si} = 0.14 \, m^2 \, K/W$.
11. Calculate the thermal transmittance, conductance and structural weight for an internal ceiling in a multi-storey office building where the heat flow is upwards, from: $R_{si} = 0.1 \, m^2 \, K/W$; 25 mm wood

blocks, $k = 0.14 \, W/m \, K$, $W = 650 \, kg/m^3$; 20 mm cement screed, $k = 0.41 \, W/m \, K$, $W = 1200 \, kg/m^3$; 180 mm cast concrete, $k = 1.13 \, W/m \, K$, $W = 2000 \, kg/m^3$; ceiling void, $R_a = 0.22 \, m^2 \, K/W$; 12 mm plasterboard, $k = 0.16 \, W/m \, K$, $W = 950 \, kg/m^3$; $R_{si} = 0.1 \, m^2 \, K/W$.

12. Calculate the thermal transmittance, conductance and structural weight for an external flat roof over a house extension from: $R_{so} = 0.02 \, m^2 \, K/W$; 20 mm roofing felt, $k = 0.5 \, W/m \, K$, $W = 1700 \, kg/m^3$; 25 mm timber boards, $k = 0.14 \, W/m \, K$, $W = 650 \, kg/m^3$; 100 mm air space, $R_a = 0.18 \, m^2 \, K/W$; 120 mm glass-fibre quilt, $k = 0.04 \, W/m \, K$, $W = 12 \, kg/m^3$; 13 mm plasterboard, $k = 0.16 \, W/m \, K$, $W = 950 \, kg/m^3$; 10 mm lightweight plaster, $k = 0.16 \, W/m \, K$, $W = 600 \, kg/m^3$; $R_{si} = 0.1 \, m^2 \, K/W$.
13. Calculate the thermal transmittance, conductance and structural weight for an external flat roof over an office from: $R_{so} = 0.09 \, m^2 \, K/W$; 20 mm white granite chippings, $k = 0.96 \, W/m \, K$, $W = 1800 \, kg/m^3$; 100 mm expanded polystyrene slab, $k = 0.035 \, W/m \, K$, $W = 25 \, kg/m^3$; 20 mm asphalt, $k = 0.5 \, W/m \, K$, $W = 1700 \, kg/m^3$; 150 mm cast concrete, $k = 1.13 \, W/m \, K$, $W = 2000 \, kg/m^3$; 13 mm dense plaster, $k = 0.5 \, W/m \, K$, $W = 1300 \, kg/m^3$; $R_{si} = 0.1 \, m^2 \, K/W$.
14. Calculate the thermal transmittance, conductance and structural weight of a tiled roof that has a pitch of 25° from the horizontal. The roof is on a single-storey house. Use only the data provided. $R_{so} = 0.07 \, m^2 \, K/W$; 20 mm concrete tiles, $k = 0.84 \, W/m \, K$, $W = 1900 \, kg/m^3$; 5 mm roofing felt, $k = 0.5 \, W/m \, K$, $W = 1700 \, kg/m^3$; roof space, $R_r = 0.18 \, m^2 \, K/W$; 150 mm mineral-fibre quilt, $k = 0.35 \, W/m \, K$, $W = 30 \, kg/m^3$; 10 mm plasterboard, $k = 0.16 \, W/m \, K$, $W = 950 \, kg/m^3$; 8 mm lightweight plaster, $k = 0.16 \, W/m \, K$, $W = 600 \, kg/m^3$; $R_{si} = 0.1 \, m^2 \, K/W$.
15. Calculate the shape factor and the hyperbolic tangent for a 40 m × 25 m solid ground floor when the surrounding wall is 350 mm thick.
16. Calculate the thermal transmittance of a solid ground floor that is 50 m long and 35 m wide when the surrounding walls have a thickness of 300 mm and the floor has four exposed edges.
17. Calculate the thermal transmittance of a solid ground floor that is 150 m long and 95 m wide

when the surrounding walls have a thickness of 300 mm and the floor has four exposed edges.

18. Calculate the thermal transmittance of a solid ground floor that is 12 m long and 8 m wide when the surrounding walls have a thickness of 300 mm, edge insulation reduces the thermal transmittance by 15%, and the floor has four exposed edges.

19. A house has floor plan dimensions of 15 m × 8 m. The floor consists of 15 mm wool carpet, 10 mm wool felt underlay, 12 mm timber floorboards that are supported on wooden joists, a ventilated floor cavity, and an over-site concrete slab. The thermal conductivities are: carpet 0.055 W/m K, underlay 0.045 W/m K, floorboard 0.14 W/m K. Use the values $R_{si} = 0.14 \, m^2 \, K/W$, $R_r = 0.2 \, m^2 \, K/W$, $R_c = 0.67 \, m^2$ K/W, $R_{t1} = 0.09 \, m^2 \, K/W$, $R_{t2} = 0.29 \, m^2 \, K/W$, $R_v = 0.63 \times B \, m^2 K/W$, B = floor width (m). Calculate the thermal transmittance of the floor using both manual and worksheet methods.

20. A suspended timber floor 22 m wide and 12 m broad in a house is to be insulated with 150 mm thick mineral-fibre slab laid onto nylon netting between the joists. The wall surrounding the floor is 300 mm wide. Calculate the thermal transmittance of the insulated floor. The floor consists of 15 mm wool carpet, 8 mm cellular rubber underlay, 12 mm timber floorboards that are supported on wooden joists, a ventilated floor cavity and an over-site concrete slab. The thermal conductivities are: carpet 0.055 W/m K, underlay 0.065 W/m K, floor board 0.14 W/m K, mineral-fibre quilt 0.035 W/m K. Use the values $R_{si} = 0.14 \, m^2 \, K/W$, $R_r = 0.2 \, m^2 \, K/W$, $R_c = 0.67 \, m^2 \, K/W$, $R_{t1} = 0.09 \, m^2$ K/W, $R_{t2} = 0.29 \, m^2 \, K/W$, $R_v = 0.63 \times B \, m^2 \, K/W$, B = floor width (m).

21. A sports hall has a floor of 100 m × 60 m and a wall thickness of 400 mm. The floor consists of 25 mm thick timber floorboards that are supported on wooden joists, 125 mm thick expanded polystyrene slab, a ventilated floor cavity, and an over-site concrete slab. The thermal conductivity of the floorboards is 0.14 W/m K, and that of the polystyrene is 0.035 W/m K. Calculate the overall thermal transmittance of the floor.

22. Calculate the thermal transmittance, conductance and structural weight of a single-glazed window in a beechwood frame. The frame occupies 12% of the wall opening. Use the data: $R_{so} = 0.08 \, m^2 \, K/W$; 18 mm beech frame, $k = 0.165 \, W/m \, K$, $W = 700 \, kg/m^3$; 6 mm plane glass, $k = 1.05 \, W/m \, K$, $W = 2500 \, kg/m^3$; $R_{si} = 0.12 \, m^2$ K/W.

23. Calculate the thermal transmittance, conductance and structural weight for a new design of housing wall from the data: $R_{so} = 0.03 \, m^2/K/W$; 8 mm cement render 0.5 W/m K, 1300 kg/m³; 200 mm medium-weight concrete block, 0.51

W/m K, 1400 kg/m³; $R_a = 0.18 \, m^2 \, K/W$; 50 mm mineral fibre slab, 0.035 W/m K, 30 kg/m³; 100 mm lightweight concrete block, 0.19 W/m K, 600 kg/m³; 10 mm lightweight plaster, 0.16 W/m K, 600 mg/m³; $R_{si} = 0.12 \, m^2 \, K/W$.

24. Calculate the thermal transmittance, conductance and structural weight for a tile hung wall from the data: $R_{so} = 0.03 \, m^2 \, K/W$; 28 mm concrete tiles, 1.1 W/m K, 2100 kg/m³; 5 mm roofing felt, 0.5 W/m K, 1700 kg/m³; $R_a = 0.18 \, m^2 \, K/W$; 100 mm phenolic foam slab, 0.04 W/m K, 30 kg/m³; 100 mm lightweight concrete block, 0.19 W/m K, 600 kg/m³; 10 mm lightweight plaster, 0.16 W/m K, 600 kg/m³; $R_{si} = 0.12 \, m^2 \, K/W$.

25. Calculate the thermal transmittance, conductance and structural weight for a tile-hung wall from the data: $R_{so} = 0.08 \, m^2 \, K/W$; 28 mm clay tiles, 0.85 W/m K, 1900 kg/m³; 5 mm roofing felt, 0.5 W/m K, 1700 kg/m³; $R_a = 0.18 \, m^2 \, K/W$; 100 mm glass-fibre slab, 0.035 W/m K, 25 kg/m³; 18 mm plasterboard, 0.16 W/m K, 950 kg/m³; 10 mm lightweight plaster, 0.16 W/m K, 600 kg/m³; $R_{si} = 0.12$ $m^2 \, K/W$.

26. Calculate the thermal transmittance, conductance and structural weight for a cavity wall from the data: $R_{so} = 0.03 \, m^2 \, K/W$; 105 mm brickwork, 0.84 W/m K, 1700 kg/m³; $R_a = 0.18 \, m^2 \, K/W$; 40 mm expanded polystyrene slab, 0.035 W/m K, 25 kg/m³; 120 mm medium concrete block, 0.51 W/m K, 1400 kg/m³; 12 mm dense plaster, 0.5 W/m K, 1300 kg/m³; $R_{si} = 0.12 \, m^2 \, K/W$.

27. Calculate the thermal transmittance, conductance and structural weight for a plasterboard cavity partition wall that separates two rooms on the first floor of a house. Use the data: $R_{si} = 0.12 \, m^2 \, K/W$; 8 mm light plaster, 0.16 W/m K, 600 kg/m³; 19 mm plasterboard, 0.16 W/m K, 950 kg/m³; 50 mm mineral fibre, 0.035 W/m K, 30 kg/m³; 19 mm plasterboard, 0.16 W/m K, 950 kg/m³; 8 mm light plaster, 0.16 W/m K, 600 kg/m³; $R_{si} = 0.12 \, m^2 \, K/W$.

28. Calculate the thermal transmittance, conductance and structural weight for an intermediate cast concrete floor from the data provided. Heat flows from the top-floor conference room downwards into an office. $R_{si} = 0.14 \, m^2 \, K/W$; 8 mm wool carpet, 0.055 W/m K, 160 kg/m³; 8 mm cellular rubber underlay, 0.065 W/m K, 270 kg/m³; 30 mm cement screed, 0.41 W/m K, 1200 kg/m³; 200 mm cast concrete, 1.13 W/m K, 2000 kg/m³; $R_a = 0.22 \, m^2 \, K/W$; 20 mm plasterboard, 0.16 W/m K, 950 kg/m³; $R_{si} = 0.14 \, m^2 \, K/W$.

29. Calculate the thermal transmittance, conductance and structural weight for an intermediate cast concrete ceiling over an office from the data provided. Heat flow is upwards. $R_{si} = 0.1$ $m^2 \, K/W$; 15 mm woodblock flooring 0.14 W/m K, 650 kg/m³; 20 mm cement screed, 0.41 W/m K,

1200 kg/m^3; 120 mm cast concrete, 1.13 W/m K, 2000 kg/m^3; $R_a = 0.17\,\text{m}^2\,\text{K/W}$; 12 mm fibreboard ceiling panel, 0.06 W/m K, 300 kg/m^3; $R_{si} = 0.1$ m^2 K/W.

30. Calculate the thermal transmittance, conductance and structural weight for an insulated external flat roof on a top-floor residence from the data provided. $R_{so} = 0.07\,\text{m}^2\,\text{K/W}$; 15 mm stone chippings, 0.96 W/m K, 1800 kg/m^3; 20 mm asphalt, 0.5 W/m K, 1700 kg/m^3; 180 mm aerated concrete slab, 0.16 W/m K, 500 kg/m^3; $R_a = 0.17\,\text{m}^2\,\text{K/W}$; 150 mm glass-fibre quilt, 0.04 W/m K, 12 kg/m^3; 12 mm plasterboard, 0.16 W/m K, 950 kg/m^3; 5 mm light plaster, 0.16 W/m K, 600 kg/m^3; $R_{si} = 0.1$ m^2 K/W.

31. Calculate the thermal transmittance, conductance and structural weight for an external floor that is over an outdoor car park, from the data: $R_{so} = 0.03\,\text{m}^2\,\text{K/W}$; 200 mm cast concrete, 1.13 W/m K, 2000 kg/m^3; $R_a = 0.17\,\text{m}^2\,\text{K/W}$; 100 mm expanded polystyrene slab, 0.035 W/m K, 25 kg/m^3; 18 mm chipboard flooring, 0.15 W/m K, 800 kg/m^3; 5 mm thermoplastic tile, 0.5 W/m K, 1050 kg/m^3; $R_{si} = 0.14\,\text{m}^2\,\text{K/W}$.

32. Calculate the thermal transmittance, conductance and structural weight of a well-insulated tiled roof that has a pitch of 65° from the horizontal. The roof is on a two-storey house. Use the data: $R_{so} = 0.07\,\text{m}^2\,\text{K/W}$; 25 mm concrete tiles, 1.1 W/m K, 2100 kg/m^3; 5 mm roofing felt, 0.5 W/m K, 1700 kg/m^3; roof space, $R_r = 0.18\,\text{m}^2\,\text{K/W}$; 18 mm chipboard, 0.15 W/m K, 800 kg/m^3; 120 mm mineral fibre, 0.035 W/m K, 30 kg/m^3; 12 mm plasterboard, 0.16 W/m K, 950 kg/m^3; 5 mm plaster, 0.16 W/m K, 600 kg/m^3; $R_{si} = 0.1$ m^2 K/W.

33. Calculate the thermal transmittance, conductance and structural weight of a double-glazed window in an aluminium thermal-break frame. The frame occupies 13% of the wall opening. Use the data provided. Frame: 5 mm outer aluminium frame, $k = 160$ W/m K, $W = 2800$ kg/m^3; 5 mm silicone rubber thermal break, $k = 0.25$ W/m K, $W = 1200$ kg/m^3; 5 mm inner aluminium frame, $k = 160$ W/m K, $W = 2800$ kg/m^3. Glass: 6 mm outer glass, $k = 1.05$ W/m K, $W = 2500$ kg/m^3; $R_a = 0.18\,\text{m}^2\,\text{K/W}$; 6 mm inner glass, $k = 1.05$ W/m K, $W = 2500$ kg/m^3. Surface films are $R_{so} = 0.03\,\text{m}^2\,\text{K/W}$ and $R_{si} = 0.12\,\text{m}^2\,\text{K/W}$.

34. Calculate the thermal transmittance, conductance and structural weight of a triple-glazed window in a rigid polyvinyl chloride (PVC) frame. The frame occupies 15% of the wall opening, and is continuous between the internal and external air surfaces. Use the data provided: 100 mm PVC frame, $k = 0.16$ W/m K, $W = 1350$ kg/m^3; three panes of 6 mm glass, $k = 1.05$ W/m K, $W = 2500$ kg/m^3; two sealed air spaces, $R_a = 0.18\,\text{m}^2\,\text{K/W}$ each; $R_{so} = 0.03\,\text{m}^2\,\text{K/W}$; $R_{si} = 0.12\,\text{m}^2\,\text{K/W}$.

Hmph!

3 Heat gain

INTRODUCTION

This worksheet allows the air conditioning cooling load to be calculated for a single room at one time of day on one date. Reference data is provided for three times during the day on two dates. This is likely to be enough for many applications, because most peak cooling load times are predictable. The user copies the required solar heat gain information into the room data section. Sufficient data is provided on the worksheet for UK conditions and typical applications. The user can enter new data onto the reference area of the worksheet for any geographical location and time. The room cooling load from outdoor and interior heat gain sources is calculated. The air conditioning system supply air volume flow rate and duct diameter are calculated from the room and supply air temperatures provided by the user. There is a space for the addition of cooling loads from additional rooms so that a zone plant capacity can be found.

LEARNING OBJECTIVES

Study and use of this chapter will enable the user to:

1. know the sources of the room cooling load for air conditioning;
2. identify the design cooling load;
3. be able to evaluate when the design cooling load should occur;
4. know the sources of heat gain;
5. understand the thermal storage effects of the building structure upon solar heat gains;
6. identify how solar heat gains occur through glazing;
7. become familiar with typical heat gain data;
8. be able to estimate the heat gain to an occupied space using manual methods;
9. use air, environmental and sol-air temperature;
10. know how to calculate the room cooling load through the opaque structure;
11. use thermal transmittance and admittance;
12. use time lag, decrement factor and surface factor;
13. calculate ventilation and electrical equipment heat gains;

14. calculate room sensible, latent and total heat gains;

15. calculate the outside air ventilation rate required for the occupants;

16. make the additions to the basic room cooling load to allow for heat gains to ductwork and air leakage from the duct system;

17. calculate the room supply air flow;

18. find the diameter of the supply and recirculation air ducts for the room;

19. know how to use solar gain correction factors for glazing;

20. use 24 h mean and cyclic heat flows;

21. calculate the room air change rate;

22. find the cooling plant load for a group of rooms.

Key terms and concepts

ROOM COOLING LOAD

The cooling load in a room is calculated from the simultaneous solar radiation heat gains and the warmer outdoor air, plus the sources of heat occurring within the building. The design cooling load is that which will maintain the specified indoor environmental conditions when it is necessary to do so. Most cooling loads are to maintain thermal comfort during the normally occupied part of the day. The design cooling load for south-facing glazed rooms in the UK is likely to occur around noon on 21 June. East-facing rooms will produce peak cooling loads in the early morning. It may be decided to design the cooling system for the east zone to meet the demand at 10 a.m. even when this is less than that at 8 a.m. The occupants can cope with an early morning warmer indoor air temperature as they acclimatize to the controlled conditions. The design cooling load for a west-facing room may be taken at 4 p.m., so that any higher gains that occur afterwards will not cause long-term discomfort for those who are vacating the room at 5 p.m. The designer can rationalize the amount of calculation to be performed in order to achieve the main objective of the cooling system. This objective is to provide an affordable and practical installation with the least time spent on the design.

The design cooling load is found from:

1. solar heat gain through the glazing;
2. convection and conduction heat gains through the glazing;
3. the infiltration of warmer outdoor air directly into the room;
4. the net heat flows into the room through the opaque structure;
5. the sensible heat gain from the occupants;
6. the electrical power input to lighting system;
7. the electrical power input to equipment;
8. ventilation air that enters from adjacent warmer rooms.

Some of these heat gains (such as the inward convection of outside air) raise the air temperature within the room immediately, and some of them warm the building structure by radiation. These structural heat inflows take time to appear as an increase in the room air temperature. Solar radiation falling upon the outside surfaces of the building warms the brick walls, concrete floors and roof, which puts the heat gain into thermal storage. The elevated temperature of the solid components leads to the release of heat into the room some hours after the solar irradiance occurs. Masonry constructions can produce a time lag of 9–12 h between the solar irradiation and the corresponding heat gain within the room as an increase in the air temperature. When the designer calculates the cooling load in a normally occupied office for noon, the net flow of heat into the room from the masonry wall is due to the outdoor conditions around 9 h earlier, or at 3 a.m. The heat flow at 3 a.m. can generate a negative heat gain at noon: that is, a flow of heat from the room into the solid wall due to the cooler conditions that occurred 9 h previously. Thermal admittance, Y W/m^2 K, is used in the assessment of these heat flows into and out from thermal storage.

Heat gains to the room through glazing occur because of the direct transmission of solar radiation, convection heat transfers from the warmed glass, and by the thermal transmittance gains. The designer takes into account the following factors:

1. latitude of the site;
2. exposure to solar radiation;
3. altitude (height above sea level);
4. transmissivity of the glass;
5. reflectance of the glass surfaces;

6. exterior shading;
7. interior shading;
8. colour of the glass.

The typical cooling load due to solar gains through glazing at a latitude of 51.7° N is given in Table 3.1. Correction factors to these cooling loads are given in Table 3.2 for body-tinted and reflective glazing. These correction factors are used when internal slatted blinds are in place, and it is assumed that the blinds are being drawn by the occupants. Other cooling load and glazing correction factor data can be used on the worksheet. The reference source is used for other latitudes, times and dates.

Table 3.1   Cooling load through vertical clear glazing (W/m^2) with intermittent use of internal blinds

Date	Orientation	Sun time (hours)		
		1000	1200	1600
21 June	N	114	138	111
	NE	81	151	124
	E	254	88	125
	SE	282	207	110
	S	179	238	89
	SW	105	183	261
	W	114	139	314
	NW	115	140	204
22 Sept.	N	76	104	73
	NE	89	105	74
	E	220	52	75
	SE	330	262	74
	S	262	333	137
	SW	78	237	274
	W	75	103	258
	NW	75	103	156

(CIBSE, 1986a)

Table 3.2   Correction factors to glazing cooling load for glass type and closed light-coloured slatted blinds

Glass type	Single glazing	Double glazing
Clear 6 mm	0.77	0.74
Tinted 6 mm	0.77	0.55
Tinted 10 mm	0.73	0.47
Reflecting	0.57	0.41
Mirrored	0.34	0.21

(CIBSE, 1986a)

EXAMPLE 3.1

An open-plan office in London has outside walls that face south-east and south. Each wall has a double-glazed window 2 m wide and 2 m high with internal slatted blinds. Body-tinted grey 6 mm thick glass is used for the outer sheet and clear 6 mm glass for the inner sheet. Calculate the cooling load on the room air conditioning system using only the information from Tables 3.1 and 3.2.

Body-tinted 6 mm double glazing and internal slatted blinds have a cooling load correction factor of 0.55. The interior sheet in a double-glazed window normally has clear glass. The time of day that produces the peak room cooling load has to be assessed. This peak load is the combination of the heat gains occurring through the two exposed glass areas of the room. Calculate the heat gain through both the windows at each time and date provided in Table 3.1.

Heat gain at 10 a.m. on 21 June through the south-east window is

$$Q_1 = 2\,\text{m} \times 2\,\text{m} \times 0.55 \times 282\,\text{W/m}^2$$
$$= 620.4\,\text{W}$$

Heat gain at 10 a.m. on 21 June through the south window is

$$Q_2 = 2\,\text{m} \times 0.55 \times 179\,\text{W/m}^2$$
$$= 393.8\,\text{W}$$

The total heat gain at 10 a.m. on 21 June through both windows is

$$Q = 620.4 + 393.8\,\text{W}$$
$$= 1014.2\,\text{W}$$

The resulting heat gains for the remaining combinations of dates and times are shown in Table 3.3.

Table 3.3 Solution to Example 3.1: room cooling load (W)

Date	Time (*hours*)		
	1000	*1200*	*1600*
21 June	1014.2	979.0	437.8
22 September	1302.4	1309.0	464.2

The peak room cooling load is 1309 W at noon on 22 September in London, latitude 51.7° N.

SOL-AIR TEMPERATURE

Sol-air temperature, t_{eo}, is the outdoor temperature that, in the absence of solar irradiance, would give the same temperature distribution and rate of heat transfer through the wall or roof as exists with the actual outside air temperature and solar irradiance (CIBSE, 1986a). Sol-air temperature is used to find the net heat flow into the opaque structure. This is done for the time of the solar gain occurrence which then appears within the room at the time of the peak solar gain through the glazing. A masonry wall having a time lag of 9 h and a peak solar gain at noon would have a net flow into the opaque structure at 3.0 a.m. The mean 24 h sol-air temperature is used to find the mean heat flow into the opaque structure.

HEAT GAINS THROUGH THE OPAQUE STRUCTURE

The worksheet calculates the 24 h mean heat flow through the structure, and the cyclic variation from the mean. Thermal transmittance is used to calculate the mean flows, and admittance is used for the periodic oscillations. The mean conduction heat gains or losses through an opaque structure, Q_u are

$$Q_u = \frac{F_u}{F_v} \times AU \times (t_{eo} - t_c)$$

where

$$F_u = \frac{18\Sigma(A)}{18\Sigma(A) + \Sigma(AU)}$$

$$F_v = \frac{6\Sigma(A)}{6\Sigma(A) + 0.33NV}$$

and $t_{eo} = 24$ mean sol-air temperature (°C); $t_c = 24$ mean internal resultant temperature (°C); A = surface area (m²); U = thermal transmittance (W/m² K); $\Sigma(A)$ = summation of the room surface areas (m²); N = room air infiltration ventilation rate (air change per hour) (h⁻¹); and V = room volume (m³).

The cyclic variation from the mean heat flow, $\tilde{Q}_u$, is

$$\tilde{Q}_u = \frac{F_y}{F_v} \times AU \times f\,\tilde{t}_{eo}$$

where

$$F_y = \frac{18\Sigma(A)}{18\Sigma(A) + \Sigma(AY)}$$

f = the decrement factor, the ratio of the heat flow through the structure to the 24 h mean heat flow. The factor for glazing is unity, while that for dense concrete walls and roofs may be as low as 0.2 because of their energy storage and release to outdoors and indoors.

$\tilde{t}_{eo}$ = the swing in the environmental temperature between that when the heat flow occurs at the external surface and the 24 h mean environmental temperature. The swing will be negative when t_{eo} at the outdoor time of the heat gain is less than the 24 h mean t_{eo} and shows a loss of heat from the room. The time difference between these events is the time lag for the structure, and this corresponds to the decrement factor reduction in the energy flows.

Y = thermal admittance (W/m² K) is the rate of heat flow into the structure. It is the same as the U value for glass but higher than the U value for insulated, dense structures.

F = the surface factor that is the ratio used for internal walls, floors and ceilings, and it relates to the variation in heat flows between rooms.

The net heat flow through the structure is the sum of the 24 h mean heat flow and the cyclic variation:

$$Q = \bar{Q}_u + \tilde{Q}_u \text{ W}$$

The heat gain through the glazing is found from

$$Q_u = \frac{F_u}{F_v} \times AU \times (t_{ao} - t_c) \text{ W}$$

where t_{ao} = outdoor air temperature (°C).

The decrement factor f for glass is 1, and the time lag is zero. The room heat gain due to the entry of outdoor air into the room is given by

$$Q_v = 0.33F_2NV \times (t_{ao} - t_c)$$

$$F_2 = 1 + \frac{F_u \Sigma(AU)}{6\Sigma(A)}$$

The other heat gains to the room come from the occupancy, electrical lighting, electrical equipment and the inflow of warmer air from other rooms.

EXAMPLE 3.2

A south-facing top-floor office in Portsmouth city centre is shown in Fig. 3.1. The exposure is normal, and the height above sea level is 10 m. The office air conditioning is to be designed to maintain a room resultant temperature of 21 °C and an air temperature of 22 °C d.b. when the outside air is at 30 °C d.b. There are four sedentary occupants and two continuously used computers having power consumptions of 200 W each. The air conditioning plant operates for 10 h per day. There are 0.25 air changes per hour due to the natural infiltration of outdoor air. The adjacent offices, corridor and office below are maintained at the same temperature as the calculated office. The constructional details of the building are:

(a) double-glazed window of clear float glass in an aluminium frame with thermal break and internal white-coloured venetian blinds;
(b) external wall of 105 mm brickwork, 10 mm cavity, 40 mm glass-fibre batts, 100 mm lightweight concrete block, 13 mm lightweight plaster, dark exterior colour;
(c) internal walls of 100 mm lightweight concrete block, 13 mm lightweight plaster both sides;
(d) 150 mm cast concrete floor, 25 mm wood block surface;
(e) 19 mm asphalt roof covering, 13 mm fibreboard, 25 mm air gap, 75 mm glass fibre, 10 mm plasterboard and dark exterior colour.

The rate of infiltration of outside air is 0.25 air changes per hour. The air conditioning system is to provide 10 litres/s per person of outside air. The minimum supply air temperature is to be 12 °C d.b. The heat emission from the occupants of the office is 90 W sensible heat and 50 W latent heat. Heat gains to the supply and recirculation air ducts are expected to contribute an additional 5% to the room heat load. The design engineer decides to provide a margin of 10% to the calculated plant heat load. The air duct system is expected to have a general leakage rate that amounts to 5% of the design air flow. The limiting duct air velocity is 5 m/s. Use the thermal data provided in the worksheet. Calculate the room cooling load and find the diameter of the supply and recirculation air ducts.

The thermal data for each surface is shown in Table 3.4.

Table 3.4 Heat transfer data for Example 3.2

Surface A	A (m²)	U (W/m²)	A × U	Y (W/m² K)	A × Y	f	Lag (h)
Glass	6	3.3	19.8	3.3	19.8	1	0
Ext. wall	9	0.45	4.1	2.4	21.6	0.42	9
Int. wall	45	1.1	0	2.3	103.5	0.88	1
Floor	25	1.5	0	2.9	72.5	0.7	1
Roof	25	0.4	10	0.7	17.5	0.99	1
	Σ(A) 110		Σ(U) 33.9		Σ(A × Y) 234.9		

Note that the internal walls and floor have no heat exchange with their adjacent rooms, so AU is zero. These internal surfaces do not have solar radiation decrement factors f or time lags; surface factor F is used.

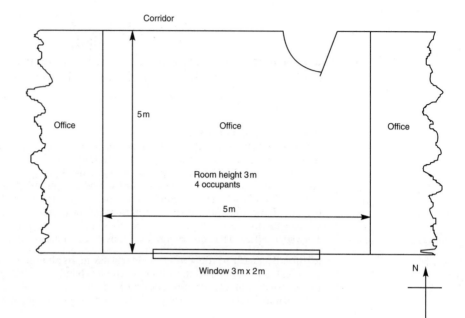

Fig. 3.1   Air conditioned office in Example 3.2

$$\text{office volume } V = 5 \times 5 \times 3 \, \text{m}^3$$
$$= 75 \, \text{m}^3$$

The cooling load due to the south-facing double clear float glazing for a light-weight building at 51.7° N latitude with the intermittent use of internal slatted blinds at noon sun time on 22 September is 333 W/m². A correction factor of 0.74 applies when the blinds are closed.

$$\text{solar gain through the window} = 333 \text{W/m}^2 \times 0.74 \times 6 \, \text{m}^2$$
$$= 1479 \, \text{W}$$

The south-facing wall has a time lag of 9 h. The solar heat gain to the south wall at the time of 9 h before noon (0300 hours) occurs when t_{eo} is 10.5 °C. The 24 h mean $\bar{t}_{eo}$ for the south-facing wall is 25°C. The swing in t_{eo} between the 0300 hours value and the 24 h mean is

$$\tilde{t}_{eo} = 10.5°\text{C} - 25°\text{C}$$
$$= -14.5°\text{C}$$

Similarly, at 1100 hours for the roof:

$$\text{at 1100 hours, } t_{eo} = 38°\text{C}$$
$$\text{24 h mean, } \bar{t}_{eo} = 20°\text{C}$$
$$\tilde{t}_{eo} = 38°\text{C} - 20°\text{C}$$
$$= 18°\text{C}$$

The window is subject to air temperature heat transfers and not sol-air temperatures, so it has zero time lag.

$$\text{at 1200 hours, } t_{ao} = 18.5°C$$
$$\text{24 h mean, } \bar{t}_{ao} = 15.5°C$$
$$\tilde{t}_{ao} = 18.5°C - 15.5°C$$
$$= 3°C$$

Table 3.5 shows the results.

Table 3.5 Sol-air and air data for Example 3.2

Surface	Lag (h)	24 h $\bar{t}_{eo}$	24 h $\bar{t}_{ao}$	Time (hours)	t_{eo}	Swing, $\tilde{t}_{eo}$
Ext. wall	9	25	15.5	0300	10.5	−14.5
Roof	1	20	15.5	1100	38	18
					t_{ao}	$\tilde{t}_{ao}$
Window	0	−	15.5	1200	18.5	3

$$F_u = \frac{18\Sigma(A)}{18\Sigma(A) + \Sigma(AU)}$$
$$= \frac{18 \times 110}{18 \times 110 + 33.9}$$
$$= 0.983$$

$$F_v = \frac{6\Sigma(A)}{6\Sigma(A) + 0.33NV}$$
$$= \frac{6 \times 110}{6 \times 110 + 0.33 \times 0.25 \times 75}$$
$$= 0.991$$

$$F_y = \frac{18\Sigma(A)}{18\Sigma(A) + \Sigma(AY)}$$
$$= \frac{18 \times 110}{18 \times 110 + 234.9}$$
$$= 0.894$$

$$F_2 = 1 + \frac{F_u\Sigma(AU)}{6\Sigma(A)}$$
$$F_2 = 1 + \frac{0.983 \times 33.85}{6 \times 110}$$
$$= 1.0504$$

Calculate the 24 h mean conduction heat gains with:

$$Q_u = \frac{F_u}{F_v} \times AU \times (\bar{t}_{eo} - t_c) \text{ W}$$

The external wall:

$$Q_u = \frac{0.983}{0.991} \times 9 \times 0.45 \times (25 - 21) \text{ W}$$
$$= 0.992 \times 9 \times 0.45 \times 4 \text{ W}$$
$$= 16 \text{ W}$$

The roof:

$$Q_u = 0.992 \times 25 \times 0.4 \times (20 - 21) \text{ W}$$
$$= -10 \text{ W}$$

The window:

$$Q_u = 0.992 A U \times (\bar{t}_{ao} - t_c) \text{ W}$$
$$= 0.992 \times 6 \times 3.3 \times (15.5 - 21) \text{ W}$$
$$= -108 \text{ W}$$

The 24 h mean conduction gain is

$$Q_u = 16 - 10 - 108 \text{ W}$$
$$= -102 \text{ W, a net loss}$$

The cyclic variation from the mean heat flow $\tilde{Q}_u$ is

$$\tilde{Q}_u = \frac{F_u}{F_v} \times A U f \tilde{t}_{eo} \text{ W}$$

The external wall:

$$\tilde{Q}_u = \frac{0.894}{0.991} \times 9 \times 0.45 \times 0.42 \times (-14.5) \text{ W}$$
$$= 0.902 \times 9 \times 0.45 \times 0.42 \times (-14.5) \text{ W}$$
$$= -22 \text{ W}$$

The roof:

$$Q_u = 0.902 \times 25 \times 0.4 \times 0.99 \times 18 \text{ W}$$
$$= 161 \text{ W}$$

The window:

$$\tilde{Q}_u = 0.902 \times A \times U \times f \times \tilde{t}_{ao} \text{ W}$$
$$= 0.902 \times 6 \times 3.3 \times 1 \times 3 \text{ W}$$
$$= 54 \text{ W}$$

The net swing in conduction gains is

$$\tilde{Q}_u = -22 + 161 + 54 \text{ W}$$
$$= 193 \text{ W}$$

The outdoor air that is 18.5 °C at noon infiltrates the conditioned office, causing a heat gain of

$$Q_v = 0.33 \times F_2 N V \times (t_{ao} - t_c) \text{ W}$$
$$= 0.33 \times 1.0504 \times 0.25 \times 75 \times (30 - 21) \text{ W}$$
$$= 58 \text{ W}$$

The four occupants each emit 90 W:

$$Q = 4 \times 90 \text{ W}$$
$$= 360 \text{ W}$$

The two computers each emit 200 W:

$$Q = 2 \times 200\,\text{W}$$
$$= 400\,\text{W}$$

The sensible heat gain Q_{ps} to the office at noon 22 September and 21 March is the sum of:

1. solar gain through the glazing, 1479 W;
2. 24 hour mean conduction through the structure, -102 W;
3. net swing in the conduction gain, 193 W;
4. ventilation air infiltration, 58 W;
5. occupancy gain, 360 W;
6. electrical equipment emission, 400 W.

$$Q_{ps} = 1479 - 102 + 193 + 58 + 360 + 400\,\text{W}$$
$$= 2388\,\text{W}$$

The air conditioning system cooling load in the room is the sensible heat gain plus the heat gain to the duct system, 5%, and the plant margin that is allowed, 10%.

$$Q_{ps} = 1.15 \times 2388\,\text{W}$$
$$= 2746\,\text{W}$$
$$= 2.746\,\text{kW}$$

$$\text{room floor area } A = 25\,\text{m}^2$$

$$\text{room volume, } V = 75\,\text{m}^3$$

$$Q_{ps} = \frac{2388\,\text{W}}{25\,\text{m}^2}$$
$$= 95.5\,\text{W/m}^2$$

$$Q_{ps} = \frac{2388\,\text{W}}{75\,\text{m}^3}$$
$$= 31.8\,\text{W/m}^3$$

$$\text{latent heat gain, } Q_{pl} = 4 \text{ people} \times \frac{50\,\text{W}}{\text{person}}$$
$$= 200\,\text{W}$$

$$\text{total heat gain, } Q_{pt} = 2388 + 200\,\text{W}$$
$$= 2588\,\text{W}$$

$$\text{outside air supply, } Q_o = 4 \text{ people} \times 10 \text{ litres/s person}$$
$$= 40\,\text{litres/s}$$

$$\text{minimum supply air temperature, } t_s = 12\,°\text{C d.b.}$$

$$\text{room air temperature, } t_r = 22\,°\text{C d.b.}$$

To find the volume flow rate, Q, of supply air into the room, the room sensible heat gain and room air temperature are entered into the equation:

$$Q = \frac{Q_{ps}\,\text{kW} \times (273 + t_s)}{351(t_R - t_S)}\,\text{m}^3/\text{s}$$

$$= \frac{2.746 \times (273 + 12)}{351 \times (22 - 12)} \, \text{m}^3/\text{s}$$

$$= 0.223 \, \text{m}^3/\text{s}$$

The leakage of air from the ductwork system is expected to be 5% of the design flow rate, so the supply air fan is to deliver

$$Q = 1.05 \times 0.223 \, \text{m}^3/\text{s}$$

$$= 0.235 \, \text{m}^3/\text{s}, \text{ to the nearest 5 litres/s}$$

The room air change rate is

$$N = Q \frac{\text{m}^3}{\text{s}} \times \frac{3600 \, \text{s}}{1 \, \text{h}} \times \frac{1 \, \text{air change}}{V \, \text{m}^3}$$

$$= 0.235 \, \frac{\text{m}^3}{\text{s}} \times \frac{3600 \, \text{s}}{1 \, \text{h}} \times \frac{1 \, \text{air change}}{75 \, \text{m}^3}$$

$$= 11.2 \, \text{air changes/h}$$

To find the supply air duct diameter:

$$Q = \frac{\pi d^2}{4} \text{m}^2 \times v \frac{\text{m}}{\text{s}}$$

$$d^2 = \frac{4Q}{\pi v} \, \text{m}^2$$

$$= \frac{4 \times 0.235}{\pi \times 5} \text{m}^2$$

$$= 0.06 \, \text{m}^2$$

$$d = 0.245 \, \text{m}$$

The standard air duct size that will maintain the air velocity below 5 m/s will be the next 50 mm diameter larger: that is, 250 mm diameter.

DATA REQUIREMENT

The user will enter new data for:

1. Your name in cell B4.
2. Job title in cell B5.
3. Job reference in cells B6 and C6.
4. Filename in cell B10.
5. Site location in cell C205.
6. Altitude, height above sea level in cell C206.
7. Type of locality in cell C207.
8. Name of room in cell C208.
9. Floor level in cell C209.
10. Activity of the occupants of the room in cell C210.
11. Design room air temperature to be maintained in cell C211.
12. Resultant temperature to be maintained in the room in cell C212.
13. Outdoor air temperature to be used for the design of the cooling load at the peak load time in cell C213. Local meteorological data is used for this

selection. It is usual to choose an outdoor air temperature that is likely to be exceeded on only three days in a year for comfort air conditioning. Where the room conditions must never be exceeded, the highest ambient temperature that is expected is entered.

14. The rate of infiltration of outside air into the air conditioned room in cell C214. The infiltration is a direct heat gain in the room. If the air conditioned room is slightly pressurized, air leakage will be outwards, and zero can be entered.

15. The supply of outside air into the room in litres per second per person in cell C215. This is normally 10 litres/s for most applications, 15 litres/s where smoking is permitted, and up to 25 litres/s for heavy smoking rooms. It may be possible to provide less than 10 litres/s if air cleaning is provided for the room, depending upon codes and regulations.

16. The rate of supply of outside air per square metre of floor area in cell C216. This is used when the number of occupants for the room is not entered.

17. The exposure of the room in cell C226.

18. The room occupancy in cell C228.

19. The sensible heat emission from each occupant in cell C229.

20. The latent heat emission from each occupant in cell C230.

21. The lighting heat emission into the cooled room in cell C231.

22. The power of all the electrical equipment that is in use during the cooling load time within the room in cell C232.

23. The operation hours for the air conditioning system per day in cell C233.

24. The date for calculation of the room cooling load in cell C235.

25. The length of floor area 1 in the room in cell C246.

26. The width of floor area 1 in cell C247.

27. The height of floor area 1 in cell C248.

28. The length, width and height of area 2 of the room in the cell range C251 to C253.

29. The thermal transmittance of the windows, door and roof lights in the cell range C266 to C268.

30. The orientation of external wall 1 in cell D286.

31. Thermal transmittance of the wall in cell D287.

32. Thermal admittance of the wall in cell D288.

33. Cooling load through the vertical glazing in cell C289.

34. Window solar gain transmission factor in cell D290.

35. Time of the cooling load in cell D291.

36. Thermal time lag for the external wall in cell D292.

37. Decrement factor for the wall in cell D294.

38. Mean 24 h sol-air temperature for the wall in cell D295.

39. Sol-air temperature at the time of heat gain for the external wall in cell C296.

40. Mean 24 h outside air temperature in cell D298.

41. Outside air temperature at the time of the room cooling load in cell D299.

42. Data for external walls 2, 3, 4 and the external roof in the cell range D306 to D379 as for wall 1.

43. Orientation of the internal wall in cell D386; enter internal.

44. Thermal transmittance of the internal wall in cell D387.

45. Thermal admittance of the internal wall in cell D388.

46. Surface factor of the internal wall in cell D389.

47. Time lag for the internal wall in cell D390.

48. Resultant temperature of the adjacent space in cell D391.
49. Length of external wall 1 in cell E403.
50. Height of external wall 1 in cell F403.
51. Length and height of the windows and doors in wall 1 in the cell range E404 to F406.
52. Dimensions for external walls 2, 3 and 4 in the cell range E411 to F436.
53. Orientation of floor 1 in cell B445; enter internal, timber, exposed or ground.
54. There would not normally be any cooling load or correction factor for floors in the cell range C445 to D446.
55. Length of floor area 1 in cell E445.
56. Width of floor area 1 in cell F445.
57. Sol-air temperature for the floor in cell H445. This does not normally apply; enter the resultant temperature of the conditioned room.
58. Thermal transmittance of the floor in cell I445.
59. Thermal admittance of the floor in cell K445.
60. Decrement factor for the floor in cell M445.
61. Time lag for the floor in cell N445.
62. Mean 24 h sol-air temperature for the floor in cell O445. Enter the average of the values for the building orientations.
63. Mean 24 h outside air temperature in cell P445.
64. Time of occurrence of the solar heat gain to the floor in cell Q445.
65. Sol-air temperature at the time of solar heat gain to the floor in cell R445. This does not apply; enter the resultant temperature of the conditioned room.
66. The data for floor area 2, the two ceiling areas and the roof are entered in the cell range B446 to R450 as for floor area 2. Enter the resultant and air temperatures of the adjacent room in place of the outdoor values for the ceilings. enter the correct sol-air temperatures for a roof.
67. The name of the current file in cells A529 and B529. The worksheet has lines for 12 files of rooms. The number of files can be extended by copying lines downwards.
68. The room sensible heat cooling load in kilowatts in cell C529.
69. The room latent heat cooling load in kilowatts in cell D529.
70. The percentage addition to the cooling load for the heat gain to the air duct system in cell E545.
71. The percentage to be added to the room cooling load for the plant margin in cell E546.
72. Surplus plant capacity to be included in cell E547.
73. Surplus latent heat plant capacity to be included in cell E548.
74. Minimum supply air temperature at the supply air outlet grille or diffuser in cell C568.
75. The percentage rate of leakage from the distribution duct system in cell C570.
76. The limiting air velocity in the air duct system in cell C574.

OUTPUT DATA

1. Current date is given in cell B7.
2. Current time is given in cell B8.
3. Outside air requirement for the occupants is given in cell E237.

4. Outside air requirement based on the floor area is given in cell E238.
5. Outside air to be used for design is given in cell E239.
6. Volume of area 1 of the room is given in cell E249.
7. Floor area 1 is given in cell E250.
8. Volume of area 2 of the room is given in cell E254.
9. Floor area 2 is given in cell E255.
10. Total volume of the room is given in cell E257.
11. Floor area of the room is given in cell E258.
12. Date used for calculation of the room cooling load is given in cell F285.
13. The time that the heat gain to wall 1 occurs is given in cell F293.
14. Swing in sol-air temperature between the mean 24 h temperature and that at the time of heat gain is given in cell F297.
15. Swing in outside air temperature between the mean 24 h temperature and that at the time of the room cooling load is given in cell F300.
16. Data for walls 2, 3, 4 and the roof is given in the cell range F305 to F380 as for wall 1.
17. Orientation of wall 1 is given in cell B403.
18. Gross area of wall 1 is given in cell G403.
19. Orientation of window 1 is given in cell B404.
20. Cooling load through window 1 is given in cell C404.
21. Glazing cooling load correction factor for window 1 is given in cell D404.
22. Area of window 1 is given in cell G404.
23. Sol-air temperature for window 1 is given in cell H404.
24. Thermal transmittance of window 1 is given in cell I404.
25. Product of area and thermal transmittance for window 1 is given in cell J404.
26. Thermal admittance of window 1 is given in cell K404.
27. Product of area and thermal admittance for window 1 is given in cell L404.
28. Decrement factor for window 1 is given in cell M404.
29. Time lag for window 1 is given in cell N404.
30. Mean 24 h sol-air temperature is given in cell O404.
31. Mean 24 h outside air temperature is given in cell P404.
32. Time of occurrence of the heat gain through window 1 is given in cell Q404.
33. Sol-air temperature at the time of heat gain through window 1 is given in cell R404.
34. Swing in sol-air temperature for window 1 is given in cell S404.
35. Mean 24 h heat gain to the room through window 1 is given in cell T404.
36. Swing in heat gain at the time of the room cooling load from the 24 h mean heat gain through window 1 is given in cell U404.
37. Cooling load through window 1 due to solar radiation is given in cell W404.
38. The heat gain data for window 2 is given in cell range B405 to W405 as for window 1
39. The heat gain data for door 1 is given in cell range B406 to W406 as for window 1.
40. Net area of wall 1 is given in cell G407.
41. Sol-air temperature for the wall is given in cell H407.
42. Thermal transmittance of the wall is given in cell I407.
43. Product of area and thermal transmittance for the wall is given in cell J407.
44. Thermal admittance of the wall is given in cell K407.
45. Product of area and thermal admittance for the wall is given in cell L407.

46. Decrement factor for the wall is given in cell M407.
47. Time lag for the wall is given in cell N407.
48. Mean 24 h sol-air temperature is given in cell O407.
49. Mean 24 h outside air temperature is given in cell P407.
50. Time of occurrence of the heat gain through the wall is given in cell Q407.
51. Sol-air temperature at the time of heat gain through the wall is given in cell R407.
52. Swing in sol-air temperature for the wall is given in cell S407.
53. Mean 24 h heat gain to the room through the wall is given in cell T407.
54. Swing in heat gain at the time of the room cooling load from the 24 h mean heat gain through the wall is given in cell U407.
55. The data for walls 2, 3 and 4 is given in the cell range B411 to W437 as for wall and window 1.
56. Area of floor 1 is given in cell G445.
57. The product of area and thermal transmittance for the floor is given in cell J445.
58. The product of area and thermal admittance for the floor is given in cell L445.
59. Swing in sol-air temperature for the roof is given in cell S449.
60. Mean 24 h heat gain through the roof is given in cell T449.
61. Swing in heat gain through the roof is given in cell U449.
62. Cooling load through the glazing in the roof is given in cell W449.
63. Data for roof area 2 is given in cell range G450 to W450 as for roof 1.
64. 24 h mean heat gain to the room is given in cell T452.
65. Swing in heat gain between the mean 24 h heat gain and that at the time of room cooling load is given in cell U452.
66. Glazing cooling load for the room is given in cell W452.
67. Total room surface area is given in cell D463.
68. Check sum for the surface area of the room is given in cell D464.
69. Any difference between the two calculations of the room surface area is given in cell D465.
70. Summation of area and thermal transmittance for the whole room is given cell D466.
71. Summation of area and thermal admittance for the whole room is given in cell D467.
72. Infiltration rate of outdoor air into the conditioned room is given in cell D468.
73. Dimensionless factor F_v is given in cell D469.
74. Dimensionless factor F_u is given in cell D470.
75. Dimensionless factor F_y is given in cell D471.
76. Dimensionless factor F_2 is given in cell D472.
77. Ratio F_u/F_v is given in cell D473.
78. Ratio F_y/F_v is given in cell D474.
79. Ratio $\Sigma(AU)/\Sigma(A)$ is given in cell D475.
80. Ratio $\Sigma(NV)/3\Sigma(A)$ is given in cell D476.
81. Solar gain through the windows is given in cell D484.
82. Sensible heat gain from the occupants is given in cell D486.
83. Lighting heat gain is given in cell D488.
84. Heat gain from electrical equipment is given in cell D490.
85. 24 h mean conduction heat gain to the room is given in cell D492.
86. Net swing in conduction heat gain is given in cell D494.
87. Heat gain due to the infiltration of outdoor air is given in cell D496.

88. Sensible heat gain to the room is given in cell D499.
89. Latent heat gain from the occupants is given in cell D502.
90. Sensible heat gain to the room (W) is given in cell B505.
91. Sensible heat gain to the room (kW) is given in cell B506.
92. Sensible heat gain (W/m^3 of room volume) is given in cell B507.
93. Sensible heat gain (W/m^3 of floor area) is given in cell B508.
94. Latent heat gain (W) is given in cell B509.
95. Latent heat gain (kW) is given in cell B510.
96. Total heat gain to the room is given in cell B511.
97. Total heat gain in (W/m^2 of floor area) is given in cell B512.
98. Total heat gain to each room is given in the cell range E529 to E540.
99. Sensible, latent and total heat gains, additions, margin and surplus capacity for the room is given in the cell range C552 to E557.
100. Name of the room is given in cell E565.
101. Air temperature in the room is given in cell E566.
102. Room volume is given in cell E567.
103. Room sensible heat gain is given in cell E569.
104. Conditioned supply air quantity (m^3/s) is given in cell E571.
105. Supply air quantity (l/s) is given in cell E572.
106. Room air change rate (air changes/h) is given in cell E573.
107. The minimum supply air duct diameter required is given in cell E575.

FORMULAE

Representative samples of the formulae are given here. Each formula can be read by moving the cursor to the cell. The equation is presented in the form in which it would normally be written, and in the format that is used by the spreadsheet.

Cell B7

@TODAY

This function produces the serial number of the current day and time. The cell is formatted to display the date.

Cell B8

@TODAY

This function produces the serial number of the current day and time. The cell is formatted to display the time.

Cell E237

+C228*C215

The outside air flow rate to be supplied into the building is calculated. The air flow rate is to meet the required standard for air quality within the air conditioned room.

$$Q_o = \text{No. occupants} \times Q \text{ l/s}$$
$$E237 = C228 \times C215 \text{ l/s}$$

Cell E238

+E258*C216

The rate of supply of outside air can be based upon the number of occupants where this is known. When only the floor area and the usage of the space are known, the outside air requirement is found from a standard air flow per square metre of floor area.

$$Q_o = A_f \, \text{m}^2 \times Q_o \, \text{litres/s} \, \text{m}^2$$
$$E238 = E258 \times C216 \, \text{litres/s}$$

Cell E239

@IF(E237⟩E238, E237, E238)

The greater of the two outside air requirements is selected for use, and it is shown in cell E239. The conditional IF statement means that when the outside air requirement calculated from the occupancy is greater than that from the floor area, the occupancy value is put into cell E239. If the occupancy air flow is not greater than the floor area calculated value, the contents of cell E238 are put into cell E239, and the floor area outside air flow is selected.

Cell E249

+C246*C247*C248

The room can be entered as two spaces. This allows an L-shaped room to be entered as one room. The volume is entered for area 1.

$$V_1 = L \, \text{m} \times W \, \text{m} \times H \, \text{m}$$
$$E249 = C246 \times C247 \times C248 \, \text{m}^3$$

Cell E250

+C246*C247

The floor area is found.

$$A_1 = L \, \text{m} \times W \, \text{m}$$
$$E250 = C246 \times C247 \, \text{m}^2$$

Cell E254

+C251*C252*C253

The volume is entered for area 2:

$$V_2 = L \, \text{m} \times W \, \text{m} \times H \, \text{m}$$
$$E254 = C251 \times C252 \times C253 \, \text{m}^3$$

Cell E255

+C251*C252

The floor area is found:

$$A_2 = L\,\text{m} \times W\,\text{m}$$
$$E255 = C251 \times C252\,\text{m}^2$$

Cell E257

+E249+E254

Total room volume is

$$V = V_1 + V_2\,\text{m}^3$$
$$E257 = E249 + E254\,\text{m}^3$$

Cell E258

+E250+E255

Total room floor area is

$$A = A_1 + A_2\,\text{m}^2$$
$$E258 = E250 + E255\,\text{m}^2$$

Cell F293

+D291−D292

The time of day, or night, when the solar heat gain occurs at the outside wall or roof surface, is found from the time of the room cooling load minus the time lag.

Time of heat gain in the room = 12.00 noon

Wall time log = 9 h

Solar gain, or loss, at the outside wall surface occurs at

$$12\ \text{noon} - 9\,\text{h} = 3\ \text{a.m.}$$
$$F293 = D291 - D292\,\text{h}$$

Cell F297

+D296−D295

The swing of the sol-air temperature from the 24 h mean value is

$$\tilde{t}_{eo} = t_{eom} - t_{eol}\ {}^\circ\text{C}$$

where $\tilde{t}_{eo}$ = sol-air swing (°C); t_{eom} = 24 h mean sol-air (°C); and t_{eol} = sol-air at time lag prior to heat gain occurring in room (°C).

$$F297 = D296 - D295\ {}^\circ\text{C}$$

Cell F300

+D299−D298

The swing of the air temperature from the 24 h mean value is

$$\tilde{t}_{ao} = t_{aom} - t_{aol} \, °C$$

where $\tilde{t}_{ao}$ = air temperature swing (°C); t_{aom} = 24 h mean air temperaure (°C); and t_{aol} = air temperature time lag prior to heat gain occurring in room (°C).

$$F300 = D299 - D298 \, °C$$

Cells F313 to F380

The time lag, sol-air and air temperatures are evaluated for the other three external walls and the roof as for wall 1.

Cell F385

+$C$235

The date for calculation of the cooling load is copied from cell C235 into cell F385 for information.

Cell B403

+$D$286

The orientation of wall 1 is copied from cell D286 into cell B403. This is repeated in cell range B404 to B406.

Cell G403

+E403*F403

The gross area of external wall 1 is

$$A = L \, m \times H \, m$$
$$G403 = E403 \, m \times F403 \, m$$

Cells C404 to C406

+$D$289

The cooling load through the vertical glazing in wall 1 is copied from cell D289.

Cells D404 to D406

+$D$290

The glazing cooling load correction factor for glass type is copied from cell D290.

Cell G404

+E404*F404

The area of window 1 is

$$A = L\,\text{m} \times H\,\text{m}$$
$$\text{G404} = \text{E404} \times \text{F404 m}^2$$

Cell G405

+E404*F405

The area of window 2 is

$$A = L\,\text{m} \times H\,\text{m}$$
$$\text{G405} = \text{E405} \times \text{F405 m}^2$$

Cell G406

+E406*F406

The area of the external door is

$$A = L\,\text{m} \times H\,\text{m}$$
$$\text{G406} = \text{E406} \times \text{F406 m}^2$$

Cell G407

+G403−G404−G405−G406

The net area of external wall 1 is

$$A = A_1 - W_1 - W_2 - D_1 \text{ m}^2$$

where A = net wall area (m^2); A_1 = window 1 area (m^2); A_2 = window 2 area (m^2); and D_1 = door area (m^2).

$$\text{G407} = \text{G403} - \text{G404} - \text{G405} - \text{G406 m}^2$$

Cells H404 to H407

+$D$298

The sol-air temperature for wall 1 at the time of the room cooling load is copied from cell D298.

Cells I404 to I407

+$C$266

The thermal transmittance for wall 1 is copied from cell C266.

Cells J404 to J407

+G404*I404

The product of area and thermal transmittance is found for each surface of wall 1.

$$A \times U = A\,\text{m}^2 \times U\,\text{W/m}^2\,\text{K}$$
$$J404 = G404 \times I404\,\text{W/K}$$

Cells K404 to K406

+$C$266

The thermal admittance of the windows and doors is copied from cell C266.

Cell K407

+$D$288

The thermal admittance of the external wall is copied from cell D288, into cell K407.

Cells L404 to L407

+G404*K404

The product of area and thermal admittance is found for the external windows, doors and wall.

$$A \times Y = A\,\text{m}^2 \times Y\,\text{W/m}^2\,\text{K}$$
$$L404 = G404 \times K404\,\text{W/K}$$

Cells M404 to M406

The decrement factor for glazing is always 1.0, because it has negligible thermal storage. The cooling load data has taken the properties of the different types of glass into account.

Cell M407

+D294

The decrement factor for the outside wall is copied from cell D294 into cell M407.

Cells N404 to N406

The time lag for glazing is always zero, because it has negligible thermal storage.

Cell N407

+D292

The time lag for the outside wall is copied from cell D292 into cell N407.

Cells O404 to O406

The mean 24 h sol-air temperature is not used for glazing as there are no thermal storage calculations.

Cell O407

+D295

The mean 24 h sol-air temperature for the outside wall is copied from cell D295 into cell O407.

Cells P404 to P407

+$D$298

The mean 24 h outside air temperature for the location is copied from cell D298.

Cells Q404 to P406

+$D$291

The time at which the heat gain takes place at the outside surface of the glazing is copied from cell D291.

Cell Q407

+$F$293

The time at which the heat gain takes place at the outside surface of the external wall is copied from cell F293 into cell Q407.

Cells R404 to R406

+$D$299

The outside air temperature at the time of heat gain through the glazing is copied from cell D299. The sol-air temperature is not used for glazing.

Cell R407

+D296

The sol-air temperature at the time of heat gain through the outside wall is copied from cell D296 into cell R407.

Cells S404 to S407

+R404−P404

The swing in outside temperature from the mean 24 h temperature and the value at the cooling load time is found. The air temperatures are used for glazing and the sol-air temperature for the wall.

$$\tilde{t}_{eo} = t_{eom} - t_{eo} \text{ }^\circ C$$
$$S404 = R404 - P404 \text{ }^\circ C$$

Cells T404 to T407

+$D$473*J204*(P404−$C$212)

The mean 24 h conduction heat gain through the glazing, door and external wall is

$$Q_u = (F_u/F_v)\Sigma(AU) \times (\bar{t}_{ao} - t_c) \text{ W}$$
$$T404 = D473 \times J204 \times (P404 - C212) \text{ W}$$

Cells U404 to U407

+$D$474*J404*M404*S404

The swing in heat gain between the mean 24 h and actual heat flow at the time of room cooling load is

$$\tilde{Q}_u = (F_y/F_v)\Sigma(AU) \times f\,\tilde{t}_{eo} \text{ W}$$
$$U404 = D474 \times J404 \times M404 \times S404 \text{ W}$$

Cells W404 to W405

+C404*D404*G404

The room cooling load due to solar radiation heat gain through the glazing is

$$Q = I\,\text{W/m}^2 \times k \times A\,\text{m}^2$$
$$W404 = C404 \times D404 \times G404 \text{ W}$$

Cells B411 to W437

The heat gain calculations for external walls 2, 3 and 4, including the glazing and doors, are conducted as for wall 1.

Cells G445 to G450

+E445*F445

The areas of the floors, ceilings and roof surfaces are calculated from

$$A = L\,\text{m} \times W\,\text{m}$$
$$G445 = E445 \times F445\,\text{m}^2$$

Cells J445 to J450

+G445*I445

The product of surface area and thermal transmittance is found for the floor, ceiling and roof surfaces from

$$(A \times U) = A\,\text{m}^2 \times U\,\text{W/m}^2\,\text{K}$$
$$J445 = G445 \times I445\,\text{W/K}$$

Cells L445 to L450

+G445*K445

The product of surface area and thermal admittance is found for the floor, ceiling and roof surfaces from

$$A \times Y = A\,\mathrm{m}^2 \times Y \ \mathrm{W/m^2\,K}$$
$$L445 = G445 \times K445 \ \mathrm{W/K}$$

Cells S445 to S450

+R445−O445

The swing in sol-air temperature for the floor, ceiling and roof is

$$\tilde{t}_{\mathrm{eo}} = t_{\mathrm{eo}} - t_{\mathrm{eom}} \ ^\circ\mathrm{C}$$
$$S445 = R445 - O445 \ ^\circ\mathrm{C}$$

Cells T445 to T450

+$D$473*J445*(O445−$C$212)

The mean 24 h conduction heat gain through the floor, ceiling and roof is

$$Q_{\mathrm{u}} = (F_{\mathrm{u}}/F_{\mathrm{v}})(AU)(t_{\mathrm{eom}} - t_{\mathrm{c}}) \ \mathrm{W}$$
$$T445 = D473 \times J445 \times (O445 - C212) \ \mathrm{W}$$

Cell T452

@SUM (T404..T450)

The sum of the mean 24 h conduction heat gains to the room is

$$\Sigma Q_{\mathrm{u}} = \Sigma(Q_{\mathrm{u}1} + Q_{\mathrm{u}2} + \ldots + Q_{\mathrm{u}22}) \ \mathrm{W}$$
$$T452 = T404 + T405 + \ldots + T450 \ \mathrm{W}$$

Cells U445 to U450

+$D$474*J445*M445*S445

The swing in heat gain through the floor, ceiling and roof is

$$\tilde{Q}_{\mathrm{u}} = (F_{\mathrm{y}}/F_{\mathrm{v}})(AU)f\tilde{t}_{\mathrm{eo}} \ \mathrm{W}$$
$$U445 = D474 \times J445 \times M445 \times S445 \ \mathrm{W}$$

Cell U452

@SUM (U404..U450)

The net swing in conduction heat gains to the room is

$$\Sigma \tilde{Q}_{\mathrm{u}} = \Sigma(\tilde{Q}_{\mathrm{u}1} + \tilde{Q}_{\mathrm{u}2} + \ldots + \tilde{Q}_{\mathrm{u}22}) \ \mathrm{W}$$
$$U452 = U404 + U405 + \ldots + U450 \ \mathrm{W}$$

Cells W447 to W450

+C447*D447*G447

The cooling load due to the roof lights is

$$Q = I \, \text{W/m}^2 \times k \times A \, \text{m}^2$$
$$\text{W447} = \text{C447} \times \text{D447} \times \text{G447} \, \text{W}$$

Cell W452

@SUM (W404..W450)

The total cooling load through the glazing in the walls and the roof is

$$Q = Q_1 + Q_2 + \ldots + Q_{20} \, \text{W}$$
$$\text{W452} = \text{W404} + \text{W405} + \ldots + \text{W450} \, \text{W}$$

Cell D463

+G404+G411+G425+G433+G445+G446+G447+G448+G449+G450

The total surface area of the room is

$$A = A_1 + A_2 + \ldots + A_{10} \, \text{m}^2$$
$$\text{D463} = \text{G404} + \text{G411} + \ldots + \text{G450} \, \text{m}^2$$

Cell D464

+G404+G405+G406+G407+G412+G413+G414+G415+G426+G427+G428
+G429+G434+G435+G436+G437+G445+G446+G447+G448+G449+G450

The total of the room surface areas is checked by adding the individual areas:

$$A = A_1 + A_2 + \ldots + A_{22} \, \text{m}^2$$
$$\text{D464} = \text{G404} + \text{G405} + \ldots + \text{G450} \, \text{m}^2$$

Cell D465

+D463−D464

Any error between the two methods of summing the room surface areas is shown as

$$\text{error} = A_1 - A_2 \, \text{m}^2$$
$$\text{D465} = \text{D463} - \text{D464} \, \text{m}^2$$

Cell D466

@SUM (J404..J450)

The sum of the surface areas and thermal transmittances for the whole room is

$$\Sigma(AU) = A_1 U_1 + \ldots + A_{22} U_{22} \, \text{W/K}$$
$$\text{D466} = \text{J404} + \ldots + \text{J450} \, \text{W/K}$$

Cell D467

@SUM(L404..L450)

The sum of the surface areas and thermal admittances for the whole room is

$$\Sigma(AY) = A_1 Y_1 + \ldots + A_{22} Y_{22} \text{ W/K}$$
$$D467 = L404 + \ldots + L450 \text{ W/K}$$

Cell D468

+C214*E257/3600

The rate of infiltration of outdoor air into the conditioned room is

$$Q = N \frac{\text{air change}}{\text{h}} \times V \, \text{m}^3 \times \frac{1 \, \text{h}}{3600 \, \text{s}}$$
$$D468 = C214 \times E257 \, \text{m}^3/\text{s}$$

Cell D469

6*D463/(6*D463+0.33*C214*E257)

Dimensionless ventilation factor:

$$F_v = \frac{6\Sigma A}{6\Sigma A + 0.33NV}$$
$$D469 = \frac{6 \times D463}{6 \times D463 + 0.33 \times C214 \times E257}$$

Cell D470

18*D463/(18*D463+D466)

Dimensionless thermal transmittance factor:

$$F_u = \frac{18\Sigma A}{18\Sigma A + \Sigma(AU)}$$
$$D470 = \frac{18 \times D463}{18 \times D463 + D466}$$

Cell D471

18*D463/(18*D463+D467)

Dimensionless thermal admittance factor:

$$F_y = \frac{18\Sigma A}{18\Sigma A + \Sigma(AY)}$$
$$D471 = \frac{18 \times D463}{18 \times D463 + D467}$$

Cell D472

1+(D470*D466/6/D463)

Dimensionless factor:

$$F_2 = 1 + \frac{F_u(AU)}{6\Sigma(A)}$$

$$\text{D472} = 1 + \frac{\text{D470} \times \text{D466}}{6 \times \text{D463}}$$

Cell D473

+D470/D469

The ratio of the dimensionless factors,

$$\frac{F_u}{F_v} = \frac{\text{D470}}{\text{D469}}$$

Cell D474

+D471/D469

The ratio of the dimensionless factors,

$$\frac{F_y}{F_v} = \frac{\text{D471}}{\text{D469}}$$

Cell D475

+D466/D463

The mean thermal transmittance for the room is

$$U = \frac{\Sigma(AU)}{\Sigma(A)} \ \text{W/m}^2\,\text{K}$$

$$\text{D475} = \frac{\text{D466}}{\text{D463}} \ \text{W/m}^2\,\text{K}$$

Cell D476

+C214*E257/3/D463

The ventilation factor is

$$\frac{NV}{3A} = \frac{\text{C214} \times \text{E257}}{3 \times \text{D463}} \ \text{m/h}$$

Cell D484

+W452

The room cooling load due to the solar heat gain through the glazing, Q_s W, is copied from cell W452 into cell D484.

Cell D486

+C228*C229

The occupancy sensible heat gain is

$$Q_o = \text{occupants} \times \text{sensible W/person}$$
$$D486 = C228 \times C229 \text{ W}$$

Cell D488

+E258*C231

The room cooling load from the artificial lighting is

$$Q_1 = \text{floor area } A \text{ m}^2 \times \text{lighting W/m}^2$$
$$D488 = E258 \times C231 \text{ W}$$

Cell D490

+C232

The heat emission from the electrical equipment within the room, Q_e W, is copied from cell C232 into cell D490.

Cell D492

@SUM(T404..T450)

The mean 24 h conduction heat gain to the air conditioned room is found from the net sum of the gains through the building surfaces:

$$Q_u = Q_{u1} + Q_{u2} + \ldots + Q_{u22} \text{ W}$$
$$D492 = T404 + T405 + \ldots + T450 \text{ W}$$

Cell D494

@SUM(U404..U450)

The swing in the conduction heat gain between the mean 24 h and heat gain at the time of the room cooling load is found from the net sum of the swing in gains:

$$\tilde{Q}_u = \tilde{Q}_{u1} + \tilde{Q}_{u2} \ldots + \tilde{Q}_{u22} \text{ W}$$
$$D494 = U404 + U405 + \ldots + U450 \text{ W}$$

Cell D496

0.33*D472*C214*E257*(C213−C212)

The room cooling load from the infiltration of outside air directly into the room is

$$Q_v = 0.33 F_2 N V \times (t_{ao} - t_c) \text{ W}$$
$$D496 = 0.33 \times D472 \times C214 \times E257 \times (C213 - C212) \text{ W}$$

Cell D499

@SUM(D484..D496)

The design sensible heat gain for the room is found from

$$Q_{ps} = Q_s + Q_o + Q_l + Q_e + Q_u + \tilde{Q}_u + Q_v \text{ W}$$
$$D499 = D484 + D485 + \ldots + D496 \text{ W}$$

Cell D502

+C228*C230

The latent heat gain from the room occupants is

$$Q_{pl} = \text{occupants} \times \text{latent W/person}$$
$$D502 = C228 \times C230 \text{ W}$$

Cell B505

+D499

The sensible cooling load, Q_{ps} W, for the room is copied from cell D499 into cell B505.

Cell B506

+B505/1000

The room design sensible cooling load is

$$Q_{ps} = Q_{ps} \text{ W} \times \frac{1 \text{ kW}}{10^3 \text{ W}}$$
$$B506 = \frac{B505}{1000} \text{ kW}$$

Cell B507

+B505/E257

The design sensible cooling load per cubic metre of room volume is

$$Q_{ps} = \frac{Q_{ps} \text{ W}}{V \text{ m}^3}$$
$$B507 = \frac{B505}{E257} \text{ W/m}^3$$

Cell B508

+B505/E258

The design sensible cooling load per square metre of room floor area is

$$Q_{ps} = \frac{Q_{ps} \text{ W}}{A \text{ m}^2}$$
$$B508 = \frac{B505}{E258} \text{ W/m}^2$$

Cell B509

+D502

The room latent heat gain, Q_{pl} W, is copied from cell D502 into cell B509.

Cell B510

+B509/1000

The room latent heat gain is

$$Q_{pl} = Q_{pl} \, \text{W} \times \frac{1 \, \text{kW}}{10^3 \, \text{W}}$$

$$\text{B510} = \frac{\text{B509}}{1000} \, \text{kW}$$

Cell B511

+B506+B510

The room total heat gain is

$$Q_p = Q_{ps} + Q_{pl} \, \text{kW}$$

$$\text{B511} = \text{B506} + \text{B510} \, \text{kW}$$

Cell B512

+B511*1000/E258

The room total heat gain per square metre of floor area is

$$Q_p = \frac{Q_p \, \text{kW}}{A \, \text{m}^2} \times \frac{10^3 \, \text{W}}{1 \, \text{kW}}$$

$$\text{B512} = \frac{\text{B511}}{\text{E258}} \times 1000 \, \text{W/m}^2$$

Cells E259 to E540

+C529+D529

The room total heat gain is

$$Q_p = Q_{ps} + Q_{pl} \, \text{kW}$$

$$\text{E529} = \text{C529} + \text{D529} \, \text{kW}$$

Cell C552

@SUM(C529..C540)

The sum of the sensible room cooling loads that have been listed in the current file is

$$Q_{ps} = Q_{ps1} + Q_{ps2} + \ldots + Q_{ps12} \, \text{kW}$$

$$\text{C552} = \text{C529} + \text{C530} + \ldots + \text{C540} \, \text{kW}$$

The current file can be used to collect the cooling loads for up to 12 rooms. Additional lines can be inserted after row number 540; the formula in this cell is then changed by the user to include the extra cells.

Cell D552

+B510

The room latent heat gain, Q_{pl} kW, is copied from cell B510 into cell D552.

Cell E552

+C552+D552

The room design total heat gain is

$$Q_p = Q_{ps} + Q_{pl} \text{ kW}$$
$$E552 = C552 + D552 \text{ kW}$$

Cell C553

+A553*C552/100

An addition to the room design sensible cooling load is made for the temperature increase by the ducted air as it passes through unconditioned spaces. Some of the supply and return air ductwork may be external to the building envelope, usually on the roof. Air ducts that pass through ceiling spaces will absorb heat from light fittings, solar gains to the spaces and heat emission from hot pipes and surfaces. Such additions affect only the sensible cooling load.

$$Q_{ps} = Q_{ps} \times \frac{\text{gain} \%}{100} \text{ kW}$$
$$C553 = C552 \times \frac{A553}{100} \text{ kW}$$

Cell E553

+C553+D553

The addition to the room total heat gain is

$$Q_p = Q_{ps} + Q_{pl} \text{ } kW$$
$$E553 = C553 + D553 \text{ kW}$$

Normally, the user will enter zero for the additional latent room load in cell D553. There may be occasions when a value needs to be entered for some additional latent cooling capacity.

Cell C554

+A554*C552/100

A percentage margin can be added to the room design sensible cooling load to create surplus plant capacity:

$$Q_{ps} = \text{margin}\% \times \frac{Q_{ps}}{100} \text{ kW}$$

$$C554 = A554 \times \frac{C552}{100} \text{ kW}$$

Cell E554

+C554+D554

The plant margin to be added to the room total heat gain is

$$Q_p = Q_{ps} + Q_{pl} \text{ kW}$$

$$E554 = C554 + D554 \text{ kW}$$

Normally, the user will enter zero for the additional latent room load in cell D554.

Cell C555

+E547

The surplus plant capacity, Q_{ps} kW, is copied from cell E547 into cell C555.

Cell D555

+E548/1000

The surplus plant latent heat capacity, Q_{pl} kW, is

$$Q_{pl} = Q_{pl} \text{ W} \times \frac{1 \text{ kW}}{10^3 \text{ W}}$$

$$D555 = \frac{E548}{1000} \text{ kW}$$

Cell E555

+C555+D555

The surplus total heat plant capacity is

$$Q_p = Q_{ps} + Q_{pl} \text{ kW}$$

$$E555 = C555 + D555 \text{ kW}$$

Cell C557

@SUM (C552..C555)

The plant sensible cooling capacity, Q_{ps} kW, is

$$Q_{ps} = Q_{ps1} + Q_{ps2} + Q_{ps3} + Q_{ps4} \text{ kW}$$

where Q_{ps1} = design room sensible heat gain (kW); Q_{ps2} = sensible heat addition for the duct system (kW); Q_{ps3} = sensible heat addition for the plant margin (kW); and Q_{ps4} = plant surplus sensible heat capacity (kW).

$$C557 = C552 + C553 + C554 + C555 \text{ kW}$$

Cell D557

@SUM(D552..D555)

The plant latent cooling capacity, Q_{pl} kW, is

$$Q_{pl} = Q_{pl1} + Q_{pl2} + Q_{pl3} + Q_{pl4} \text{ kW}$$

where Q_{pl1} = design room latent heat gain (kW); Q_{pl2} latent heat addition for the duct system (kW); Q_{pl3} latent heat addition for the plant margin (kW); and Q_{pl4} plant surplus latent heat capacity (kW).

$$D557 = D552 + D553 + D554 + D555 \text{ kW}$$

Cell E557

+C557+D557

The plant total cooling capacity is

$$Q_p = Q_{ps} + Q_{pl} \text{ kW}$$

where Q_{ps} = plant sensible cooling capacity (kW), and Q_{pl} = plant latent cooling capacity (kW).

$$E557 = C557 + D557$$

Cell E565

+C208

The name of the air conditioned room is copied from cell C208 into cell E565.

Cell E566

+C211

The room design air temperature, t_{ai} °C, is copied from cell C211 into cell E566.

Cell E567

+E257

The room volume, V m³, is copied from cell E257 into cell E567.

Cell E569

+C557

The plant sensible cooling load for the room, Q_{ps} kW, is copied from cell C557 into cell E569.

Cell E571

(100+C570)*E569*(273+C568)/351/(E566−C568)/100

The supply air volume flow rate that is required to meet the plant sensible cooling load for that room is

$$Q = \frac{100 + L\%}{100} \times \frac{Q_{ps}\,\text{kW}}{(t_{ai} - t_s)} \times \frac{273 + t_s}{351} \frac{\text{m}^3}{\text{s}}$$

where Q = supply air (m^3/s); L = duct leakage (%); Q_{ps} = plant sensible load (kW); t_{ai} = room air temperature (°C d.b.); and t_s = supply air temperature at room outlet (°C d.b.).

$$E571 = \frac{100 + C570}{100} \times \frac{E569\,\text{kW}}{E566 - C568} \times \frac{273 + C568}{351} \frac{\text{m}^3}{\text{s}}$$

Note that plant additional capacity and margin are included in the value for Q_{ps}. The user may enter zero for the plant margin percentage and the surplus capacity in kW to evaluate the supply air flow. Additions can then be made to the refrigeration plant capacity to facilitate the selection of a suitable machine from catalogues.

Cell E572

+E571*1000

The supply air flow rate Q litres/s is

$$Q = Q\,\frac{\text{m}^3}{\text{s}} \times \frac{10^3\,\text{litres}}{1\,\text{m}^3}$$

$$E572 = E571\,\frac{\text{m}^3}{\text{s}} \times \frac{10^3\,\text{litres}}{1\,\text{m}^3}$$

Cell E573

+E571*3600/E567

The room air change rate is

$$N = Q\,\frac{\text{m}^3}{\text{s}} \times \frac{3600\,\text{s}}{1\,\text{h}} \times \frac{1}{V\,\text{m}^3}$$

where N = air changes per hour (h^{-1}); Q = supply air flow rate (m^3/s); and V = room volume (m^3).

$$E573 = E571\,\frac{\text{m}^3}{\text{s}} \times \frac{3600\,\text{s}}{1\,\text{h}} \times \frac{1}{E567\,\text{m}^3}$$

Cell E575

(1000*@SQRT((4*E571)/@PI/C574)

The internal diameter of the supply air duct required is

$$d = 10^3 \times \sqrt{\left(\frac{4Q\,\text{m}^3/\text{s}}{\pi V\,\text{m/s}}\right)}\,\text{mm}$$

where d = duct internal diameter (mm); Q = supply air flow rate (m^3/s); and V = limiting air velocity in the duct (m/s).

$$E575 = 10^3 \times \sqrt{\left(\frac{4 \times E571\,\text{m}^3/\text{s}}{\pi \times C574\,\text{m/s}}\right)}\,\text{mm}$$

Questions

Data from the *CIBSE Guide A* (CIBSE, 1986a) is provided for these questions, and the reader should have access to it to verify the information and to become familiar with its use.

1. State the sources of heat gain that will affect the internal thermal environment in the following applications: residences; offices; retail premises; an atrium in a shopping concourse; pharmaceutical manufacturing building; air conditioning plant room; railway passenger carriage; motor vehicle; aeroplane; entertainment theatre.
2. Explain how heat gains to an occupied building have intermittent characteristics.
3. Explain what is meant by the following terms: 24 h mean conduction gain; net conduction heat flow; direct solar transmission; cyclic variation in heat flow; decrement factor; time lag; surface factor; thermal admittance; net swing in gains.
4. Discuss the personal involvement factors that must be taken into account during the design of an air conditioned room. The term 'personal involvement factors' refers to the preferences for the architectural design, room layout and systems of mechanical air conditioning by the users of the rooms.
5. List the ways in which an office with south-facing glazing in a city in the northern hemisphere could be made comfortable by architectural design and with mechanical cooling methods.

6. Describe, with the aid of sketches, how shading devices and glass types are used to assist in the provision of thermal comfort within buildings, and how they affect the natural illumination and cooling plant loads.
7. Explain why thermal admittance, decrement factor, surface factor, structural time lag and environmental temperature swing are used in preference to thermal transmittance in the assessment of summer internal conditions.
8. List the advice that can be given to the owner of a building that suffers from summer overheating. Explain why the use of additional outdoor air mechanical ventilation may be an unsuitable solution for some applications but correct for some buildings.
9. A 30 m × 20 m × 3 m top-floor office building in a rural location near Guildford is shown in Fig. 3.2. The exposure is normal, and the height above sea level is 50 m. The office air conditioning is to be designed to maintain a room resultant temperature of 21 °C and an air temperature of 24 °C d.b. when the outside air is at 28 °C d.b. The office has computer workstations for 40 people. The maximum occupancy will not exceed 55 people when additional people are included. The computers have power consumptions of 100 W each. There are two photocopiers of 400 W each. The surface-mounted fluorescent luminaires are expected to be used continuously. An illumination level of 250 lux is to be provided by lamps having an efficacy of 85 lm/W; 1 lux is 1 lumen/m. The air condi-

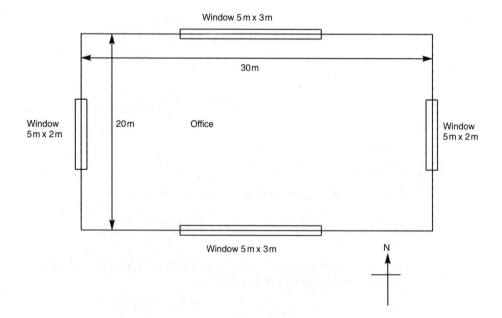

Fig. 3.2   Air conditioned office in question 9.

tioning plant operates for 10 h per day. There are 0.5 air changes per hour due to the natural infiltration of outdoor air. The office below is maintained at the same temperature as the calculated office. The constructional details of the building are:

(a) double-glazed window of body-tinted green heat-absorbing glass in an aluminium frame with thermal break and internal cream-coloured venetian blinds;
(b) external wall of 105 mm brickwork, 10 mm cavity, 40 mm glass-fibre batts, 100 mm lightweight concrete block, 13 mm lightweight plaster, light exterior colour;
(c) internal walls of 100 mm lightweight concrete block, 13 mm lightweight plaster both sides;
(d) 25 mm wood block floor, 50 mm cement screed, 150 mm cast concrete;
(e) 19 mm asphalt roof covering, 13 mm fibreboard, 25 mm air gap, 75 mm glass fibre, 10 mm plasterboard and dark exterior colour.

The air conditioning system is to provide 10 litres/s per person of outside air. The minimum supply air temperature is to be 13°C d.b. The heat emission from the occupants of the office is 80 W sensible heat and 60 W latent heat. The supply and recirculation air ducts are on the roof of the building in the external air. The air ducts are internally lined with thermal and acoustic insulation. Heat gains to the ducts are estimated to be equivalent to 6% of the room cooling load. The design engineer decides to provide a margin of 15% to the calculated plant heat load. The air duct system is expected to have a general leakage rate that

amounts to 5% of the design air flow. The limiting duct air velocity is 5 m/s. Use the thermal data provided in the worksheet. Calculate the room cooling load for noon on 22 September, and find the diameter of the supply and recirculation air ducts.

10. Explain how the air conditioning design engineer can rationalize the range of combinations of heat gains to be able to assess the room cooling load with the minimum of calculation.

11. State the meaning of the following terms: air temperature; environmental temperature; sol-air temperature.

12. Sketch a graph of the heat gain through a south-facing masonry wall to illustrate the meaning of the following terms: time of heat gain; time lag; decrement factor; 24 h mean conduction heat gain; cyclic heat gain; peak heat gain.

13. A 20 m × 20 m × 3 m office in a 20-year-old multi-storey building in London is shown in Fig. 3.3. The office is on the third floor of a six-storey building. The building is to be refurbished with new interior surface materials and replacement windows, and is to have air conditioning installed. The exposure is normal, and the height above sea level is 25 m. The office air conditioning is to be designed to maintain a room resultant temperature of 21°C and an air temperature of 22°C d.b. The office has computer workstations for 12 people. The computers have power consumptions of 120 W each. The recessed fluorescent luminaires are expected to be used continuously to provide 250 lx. The lamps having an efficacy of 65 lm/W. The air conditioning plant operates for 10 h per day. The building is positively pressurized by

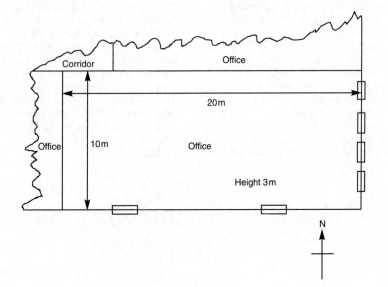

Fig. 3.3 Air conditioned office in question 13.

mechanically extracting less air than the outside air supply. Do not allow for any natural infiltration of outdoor air. The surrounding rooms are maintained at the same temperature as the office. Take the sol-air and outside air temperatures from the worksheet at the nearest time to that required. The constructional details of the building are:

(a) single-glazed window of body-tinted grey glass in an aluminium frame with thermal break and internal cream-coloured venetian blinds;

(b) external wall of 200 mm heavyweight concrete block, 25 mm cavity, 10 mm foil-backed plasterboard, light exterior colour;

(c) internal walls of 105 mm brickwork, 13 mm plaster both sides;

(d) 25 mm wood block, 50 mm cement screed and 150 mm cast concrete floor and ceiling.

The air conditioning system is to provide 10 litres/s per person of outside air. The minimum supply air temperature is to be 14°C d.b. The heat emission from the occupants of the office is 80 W sensible heat and 60 W latent heat. Heat gains to the ducts are estimated to be equivalent to 2% of the room cooling load. The design engineer decides to provide a margin of 12% to the calculated plant heat load. The air duct system is expected to have a general leakage rate that amounts to 3% of the design air flow. The limiting duct air velocity is 5 m/s. Use the thermal data provided on the worksheet. Calculate the room cooling load for 21 June at 4 p.m. and find the diameter of the supply and recirculation air ducts.

14. A single-storey retail furniture warehouse in Basingstoke is 100 m × 50 m in plan and has a 4 m ceiling height. The exposure is normal, and the height above sea level is 80 m. The air conditioning is to be designed to maintain a room resultant temperature of 21°C and an air temperature of 24°C d.b. when the outside air is at 27°C d.b. at 12 noon in June. One long side of the showroom faces west and one short side faces south. The east long side has five windows of 4 m × 2 m. The short south side has three windows of 4 m × 2 m. The other two walls are external and face north and west. The showroom has computer workstations for 10 people. The maximum occupancy will not exceed 100 people when customers are included. The computers have power consumptions of 120 W each. There is a photocopier of 400 W. The recessed fluorescent luminaires are expected to be used continuously. An illumination level of 300 lx is to be provided by lamps having an efficacy of 70 lm/W. The air conditioning plant operates for 10 h per day. There are 0.5 air changes per hour due to the natural infiltration of outdoor air through the entrance doors.

The constructional details of the building are:

(a) single-glazed windows of body-tinted bronze 10 mm thick plate glass in an aluminium frame with light-coloured internal slatted blinds;

(b) external walls of 220 mm brickwork, 25 mm cavity, 10 mm foil-backed plaster board, light exterior colour;

(c) thermoplastic tiles on a cast concrete floor, $U = 0.6 \, \text{W/m}^2 \, \text{K}$, $Y = 6.0 \, \text{W/m}^2 \, \text{K}$, surface factor 0.7, time lag 1 h;

(d) corrugated sheet steel roof, cavity, 100 mm insulation batts and 10 mm plasterboard ceiling.

The air conditioning system is to provide 10 l/s per person of outside air. The minimum supply air temperature is to be 13°C d.b. at the supply air diffusers. The heat emission from the occupants is 80 W sensible heat and 60 W latent heat. The supply and recirculation air ducts are within the ceiling space, and are supplied from an air-handling plant on the roof deck above the showroom. Heat gains to the ducts are estimated to be equivalent to 2% of the room cooling load. The design engineer decides to provide a margin of 5% to the calculated plant heat load. The air duct system is expected to have a general leakage rate that amounts to 3% of the design air flow. The limiting duct air velocity is 5 m/s. Use the thermal data provided in the question and in the worksheet. Take the sol-air and air temperatures from the nearest appropriate time on the worksheet. Calculate the room cooling load for noon on 21 June, and find the diameter of the supply and recirculation air ducts.

Most people.

4 Combustion of a fuel

INTRODUCTION

Analysis of the combustion of a fuel requires a knowledge of its chemical constituents and the excess air that is to be supplied. Sufficient information is provided to allow the analysis of a wide variety of fuels without recourse to further references. The chemicals in fuels, and their symbols, are listed. The equations of combustion in air are given for each chemical element and compound. The lists are not intended to be exhaustive, but they do cover a very large number of normal cases. The user may add further fuels once their chemical compositions are known.

The products of combustion are given for each fuel. The stoichiometric and actual air supply is calculated. The flue gas analysis for the fuel is converted from units of mass into percentages by volume to allow comparison with on-site measurements during commissioning and maintenance. The flue gas composition from site measurements can be entered. The program then calculates the amount of excess air that was supplied. The gross and net calorific values of the fuel are assessed.

The quantity of data that is to be entered is small, and all calculations and output data are generated automatically. The program acts as an on-screen tuition package. The spreadsheet has a large amount of data and some basic explanations.

LEARNING OBJECTIVES

Study and use of this program will enable the user to:

1. calculate flue gas composition;
2. calculate stoichiometric and actual air requirement;
3. calculate the excess air that was supplied;
4. calculate the volumetric constituents of flue gas;
5. identify the different types of fuel used;
6. know the chemical constituents of fuels;
7. know the chemical equations of combustion;
8. know the chemical names and symbols of fuels;
9. know the chemical symbols of fuel elements and compounds;
10. know the constituents of flue gas;

11. know the flue products from fuels;

12. calculate the gross and net calorific value of a fuel;

13. know the molecular mass of elements and compounds;

14. understand stoichiometric, actual and excess air;

15. use mass analysis of fuels.

Key terms and concepts

actual air supply	109	heat content	115
calorific value	115	hydrocarbon	105
chemical composition	106	mass analysis	106
chemical compounds	106	mass percentage	112
chemical elements	106	net calorific value	115
chemical symbols	106	nitrous oxide	110
combustion equation	108	products of combustion	107
condensation	116	relative atomic mass	106
constituents	107	relative molecular mass	106
dry flue gas	107	stoichiometric air	107
excess air	114	stoichiometric oxygen	107
fuel	105	volume percentage	112
gross calorific value	115		

COMBUSTION OF A FUEL

The equations describing the combustion of hydrocarbon fuels in air are listed here and in the spreadsheet. These are for liquid, solid and gaseous fuels. In the unlikely event that the particular fuel to be analysed is not listed, the combustion equation may be evaluated from basic principles. Further information on the chemical equations can be gained from the references.

It is important to note that the chemical symbols displayed on the spreadsheet are not in the correct form. It is not possible to show subscripts on the spreadsheet as it is on a printed document or on a word processor screen. For example, carbon dioxide is written as CO_2, but as the number two cannot be shown as a subscript, it is displayed as CO2 on the spreadsheet screen and printed output. A complete list of the correct symbols and their counterparts on the spreadsheet is shown in Table 4.1. Cells Q50 to T76 on the spreadsheet contain the same information as in Table 4.1.

DATA REQUIREMENT

The user will enter new data for:

1. Your own name in cells B6 and B150.
2. The name of the job or the college assignment in cells B7 and C7; further cells along the row can be used. Repeat this information in cells B151 and C157.

Table 4.1 Chemical symbols and relative molecular mass of common elements and compounds

Chemical	Symbol	Spreadsheet	Relative molecular mass
Acetylene	C_2H_2	C2H2	26
Benzene	C_6H_6	C6H6	78
Butane	C_4H_{10}	C4H10	58
Carbon	C	C	12
Carbon dioxide	CO_2	CO2	44
Carbon monoxide	CO	CO	28
Ethane	C_2H_6	C2H6	30
Ethyl alcohol	C_2H_6O	C2H6O	46
Ethylene	C_2H_4	C2H4	28
Hydrogen	H_2	H2	2
Hydrogen sulphide	H_2S	H2S	34
Methane	CH_4	CH4	16
Nitrogen	N_2	N2	28
Octane	C_8H_{18}	C8H18	114
Oxygen	O_2	O2	32
Pentane	C_5H_{12}	C5H12	72
Propane	C_3H_8	C3H8	44
Propylene	C_3H_6	C3H6	42
Sulphur	S	S	32
Sulphur dioxide	SO_2	SO2	64
Sulphur trioxide	SO_3	SO3	80
Toluene	C_7H_8	C7H8	92
Water, water vapour	H_2O	H2O	18

3. The current date in cells B8 and B152.
4. The file name in cells B10 and B154.
5. The fuel type in cells D29, E29 and F29. These need to be repeated into cells C159, D159 and E159.
6. The excess air percentage to be supplied in cell C31.
7. Select the fuel type from cells Q3 to Q44. Read the percentage constituents of the fuel from columns S to AC. For light fuel oil, go to cell S12 for the percentage carbon. Either write this percentage onto paper and then type it into the % constituent cell, D38, or COPY the contents of cell S12 into D38. Repeat this operation for the other percentage constituents of the fuel. Make sure that there is a zero in all the percentage constituent cells for chemicals that are not present in the fuel. The total of the constituent cells D38 to D61 must be 100% in cell D62.
8. Move to cell A120. If a combustion test has been made on site for the fuel that has been entered, the actual amount of excess air that was supplied can be calculated. Enter the percentages of carbon dioxide, % CO_2 from the measurements into cell D130, carbon monoxide CO in D131, oxygen O_2 in D132 and sulphur dioxide SO_2 in D133. The remainder of the dry flue gas will be nitrogen N_2, and this is typed into cell D134. The total of the flue gas constituents must be 100% in cell D135.
9. If the fuel is not listed, its chemical composition must be known. For example, if C_4H_8 is to be combusted, it can be entered into the spreadsheet as carbon plus hydrogen; to do this, calculate the molecular mass from the symbols. In this case, 1 kg of C_4H_8 contains $100 \times (4 \times 12)/(4 \times 12 + 8 \times 1) = 85.7\%$ carbon plus 14.3% hydrogen.

Save this file on the hard disk in drive C and on your own floppy disk in drive A now. This terminates the data entry for the fuel type chosen. Space has been left for the user to enter more types of fuel to the list by typing them onto the lines where new oil, new solid and new gas fuel types are located. Additional space can be formed by moving blocks of cells downwards. Enter the name and chemical compositions of the new fuels into the correct columns. Ensure that they total 100%.

OUTPUT DATA

1. Stoichiometric oxygen O_2 for each constituent of the fuel is listed in cells F38 to F61.
2. Stoichiometric oxygen O_2 for the fuel is given in cell D64.
3. Stoichiometric air for the fuel is given in cell D65.
4. Actual air for the fuel is given in cell D66.
5. Flue gases produced are listed in the cell range A76 to A80.
6. The percentage of each constituent of the dry flue gas is given in the cell range J76 to J80.
7. The combustion equation for each constituent of the fuel commences in cell C94. This list extends to cell C117. The products of combustion are listed in cells H94 to L115. This information is provided for reference and learning, because all the equations have been programmed into the calculation cells.
8. Mass and percentage of excess air are given in cells D145 and D147.
9. The gross calorific value of the fuel is displayed in cell D67, and the net calorific value is in cell J67.
10. Input and output data are repeated in a range of cells from A150 to G223. This block of cells has been identified in the print menu so that an output report can be printed.

FORMULAE

Representative samples of the formulae are given here. Each formula can be read on the spreadsheet by moving the cursor to the cell. The equation is presented in the form in which it would normally be written and in the format that is used by the spreadsheet.

Cell E38

(F3) +$F$94

This states that the contents of cell E38 are to be replaced with those from cell F94, or E38 = F94. F94 is specified as an absolute cell reference by the use of $ symbols. The $ before the column reference fixes the column and that in front of the row fixes the row. The absolute cell reference $F$94 will not change if additional rows or columns are inserted. Lotus 1-2-3 uses the + symbol to denote a formula. +$F$94 in E38 means E38 = $F$94. The (F3) after the cell reference is the format of the number that is displayed. (F3) is for three fixed decimal places.

The other display formats can be seen on the screen. F94 contains the stoichiometric oxygen required to completely burn 1 kg of the fuel. The fuel in this case is 1 kg of carbon. F94 contains 2.67. This is found from consideration of the mass analysis equation of the fuel:

$$C + O_2 = CO_2$$

The molecular masses are 12 for carbon and 32 for oxygen. The mass equation for the complete combustion of carbon is

$$12 + 32 = 44$$

Divide through by 12 to find the oxygen required to burn 1 kg of fuel:

$$1 + \frac{32}{12} = \frac{44}{12}$$

$$1 + 2.67 = 3.67$$

The complete combustion equation becomes

1 kg carbon reacted with 2.67 kg oxygen during complete combustion to produce 3.67 kg carbon dioxide.

$$1\,kg\,C + 2.67\,kg\,O_2 = 3.67\,kg\,CO_2$$

The range of cells from E38 to E61 contains a similar formula, which reads the stoichiometric oxygen for 1 kg of fuel from cells F94 to F115. The reader can check the numbers stored in these cells from the combustion equations listed, or consult the references. Notice that oxygen in the fuel does not have to be supplied by the combustion air, and is a negative quantity. Incombustible elements in the fuel, such as ash, sediment and water, do not require oxygen.

Cell F38

+D38*E38/100

This calculates the stoichiometric oxygen that is needed for the amount of the constituent in the fuel. If the fuel has 86.3% carbon, then it needs only 86.3% of the oxygen requirement for 1 kg carbon:

$$\frac{kg\ stoichiometric\ O_2}{kg\ fuel} = \frac{D38\%\,C}{100} \times \frac{E38\ kg\ stoichiometric\ O_2}{kg\ C}$$

$$= \frac{86.3}{100} \times \frac{2.67\ kg\ stoichiometric\ O_2}{kg\ C}$$

$$= 2.304\,kg\,O_2/kg\ fuel$$

The range of cells from F38 to F58 contains similar formulae for the other constituents of the fuel.

Cell G38

+D38*H94/100

The complete combustion of 1 kg carbon produces 3.67 kg of carbon dioxide. The products of combustion for each fuel are listed in the range of cells from H94 to

L115. The appropriate products are automatically calculated for each constituent of the fuel, and these appear in the range of cells from G38 to K58.

Cell F62

@SUM (F38..F59)

This is the total of the cells that contain the stoichiometric oxygen requirement for each constituent of the fuel.

Cell D64

+$F$62

The stoichiometric oxygen required for the fuel is copied into cell D64 for the output data.

Cell D65

100*D64/$B$14

This calculates the stoichiometric air for the fuel. The percentage of oxygen in air is in cell B14. This is normally 23.3% by mass:

$$\frac{D65 \text{ stoich. air}}{\text{kg fuel}} = \frac{D64 \text{ kg stoichiometric } O_2}{\text{kg fuel}} \times \frac{100}{B14\% \, O_2 \text{ air}}$$

$$= 3.260 \times \frac{100}{23.3}$$

$$= 13.992 \text{ kg stoichiometric air/kg fuel}$$

Cell D66

(100+$C$31)*$D$65/100

This finds the actual air supply from the stoichiometric air that has just been calculated and the percentage excess air that has been specified in cell C31. For 25% excess air, the air required is 1.25 times the stoichiometric air supply:

$$\frac{D66 \text{ kg air}}{\text{kg fuel}} = \frac{D65 \text{ kg stoichiometric air}}{\text{kg fuel}} \times \frac{(100 + \text{excess air\%})}{100}$$

$$= 13.992 \times \frac{(100 + 25)}{100}$$

$$= 17.490 \text{ kg air/kg fuel}$$

Cell K60

+$D$14*$D$65/100

The air supplied for combustion comprises 23.3% oxygen and 76.7% nitrogen. The oxygen is either consumed during combustion and appears combined with elements of the fuel such as carbon dioxide, or it is not needed and remains uncombined. This excess oxygen is present in the flue gas. The nitrogen in the combustion air passes through the combustion process and appears in the form of heated gas. A minor part of the nitrogen may be combined with oxygen to form nitrous oxides, NO_x, but these are of the order of 200 parts per million, or 0.020%, and they will be ignored for calculation purposes. The nitrogen associated with the stoichiometric air is

$$\frac{K60 \text{ kg } N_2}{\text{kg fuel}} = \frac{D14\%}{100} \times \frac{D65 \text{ kg stoichiometric air}}{\text{kg fuel}}$$

$$= \frac{76.7\%}{100} \times \frac{13.992 \text{ kg stoichiometric air}}{\text{kg fuel}}$$

$$= 0.767 \times 13.992$$

$$= 10.732 \text{ kg } N_2/\text{kg fuel}$$

This is calculated from the stoichiometric air.

Cell K61

(D14*D66/100)−K60

This calculates the mass of nitrogen that accompanies the excess air and then appears in the flue gas:

$$\frac{K61 \text{ kg } N_2}{\text{kg fuel}} = \frac{D14\%}{100} \times \frac{D66 \text{ kg air}}{\text{kg fuel}} - \frac{K60 \text{ kg } N_2 \text{ from stoich. air}}{\text{kg fuel}}$$

$$= \frac{76.7 \times 17.49}{100} - 10.732$$

$$= 2.683 \text{ kg } N_2/\text{kg fuel from the excess air}$$

Cell I61

($B$14*$D$66/100)−$D$64

This calculates the mass of oxygen that accompanies the excess air and then appears in the flue gas as excess oxygen:

$$\frac{I61 \text{ kg } O_2}{\text{kg fuel}} = \frac{B14\%}{100} \times \frac{D66 \text{ kg air}}{\text{kg fuel}} - \frac{D64 \text{ kg stoich. } O_2}{\text{kg fuel}}$$

$$= \frac{23.3 \times 17.49}{100} - 3.26$$

$$= 0.815 \text{ kg } O_2/\text{kg fuel from the excess air}$$

Cell G62

@SUM (G38..G58)

Cell G62 contains the sum of the carbon dioxide amounts that are present in the flue gas from all the constituents of the fuel. These are all within the range of cells from G38 to G58. There are similar formulae in G62 to K62 for the totals of each flue gas constituent.

Cell D76

+G62

The total carbon dioxide content is copied into D76 to facilitate conversion of the mass analysis of the flue gas into a volumetric analysis. This is so that a comparison can be made with on-site measurements, which will normally be based on volumetric data.

Cell F76

+D76/E76

Divide the mass of carbon dioxide by its relative molecular mass:

$$F76 = \frac{3.167 \, \text{kg} \, CO_2}{44 \, \text{kg fuel}}$$
$$= 0.072$$

A similar calculation is made for each of the other flue gas constituents: water vapour, oxygen, sulphur dioxide and nitrogen. The results are put into the cell range F76 to F80.

Cell F81

@SUM (F76..F80)

This is the total of the range of cells from F76 to F80.

Cell F82

−F77

Subtract the water vapour content of the flue gas.

Cell F83

+F81+F82

Calculate the total for the dry flue gas by removal of the water vapour content.

Cell H76

(F76/$F$81)*100

Divide the CO_2 relative volume by the total of that column and multiply by 100 to form the percentage carbon dioxide in the wet flue gas by volume. Repeat the calculation for the range of cells from H76 to H80 for all the products.

Cell H81

@SUM(H76..H80)

This checks that all the flue gas products have been accounted for, and the total must be 100%.

Cell J76

(F76/$F$83)*100

This is similar to the formula in H76, and calculates the percentage constituents of the dry flue gas. These are the figures to be compared with site measurements, because the water vapour in the hot flue gas condenses at around 50 °C. The measurement of the flue gas constituents will have taken place at near to 20 °C.

Cell I130

+D130*G130

The measured flue gas constituents are to be converted from a volume to a mass percentage. Multiply the volume percentage carbon dioxide by its relative molecular mass. Repeat this for carbon monoxide, oxygen, sulphur dioxide and nitrogen:

$$I130 = D130 \; CO_2 \text{ \% volume} \times G130 \text{ molecular mass}$$
$$= 9 \times 44$$
$$= 396$$

The formulae in the cell range I130 to I134 are similar.

Cell I135

@SUM(I130..I134)

Find the total of the cell range I130 to I134.

Cell L130

(I130/I135)*100

Divide the content of cell I130 by the total for the range that is held in I135 and multiply by 100 to calculate the percentage mass of carbon dioxide in the dry flue gas. Repeat for the other flue gas constituents.

Cell L135

@SUM(L130..L134)

The sum of the percentage mass analysis must be 100% in L135.

Cell F137

+D38

This retrieves the percentage carbon in the fuel from cell D38 and displays it in F137.

Cell F138

+D65

This retrieves the stoichiometric air mass from the fuel analysis that was carried out.

Cell D140

(3*L130/11/100)+(3*L131/7/100)

L130 contains the mass percentage of carbon dioxide and L131 has the mass percentage of carbon monoxide in the dry flue gas. Each kg of carbon dioxide and carbon monoxide contain a fixed amount of carbon. The relative molecular mass of carbon dioxide is 44. This is made up from 12 for the carbon and 32 for the oxygen. The carbon atom weighs 12, so the proportion of carbon in the total is 12 divided by 44, or 3/11. Each kg of carbon dioxide in the dry flue gas contains 3/11 kg carbon. Carbon monoxide has a relative mass of 28. This is from the carbon 12 and half a molecule of oxygen 16. The carbon content is 12 divided by 28, or 3/7. Each kg of carbon monoxide in the dry flue gas contains 3/7 kg carbon.

$$\frac{D140 \text{ kg carbon}}{\text{kg DFG}} = \frac{3}{11} \times \frac{L130\% \text{ CO}_2}{100} + \frac{3}{7} \times \frac{L131\% \text{ CO}}{100}$$
$$= \frac{3}{11} \times \frac{13.36\% \text{CO}_2}{100} + \frac{3}{7} \times \frac{0.94\% \text{ CO}}{100}$$
$$= 0.0405 \text{ kg C/kg DFG}$$

Cell D141

+F137/100

F137 contains the percentage carbon in the fuel. Divide this by 100 to find the mass of carbon per kg fuel:

$$\frac{D141 \text{ kg C}}{\text{kg fuel}} = \frac{F137\% \text{ C}}{100}$$
$$= \frac{86.3\% \text{ C}}{100}$$
$$= 0.863 \text{ kg C/kg fuel}$$

Cell D142

+D141/D140

This uses the fact that all the carbon in the dry flue gas can only have come from the fuel during steady-state combustion conditions. Unless someone is shovelling lumps of carbon into the flue pipe, or soot is being dislodged from dirty flue passageways during the test, this must be true. As the carbon in the dry flue gas is the same as that in the fuel, a direct calculation can be made between the two:

$$\frac{D142 \text{ kg DFG}}{\text{kg fuel}} = \frac{D141 \text{ kg C}}{\text{kg fuel}} \times \frac{\text{kg DFG}}{D140 \text{ kg C}}$$
$$= \frac{0.863 \text{ kg C}}{\text{kg fuel}} \times \frac{\text{kg DFG}}{0.0405 \text{ kg C}}$$
$$= 21.3161 \text{ kg DFG/kg fuel}$$

Cell D144

+$D$142*$L$132/100

D142 contains the mass of dry flue gas that was produced from a kilogram of fuel. The excess oxygen has been expressed as percentage excess oxygen per kg of dry flue gas in L132. Multiply these cells to produce the kg of excess oxygen per kg fuel that were found from the flue gas analysis:

$$\frac{\text{D144 kg excess } O_2}{\text{kg fuel}} = \frac{\text{D142 kg DFG}}{\text{kg fuel}} \times \frac{\text{L132\% excess } O_2}{\text{kg DFG}} \times \frac{1}{100}$$

$$= \frac{21.3161 \times 5.4}{100}$$

$$= 1.507 \text{ kg excess } O_2/\text{kg fuel}$$

Cell D145

($D$144/$B$14)*100

The excess oxygen was contained in the excess air supply of

$$\frac{\text{D145 kg excess air}}{\text{kg fuel}} = \frac{\text{D144 kg excess } O_2}{\text{kg fuel}} \times \frac{\text{kg air}}{\text{B14\% } O_2} \times 100$$

$$= \frac{1.1507}{23.3} \times 100$$

$$= 4.9385 \text{ kg excess air/kg fuel}$$

Cell D146

+$F$138+$D$145

The total mass of air supplied to the combustion process was the stoichiometric air from F138 plus the excess air from D145:

$$\frac{\text{D146 total kg air}}{\text{kg fuel}} = \frac{\text{F138 kg stoichiometric air}}{\text{kg fuel}} + \frac{\text{D145 kg excess air}}{\text{kg fuel}}$$

$$= 13.99 + 4.9385$$

$$= 18.9304 \text{ kg air/kg fuel}$$

Cell D147

($D$145/$F$138)*100

The percentage of excess air, D147, is found by dividing the excess air D145 by the stoichiometric air F138 and multiplying by 100:

$$\frac{\text{D147 ex. air}}{\text{kg fuel}} = \frac{\text{D145 kg excess air}}{\text{kg fuel}} \times \frac{\text{kg fuel}}{\text{F138 kg stoichiometric air}} \times 100$$

$$= \frac{4.9385}{13.99} \times 100$$

$$= 35.3\%$$

Cell M38

+D38*N94/100/1000

Calculates the higher heat content (calorific value) of the constituent of the fuel. In the first case, this is the carbon content:

$$\frac{M38 \text{ MJ}}{kg} = \frac{D38 \text{ carbon } \%}{100} \times \frac{N94 \text{ kJ}}{kg} \times \frac{1 \text{ MJ}}{1000 \text{ kJ}}$$
$$= \frac{86.3 \times 32793}{100 \times 1000}$$
$$= 46.2 \text{ MJ/kg}$$

Cell N38

+D38*094/100/1000

Calculates the lower heat content (calorific value) of the constituent of the fuel. In the first case, this is the carbon content:

$$\frac{N38 \text{ MJ}}{kg} = \frac{D38 \text{ carbon } \%}{100} \times \frac{094 \text{ kJ}}{kg} \times \frac{1 \text{ MJ}}{1000 \text{ kJ}}$$
$$= \frac{86.3 \times 32793}{100 \times 1000}$$
$$= 46.2 \text{ MJ/kg}$$

The higher and lower heat contents of carbon are the same because of the absence of hydrogen. Water vapour is not formed during the combustion of carbon alone.

Cell M62

@SUM(M38..M57)

This is the sum of the components of the gross calorific value of the fuel from the range of cells M38 to M57.

$$M62 = SUM(28.3 + 17.8 + 0.1)$$
$$= 46.2 \text{ MJ/kg}$$

Cell N62

@SUM(N38..N57)

This is the sum of the components of the net calorific value of the fuel from the range of cells N38 to N57:

$$N62 = SUM(28.3 + 15.0 + 0.1)$$
$$= 43.4 \text{ MJ/kg}$$

Cell D67

+M62

The gross calorific value of the fuel is copied from all M62 into cell D67 for display with its units.

Cell J67

+N62

The net calorific value of the fuel is copied from cell N62 into cell J67 for display with its units.

Cells N94 to N115 contain the higher heat value that is released during the complete combustion of that constituent of the fuel. Any water vapour that is formed from the presence of hydrogen in the constituent is taken as being condensed to liquid at the air entry temperature of 25 °C. This means that all the potential heat that is contained within the fuel is released, and can be recovered. This potential heat includes the latent heat content of the water vapour in the flue gas. A condensing boiler is required to utilize this latent energy. Condensing boilers may only lower the condensed water to around 50 °C rather than the 25 °C of the combustion air. The latent heat has been recovered, but a small part of the sensible heat remains unrecoverable. It may not be possible to gain latent heat from the flue gas when the water returning to the boiler exceeds 50 °C during cold weather.

Cells O94 to O115 contain the lower heat value of the fuel constituents, NCV. This is the higher heat value, GCV, less the latent heat of condensation of the water vapour in the flue gas. At 25 °C this is calculated from:

$$NCV = GCV - h_{fg} \times 9 \times m\,H_2\ MJ/kg$$

where h_{fg} of water vapour at 25 °C = 2442 kJ/kg; $m\,H_2$ = mass of H_2 in 1 kg of the fuel; and 9 = mass of H_2O vapour produced from 1 kg of hydrogen. The molecular mass of methane (CH_4) is C 12 plus H_4 4, = 16. So, in 1 kg CH_4 there is (4/16) kg of hydrogen atoms (H), or 0.25 kg/kg CH_4. The gross and net calorific values of methane are listed in cells N103 and O103, and are calculated from

$$NCV = GCV - h_{fg} \times 9 \times m\,H_2\ HJ/kg$$
$$= 55\,643 - 2442 \times 9 \times 0.25\ kJ/kg$$
$$= 50\,149\ kJ/kg$$
$$= 50.2\ MJ/kg$$

The gross and net calorific values of the fuels used in the following questions are not specifically requested. However, the spreadsheet will automatically calculate them, and the reader should be aware of their values.

Questions

1. 1 kg of the following chemical elements and compounds are completely burnt in air. Write the chemical equation that describes the complete combustion process for each example. Manually calculate the stoichiometric air required. Enter the fuel into the spreadsheet, and find the stoichiometric air for each one. Make sure that you understand where all the calculated numbers come from, and that you can reproduce them on paper with the aid of a calculator.

   (a) carbon, C;

   (b) hydrogen, H_2;
   (c) sulphur, S:
   (d) carbon monoxide, CO;
   (e) methane, CH_4;
   (f) octane, C_8H_{18}.

2. A fuel oil consists of 76.5% carbon, 20% hydrogen, 0.5% sediment, 1% sulphur and 2% oxygen by mass. Calculate the mass of stoichiometric air required.

3. Calculate the stoichiometric mass of air that is required to completely burn 1 kg of benzene (C_6H_6).

4. Calculate the mass of stoichiometric air and the percentage volumetric dry flue gas analysis for a

fuel oil that consists of 82% carbon, 12% hydrogen, 2% oxygen, 1% sulphur and 3% nitrogen.

5. A gas that has been produced from the distillation of crude oil has the volumetric composition 14% hydrogen, 2% methane, 22% carbon monoxide, 1% ethane, 4% oxygen and 57% nitrogen. The gas is to be burnt with 30% excess air.

   (a) Manually convert this volume analysis into an analysis by mass. This is done by multiplying the volume percentages by the relative molecular mass of the constituent, summing these figures, then expressing each as a percentage of the total.
   (b) Enter the mass percentages of the constituents of the gas into the spreadsheet and find the stoichiometric air, actual air-to-fuel ratio, and the expected volume analysis of the dry flue gas.

6. The liquid hydrocarbon decane ($C_{10}H_{22}$) is fully burnt in an industrial process heat exchanger with excess air. The volumetric flue gas analysis shows carbon dioxide 11.97%, carbon monoxide 0.48%, oxygen 3.32% and nitrogen 84.23%.

   (a) Find the combustion equation from first principles: that is, by manually evaluating the number of molecules in the combustion equation.
   (b) Manually calculate the stoichiometric air mass per kg of decane and enter it into the correct cell on the spreadsheet. Ignore the other fuel data.
   (c) Read the percentage excess air that was supplied to the combustion process from the spreadsheet. Before quitting the work sheet or saving it, type the correct formula back into cell D65. This is +100*(D64/B14), otherwise the number that was typed into cell D65 becomes a constant, and it will invalidate the worksheet.

7. The decomposition of sewage in a treatment plant produces a gas that comprises 45% hydrogen, 40% carbon monoxide and 15% methane by mass. This gas is burnt with 50% excess air in a hot-water heat generator. Find the air-to-fuel ratio that is supplied to the furnace and the expected wet flue gas analysis from it.

8. Town gas, which was produced from the heating of coal, had a composition of 14.2% methane, 39.9% carbon monoxide, 40.5% hydrogen, 0.5% oxygen and 4.9% nitrogen by volume. Typical poorly controlled gas appliances may have had 100% excess air.

   (a) Manually calculate the percentage mass analysis of the gas by multiplying the volume percentages by the relative molecular mass of

the constituent, summing these figures, then expressing each as a percentage of the total.
   (b) Enter the mass percentages of the fuel constituents of the gas into the spreadsheet, and find the stoichiometric air, actual air-to-fuel ratio, and the expected volume analysis of the dry flue gas.

9. Diesel oil that was supplied to an engine had the mass composition of 84% carbon, 14% hydrogen and 2% oxygen. Measurements of the flue gas revealed that the dry flue gas composition was 8.85% carbon dioxide, 1.2% carbon monoxide, 6.8% oxygen and 83.15% nitrogen by volume. Enter the test data into the spreadsheet and find the stoichiometric mass of air that was supplied to the engine for each kilogram of diesel oil, and percentage excess air.

10. A petrol has a mass composition of 86% carbon and 14% hydrogen. It is burnt with 20% excess air in a four-stroke engine. Find the air-to-petrol ratio supplied to the engine, and the percentage volume composition of the wet exhaust gas.

11. Coal that was burnt in a steam boiler contained 78% carbon, 6% hydrogen, 9% oxygen and 7% ash. 50% excess air was supplied. Find the air-to-coal ratio, the volume composition of the dry flue gas, and the gross and net calorific values of the fuel.

12. Pulverized coal is blown into an electrical power station steam boiler. The coal mass analysis is 84% carbon, 4% hydrogen, 5% oxygen with 7% ash and incombustibles. The excess air is controlled to 25% of the stoichiometric air requirement by variable-speed centrifugal fans. Find the actual air mass supplied per kilogram of coal and the percentage volumetric dry flue gas composition.

13. Calculate the stoichiometric and actual air quantities supplied and the volumetric analysis that is expected of the dry flue gas in a coal-fired boiler plant when 60% excess air is provided. The coal had a composition of 89% carbon, 4% hydrogen, 3% oxygen, 1% sulphur. The remainder of the coal sample were incombustibles.

14. Calculate the stoichiometric air requirement and volumetric wet flue gas composition for a coal that has 72% carbon, 12% hydrogen, 7% oxygen and the remainder being ash. The coal is burnt with 45% excess air.

15. A fuel oil has a composition of 85% carbon and 15% hydrogen. It is burnt with an unknown quantity of excess air in a pressure-jet burner. A test of the flue gas revealed the volumetric composition of the dry flue gas as 13.1% carbon dioxide, 2.4% oxygen and 84.5% nitrogen. Calculate the percentage of excess air that was supplied.

16. Anthracite is burnt with 30% excess air in a hot-water boiler serving a housing estate heating

system. Analysis of the fuel revealed that it contained 89% carbon, 4% hydrogen, 2.5% oxygen, 1% nitrogen, 0.4% sulphur and 3.1% ash. Calculate the air-to-fuel ratio and the percentage volumetric composition of the dry flue gas.

17. The flue gas oxygen content from an oil-fired boiler is continuously measured and used to modulate a damper in the combustion air inlet duct to maximize efficiency. The fuel oil consists of 86% carbon, 13% hydrogen and 1% sulphur. Calculate the percentage excess air that was being supplied at a time when the dry flue gas contained 14.6% carbon dioxide, 1.1% oxygen, 1% carbon monoxide, 0.06% sulphur dioxide and 83.2% nitrogen from a volumetric analysis.

18. Gas that was produced from the heating of coal had a composition of 49.4% H_2, 18% CO, 20% CH_4, 2% C_4H_{10}, 0.4% O_2, 6.2% N_2 and 4% CO_2. The gas is burnt with 20% excess air. Calculate the stoichiometric and actual air-to-fuel ratios, the expected volumetric wet flue gas analysis, and the gross and net calorific values. Compare the calorific value with that for natural gas (methane). Comment upon the effects that the difference made when gas-fired appliances were converted from burning manufactured to natural gas.

19. A petrol engine that is to be adapted for racing is to be fed with ethyl alcohol (C_2H_6O) and 25% excess air. Calculate the percentage mass analysis of ethyl alcohol manually. Use the spreadsheet to calculate the stoichiometric and actual air supply rates, the expected volumetric dry flue gas analysis, and the gross calorific value of the fuel.

20. Electricity and heat are generated from a diesel engine that is supplied with benzene having constituents of 92.3% carbon and 7.7% hydrogen. The volumetric dry flue gas contained 13.9% carbon dioxide, 1.2% carbon monoxide and 4.35% oxygen; the remainder was assumed to be nitrogen. Calculate the percentage excess air that was supplied to the engine. Find the oxygen content of the dry flue gas if 100% excess air were to be provided.

Now, that is a really useful spreadsheet!

5 Building heat loss

INTRODUCTION

The calculation of the capacity of the heating plant to offset the loss of heat from rooms and buildings is described. The Chartered Institution of Building Services Engineers steady-state method is followed. All the relevant dimensionless ratios and overall values are calculated for the user to validate against either tabulated data or expected answers. The worksheet accommodates ten rooms. The user can duplicate the file where the building has more rooms. The total heat loss from each of these extra files is typed into the accumulation area at the end of the worksheet. This provides for ten worksheets of ten rooms each, a total of 100 rooms. This should be sufficient for most applications.

The user enters data for the room dimensions, indoor comfort temperature, outdoor air temperature, thermal transmittances and outdoor air infiltration rate, directly from the plans and elevations, exactly as is done for manual calculations. A laptop or palmtop computer can be placed onto the architect's drawings while data is typed. There is no need for data preparation sheets as with some dedicated software. The U values may have been calculated from the UVAL spreadsheet or read from tables in the references. All the data is clearly visible, and any numbers can easily be changed. The effect of increasing some thermal insulation or changing from a convective to a radiative heating system can be investigated in seconds. All the formulae are open for the user to inspect. Each formula is fully explained in the text. The heat loss file is A:HLOSS.WK1.

A section is devoted to the calculation of the air temperature that is expected to be created within unheated rooms or spaces. This allows designers to assess the likely space temperature realistically in preference to making a guess. The unheated room temperature file is A:UNHEAT.WK1.

LEARNING OBJECTIVES

Study and use of this chapter will enable the user to:

1. be aware of the difference between steady-state and intermittent heat loss calculation methods;
2. understand the terms used in the calculation of heat loss and plant power;
3. calculate heater plant energy output for rooms and buildings;
4. know how the volumetric specific heat of air is used;

5. modify the thermal transmittance of a partition wall, floor or ceiling where it has an adjacent warm space;

6. calculate the dimensionless factors F_u, F_v, F_1 and F_2;

7. understand how the surface temperature of a structure can be calculated and measured;

8. calculate the area-weighted mean surface temperature of a room;

9. calculate the environmental, air and mean radiant temperatures that are produced in a room as a result of the selection of a resultant temperature for the comfort condition;

10. know how the different types of heating system are incorporated into the calculation of heat loss;

11. validate the calculated values for factors F_1 and F_2 with published data;

12. gain experience of typical values for the heat loss per unit floor area and room volume;

13. evaluate heat exchanges between spaces that are at different temperatures;

14. calculate the temperature of an unheated room or space.

Key terms and concepts

HEAT LOSS CALCULATION

Raising the indoor air temperature of a building above that of the outdoor air by means of a heating system causes a loss of heat from the enclosed space. Heat flows through the structure, and is removed as the result of natural and mechanical ventilation with outdoor air. The heating system provides a heater output power of Q_p W to balance the rate of heat loss from the building. This is the plant heat output. It is usual to calculate thermal transmittance U, and heat loss for steady-state heat flow from the building. 'Steady state' means a rate of heat flow that does not vary with time, so that a heat loss rate of 3 kW requires a continuous heater output of 3 kW. When varying rates of heat loss are being considered, such as for the highly intermittent heating of a building used only one day per week, then admittance (Y values) are used. These generate the need for heater output powers of up to double the steady-state figure, because a large part of the heater output power is used to raise the temperature of the structure. Only the steady-state calculation is used here, as it will apply to most cases. Steady-state calculations assume that the building structure is warm, even though there are always cyclic variations in the temperature of the bricks and concrete. Such variations result from the daily cycle of use of the building from its heating, lighting and occupancy and that of the weather. Admittance values are used in the calculation of heat gains to buildings in Chapter 3 (p. 65). The heater power output that is required to offset the steady state heat loss from a room, is given by

$$Q_p = [F_1\Sigma(AU) + 0.33F_2NV] \times (t_c - t_{ao})$$

where, Q_p = plant energy output (W); A = area of heat transfer surface, (m^2); U = thermal transmittance of the structure (W/m^2 K); N = number of air changes per hour from the infiltration of unheated outdoor air (h^{-1}); V = volume of the room (m^3); t_c = dry resultant temperature at the centre of the room (°C); t_{ao} = outdoor air temperature (°C); F_1 = environmental temperature ratio; and F_2 = indoor air temperature ratio. F_1 and F_2 are read from Tables A9.1–A9.7 in the *CIBSE Guide A* (CIBSE, 1986a), or calculated from the equations that are demonstrated later in this chapter.

The ventilation coefficient 0.33 is the volumetric specific heat capacity of air when expressed in W/m^3 K. It comes from

$$\text{coefficient} = 1012\,\frac{\text{J}}{\text{kg K}} \times 1.1906\,\frac{\text{kg}}{\text{m}^3} \times N\,\frac{\text{air change}}{\text{h}} \times \frac{1\,\text{h}}{3600\,\text{s}} \times \frac{1\,\text{W s}}{1\,\text{J}}$$

$$= \frac{1012 \times 1.1906}{3600}\,\frac{\text{W}}{\text{m}^3\,\text{K}}$$

$$= 0.33\,\text{W/m}^3\,\text{K}$$

The specific heat capacity of air, 1012 J/kg K, and density, 1.1906 kg/m^3, are standard values that remain essentially constant for this application. The ventilation coefficient is multiplied by the hourly rate of air infiltration N, the room volume V, the indoor air temperature ratio F_2 and the difference between the room resultant and outdoor air temperatures, to find the ventilation heater power Q_v. Q_v is included in the equation for Q_p, and it is not normally calculated alone.

When the cooler side of a structure is not at the outdoor air, t_{ao}, it is at a resultant temperature, t_{c2}. The thermal transmittance U of the wall, ceiling or floor is calculated in the normal manner. It is recommended that this U value is adjusted by a temperature ratio so that the summation of $A \times U$ is still multiplied by the whole temperature difference $(t_c - t_{ao})$. This simplifies the way in which intermediate temperature differentials are included. The U value of such a partition is modified thus:

$$U_2 = U_1 \frac{t_{c1} - t_{c2}}{t_{c1} - t_{ao}} \text{ W/m}^2 \text{ K}$$

where $U_1 =$ the original value for the structure (W/m^2 K); $U_2 =$ the modified value (W/m^2 K); $t_{c1} =$ dry resultant temperature at the centre of the room (°C); and $t_{c2} =$ dry resultant temperature in the other room (°C).

This modification applies just as well when the adjacent room, or space, is at a resultant temperature t_{c2} that is higher than that of the subject room. A heat gain to the subject room is calculated in this case. The heat lost from the warmer room into the cooler space will equal the heat gained by the cooler room.

EXAMPLE 5.1

Calculate the thermal transmittance U_1, the modified value U_2 and the heat flow Q_p through a partition wall of 8 m^2, which separates two rooms. The rooms are at resultant temperatures of 20 °C and 12 °C when the outdoor air temperature is -2 °C. The wall consists of 100 mm lightweight concrete blockwork and 13 mm plaster on both sides. The thermal conductivity of concrete blockwork is 0.2 W/m K and that of plaster is 0.5 W/m K. The surface film resistance is 0.12 m^2 K/W on each side. F_1 was found to be 0.98.

Thermal resistance of the wall $= R$ m^2 K/W.

$$R = 0.12 + \frac{0.013}{0.5} + \frac{0.1}{0.2} + \frac{0.013}{0.5} + 0.12 \text{ m}^2 \text{ K/W}$$
$$= 0.792 \text{ m}^2 \text{ K/W}$$

$$U_1 = \frac{1}{R} \frac{\text{W}}{\text{m}^2 \text{ K}}$$
$$= \frac{1}{0.792} \frac{\text{W}}{\text{m}^2 \text{ K}}$$
$$= 1.263 \text{ W/m}^2 \text{ K}$$

The modified U value is

$$U_2 = U_1 \frac{t_{c1} - t_{c2}}{t_{c1} - t_{ao}} \text{ W/m}^2 \text{ K}$$
$$= 1.263 \times \frac{20 - 12}{20 - -2} \text{ } W/m^2 \text{ } K$$
$$= 0.459 \text{ W/m}^2 \text{ K}$$

$$Q_p = [F_1 \Sigma (A \times U) + 0.33 NVF_2] \times (t_c - t_{ao}) W$$

In this case, the ventilation heat loss is not being found, so $(0.33 \times N \times V \times F_2)$ is ignored.

$$Q_p = [F_1 \Sigma (AU)] \times (t_c - t_{ao}) W$$
$$= (0.98 \times 8 \text{ m}^2 \times 0.459 \text{ W/m}^2 \text{ } K) \times (20 - -2)°C$$
$$= 79.2 \text{ W}$$

If the unmodified U value is used, then the actual temperature difference $(t_{c1} - t_{c2})$ is applied:

$$Q_p = [F_1 \Sigma(AU)] \times (t_{c1} - t_{c2})\text{W}$$
$$= (0.98 \times 8\,\text{m}^2 \times 1.263\,\text{W/m}^2\,K) \times (20 - 12)°\text{C}$$
$$= 79.2\,\text{W}$$

Notice that the first decimal place in the solution for Q_p remains the same for both calculations, but that subsequent numbers vary because of the choice of significant decimal places written. The nearest watt may be considered of sufficient accuracy, and this is found from rounding the first decimal place to 79 W.

The dimensionless factors F_1 and F_2 are calculated from data from the room, a U value factor F_u, a ventilation factor F_v, the infiltration ventilation rate, and the type of heating system. The type of heating system is characterized by the proportions of radiated and convected heat input. F_u and F_v are calculated from

$$F_u = \frac{18\Sigma(A)}{\Sigma(AU) + 18\Sigma(A)}$$

$$F_v = \frac{6\Sigma(A)}{1012 \times 1.1906Q + 6\Sigma(A)}$$

This allows F_1 to be found from

$$F_1 = \frac{E \times [F_v \times 1012 \times 1.1906Q + F_u \times \Sigma(AU)]}{\Sigma(AU) + 18\Sigma(A)(KF_v + EF_u)} + F_u$$

and F_2 to be found from

$$F_2 = \frac{K \times [F_v \times 1012 \times 1.1906Q + F_u \times \Sigma(AU)]}{1012 \times 1.1906Q + 6 \times \Sigma(A)(KF_v + EF_u)} + F_v$$

In order to read values for F_1 and F_2 from the reference, it is necessary to calculate the ratios

$$\frac{\Sigma(AU)}{\Sigma(A)} \quad \text{and} \quad \frac{NV}{3\Sigma(A)}$$

When F_1 and F_2 have been read from tables, or calculated, they are used to find the environmental and air temperatures that will be generated within the heated space. F_1 and F_2 are the temperature ratios:

$$F_1 = \frac{t_{ei} - t_{ao}}{t_c - t_{ao}}$$

$$F_2 = \frac{t_{ai} - t_{ao}}{t_c - t_{ao}}$$

These equations are used to find the environmental and air temperatures. Rearrange the ratio equation:

$$F_1 = \frac{t_{ei} - t_{ao}}{t_c - t_{ao}}$$

$$F_1 \times (t_c - t_{ao}) = t_{ei} - t_{ao}$$

$$t_{ei} = t_{ao} + F_1(t_c - t_{ao})\,°\text{C}$$

and similarly:

$$F_2 = \frac{t_{ai} - t_{ao}}{t_c - t_{ao}}$$

$$F_2 \times (t_c - t_{ao}) = t_{ai} - t_{ao}$$
$$t_{ai} = t_{ao} + F_2(t_c - t_{ao}) \,°C$$

When the resultant and air temperatures are known for a space, mean radiant temperature t_r can be found from

$$t_c = 0.5t_r + 0.5t_{ai} \,°C$$

Rearrange to

$$0.5t_r = t_c - 0.5t_{ai}$$
$$t_r = 2(t_c - 0.5t_{ai})$$
$$= 2t_c - t_{ai} \,°C$$

The terms and constants used are as follows: 18 and 6 are heat transfer coefficients in $W/m^2\,K$; 1012 is the specific heat capacity of air in $J/kg\,K$; 1.1906 is the density of air at 20 °C in kg/m^3; $Q =$ infiltration rate of outdoor air (m^3/s); $E =$ percentage radiant heat input 0–100% in 10% steps; and $K =$ percentage convective heat input 0–100% in 10% steps. Table 5.1 gives the range of values for radiative and convective heat input proportions.

Table 5.1　Radiant and convective heat input proportions

Type of heating system	Convection, K (%)	Radiation, E (%)
Forced warm air or ducted	100	0
Natural convector or convector radiator	90	10
Multi-column radiator	80	20
Double and triple panel radiator	70	30
Single column, warmed floor, storage heater	50	50
Vertical and ceiling panels	33	67
High-temperature radiant system	10	90

(Source CIBSE 1986a)

The heat loss calculation is based on the maintenance of a desired dry resultant temperature at the centre of the room, $t_c\,°C$, for a determined outdoor air temperature, $t_{ao}\,°C$. When the values of F_1 and F_2 have been found by calculation or from tables, the mean surface, environmental, mean radiant and room air temperatures can be calculated. These are the temperatures that will be created within the room by the heating system under the stated operating conditions. It is important to realize that stable temperatures are unlikely to be found on a day of test with a thermocouple instrument. It is more likely that the external room surface temperatures are rising or falling in accordance with the weather variations and the time from the switching on of the heating plant. Figure 5.1 shows the data needed to find the internal surface temperature t_s of a structure.

The heat flow Q_p through the wall is

$$Q_p = F_1 UA(t_c - t_{ao})$$

The thermal resistance from node 1 to node 2 is R_{si}. The total thermal resistance for the wall is R, and

$$R = \frac{1}{U}$$

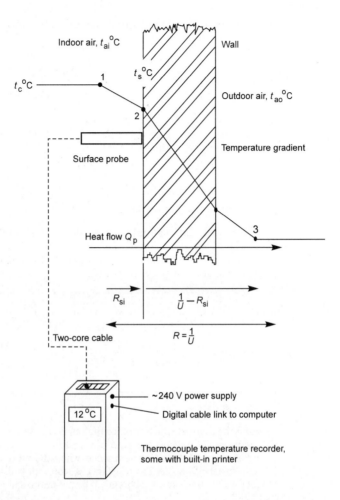

Fig. 5.1 Surface temperature of a wall.

The thermal resistance from node 2 to node 3 is $R - R_{si}$. The linear simultaneous equations that describe the heat flow through the wall are (Chadderton, 1995, ch. 10)

$$Q_p = F_1 UA(t_c - t_{ao})$$

$$= F_1 \times \frac{1}{R_{si}} \times A \times (t_c - t_s)$$

$$= \frac{1}{\dfrac{1}{U} - R_{si}} \times A \times (t_s - t_{ao})$$

F_1 does not apply to the third equation, because this is concerned only with heat flow through the structure and not what caused it in the room. Q_p is equal to both the first and third of these equations, so

$$F_1 UA(t_c - t_{ao}) = \frac{1}{\dfrac{1}{U} - R_{si}} \times A \times (t_s - t_{ao})$$

This can be rearranged to find the unknown surface t_s °C:

$$F_1 U \left(\frac{1}{U} - R_{si} \right) = \frac{t_s - t_{ao}}{t_c - t_{ao}}$$

$$t_s - t_{ao} = (t_c - t_{ao}) \times F_1 U \times \left(\frac{1}{U} - R_{si} \right)$$

$$t_s = t_{ao} + (t_c - t_{ao}) \times F_1 U \times \left(\frac{1}{U} - R_{si} \right)$$

$$t_s = t_{ao} + (t_c - t_{ao}) \times F_1 \times (1 - U \times R_{si}) °C$$

EXAMPLE 5.2

Calculate the internal surface temperature of a wall that has a thermal transmittance U of 0.96 W/m² K, an internal surface film resistance R_{si} of 0.12 m² K/W, and an outdoor air temperature of -3 °C when the room dry resultant temperature is 21 °C. The overall room factor F_1 is 1.02.

$$t_s = t_{ao} + (t_c - t_{ao}) \times F_1 (1 - UR_{si}) °C$$
$$= -3 + (21 - -3) \times 1.02 \times (1 - 0.96 \times 0.12) °C$$
$$= 18.7 °C$$

EXAMPLE 5.3

Find the thermal transmittance that is required to avoid surface condensation on the indoor surface of a brick and block cavity wall. A minimum internal surface temperature of 13 °C is to be provided when the outdoor air remains steady at -3 °C. The surface film thermal resistance of the wall is 0.12 m² K/W, the air temperature factor F_1 is 0.97, and the room will be maintained at a resultant temperature of 16 °C.

Using the equation for surface temperature:

$$t_s = t_{ao} + (t_c - t_{ao}) \times F_1 \times (1 - UR_{si}) °C$$

Rearrange to find the U value:

$$\frac{t_s - t_{ao}}{t_c - t_{ao}} \times \frac{1}{F_1} = 1 - UR_{si}$$

$$UR_{si} = 1 - \frac{t_s - t_{ao}}{t_c - t_{ao}} \times \frac{1}{F_1}$$

$$U = \frac{1}{R_{si}} \times \left(1 - \frac{t_s - t_{ao}}{t_c - t_{ao}} \times \frac{1}{F_1} \right)$$

Insert the known data and calculate the U value that will satisfy the criteria:

$$U = \frac{1}{0.12} \times \left(1 - \frac{13 - -3}{16 - -3} \times \frac{1}{0.97} \right) \frac{W}{m^2 K}$$
$$= 1.1 \, W/m^2 \, K$$

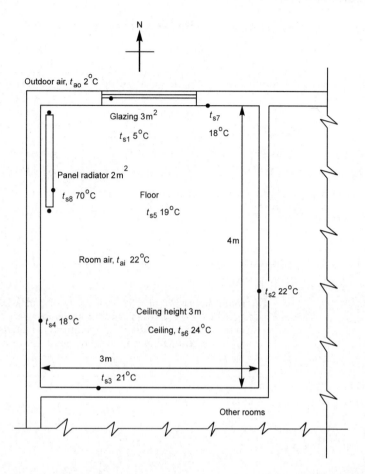

Fig. 5.2 Data for Example 5.4 (mean surface temperature) and Example 5.6 (heat loss).

EXAMPLE 5.4

Calculate the area weighted mean surface temperature of the room shown in Fig. 5.2.

$$\text{Area weighted mean } t_s = \frac{\Sigma(A t_s)}{\Sigma(A)} \, ^\circ C$$

The data are presented in Table 5.2.

Table 5.2 Mean surface temperature data for Example 5.4

Surface	Area, A (m^2)	t_s (°C)	$(A \times t_s)$ (m^2 °C)
Glazing surface 1	3	5	15
Wall surface 2	12	22	264
Wall surface 3	9	21	189
Wall surface 4	10	18	180
Floor surface 5	12	19	228

Table 5.2 (*contd.*)

Surface	Area, A (m²)	t_s (°C)	$(A \times t_s)$ (m² °C)
Ceiling surface 6	12	24	288
Wall surface 7	6	18	108
Radiator surface 8	2	70	140
$\Sigma(A) = 66$		$\Sigma(At_s) = 1412$	

$$\text{area} - \text{weighted mean } t_s = \frac{\Sigma(At_s)}{\Sigma(A)} {}^{\circ}\text{C}$$

$$= \frac{1412}{66} {}^{\circ}\text{C}$$

$$= 21.4\,{}^{\circ}\text{C}$$

EXAMPLE 5.5

A room has an F_1 of 0.98 and an F_2 of 1.05 when the resultant temperature is designed to be 19 °C at an outdoor air temperature of −4°C. Calculate the environmental, air and mean radiant temperatures for the room that will be produced in the room if the heating plant is correctly sized. Deduce what type of heating system is being used.

$$t_{ei} = t_{ao} + F_1(t_c - t_{ao})\,{}^{\circ}\text{C}$$
$$= -4 + 0.98 \times (19 - -4)\,{}^{\circ}\text{C}$$
$$= 18.5\,{}^{\circ}\text{C}$$
$$t_{ai} = t_{ao} + F_2(t_c - t_{ao})\,{}^{\circ}\text{C}$$
$$= -4 + 1.05 \times (19 - -4)\,{}^{\circ}\text{C}$$
$$= 20.2\,{}^{\circ}\text{C}$$
$$t_r = 2t_c - t_{ai}\,{}^{\circ}\text{C}$$
$$= 2 \times 19 - 20.2\,{}^{\circ}\text{C}$$
$$= 17.8\,{}^{\circ}\text{C}$$

The mean radiant temperature is a little lower than the resultant and environmental temperatures, so a minor proportion of the plant heat emission is by radiation. The room air temperature is greater than the resultant temperature, so there is a considerable amount of convective heat output. It is likely that the heat emitter is a panel or column radiator.

EXAMPLE 5.6

Calculate the heating plant output that is required for the office shown in Fig. 5.2. The room is to be maintained at a resultant temperature of 22 °C when the outdoor air is at 2 °C by a ducted air heating system. The rate of infiltration of outdoor air is one air change per hour. There is an identical room below. The adjacent rooms are at a resultant temperature of 18 °C. The thermal transmittances are 5.7 W/m² K for the glazing, 0.6 W/m² K for the external walls, 0.35 W/m² for the flat roof, and 1.2 W/m² for the interior walls and floor. Calculate F_u, F_v, F_1, F_2, $\Sigma(AU)/\Sigma(A)$ and $(NV)/[3\Sigma(A)]$. Compare the calculated values for F_1 and F_2 with those in Tables A9.1–A9.7 in the *CIBSE Guide A* (CIBSE, 1986a). Calculate the environmental, air, mean surface and mean radiant temperatures and the plant heat output per m² of floor area and per m³ of volume for the room.

The solution is shown as room 1 on the data disk in the file A:HLOSS.WK1.

$$Q_p = [F_1 \Sigma(AU) + 0.33NVF_2] \times (t_c - t_{ao}) \, \text{W}$$

The modified U value for the internal wall is

$$U_2 = U_1 \frac{t_{c1} - t_{c2}}{t_{c1} - t_{ao}} \, \text{W/m}^2 \, \text{K}$$

$$= 1.2 \times \frac{22 - 18}{22 - 2} \, \text{W/m}^2 \, \text{K}$$

$$= 0.24 \text{W/m}^2 \, \text{K}$$

Find $\Sigma(A)$ and $\Sigma(AU)$ from Table 5.3.

Table 5.3 Area and thermal transmittance data for Example 5.6

Surface	Area, A (m²)	U (W/m² K)	$A \times U$ (W/K)
Glazing	3	5.7	17.1
North external wall	6	0.6	3.6
West external wall	12	0.6	7.2
South internal wall	9	0.24	2.16
East internal wall	12	0.24	2.88
Floor (no heat loss)	12	0	0
Flat roof	12	0.35	4.2
$\Sigma(A) = 66$			$\Sigma(AU) = 37.14$

$$F_u = \frac{18\Sigma(A)}{\Sigma(AU) + 18\Sigma(A)}$$

$$= \frac{18 \times 66}{37.14 + 18 \times 66}$$

$$= 0.9697$$

$$F_v = \frac{6\Sigma(A)}{1012 \times 1.1906Q + 6\Sigma(A)}$$

$$Q = NV/3600 \, \text{m}^3/\text{s}$$

$$= 1 \times 4 \times 3 \times 3/3600 \, \text{m}^3/\text{s}$$

$$= 0.01 \, \text{m}^3/\text{s}$$

$$F_v = \frac{6 \times 66}{1012 \times 1.1906 \times 0.01 + 6 \times 66}$$

$$= 0.9705$$

For a forced warm air heating system, K is 100%, so $K = 1$ and $E = 0$. F_1 and F_2 can be found:

$$F_1 = \frac{E \times (F_v \times 1012 \times 1.1906Q + F_u\Sigma(AU))}{\Sigma(AU) + 18\Sigma(A) \times (KF_v + EF_u)} + F_u$$

$$= \frac{0 \times (0.9705 \times 1012 \times 1.1906 \times 0.01 + 0.9697 \times 37.14)}{37.14 + 18 \times 66 \times (1 \times 0.9705 + 0 \times 0.9697)} + 0.9697$$

It can be seen that when the radiant fraction E is zero, the numerator is always zero, and F_1 will always be equal to F_u. So,

$$F_1 = F_u$$
$$= 0.9697$$

$$F_2 = \frac{K \times (F_v \times 1012 \times 1.1906Q + F_u\Sigma(AU))}{1012 \times 1.1906Q + 6\Sigma(A) \times (KF_v + EF_u)} + F_v$$

$$= \frac{1 \times (0.9705 \times 1012 \times 1.1906 \times 0.01 + 0.9697 \times 37.14)}{1012 \times 1.1906 \times 0.01 + 6 \times 66 \times (1 \times 0.9705 + 0 \times 0.9697)} + 0.9705$$

$$= \frac{47.708}{396.367} + 0.9705$$

$$= 1.0909$$

In order to read values for F_1 and F_2 from the *CIBSE Guide* A (CIBSE, 1986a) Table A9.1, it is necessary to calculate the ratios

$$\frac{\Sigma(AU)}{\Sigma(A)} = \frac{37.14}{66}$$

$$= 0.5627$$

$$\text{Room volume } V = 4 \times 3 \times 3\,\text{m}^3$$

$$= 36\,\text{m}^3$$

$$\frac{NV}{3\Sigma(A)} = \frac{1 \times 36}{3 \times 66}$$

$$= 0.1818$$

CIBSE Table A9.1 $F_1 = 0.97$, calculated $F_1 = 0.9697 = 0.97$. CIBSE Table A9.1 $F_2 = 1.1$, calculated $F_2 = 1.0614 = 1.1$. The calculated values agree with those from the table to the nearest significant decimal place.

$$Q_p = [F_1\Sigma(AU) + 0.33NVF_2] \times (t_c - t_{ao})\,\text{W}$$
$$= (0.9697 \times 37.14 + 0.33 \times 1 \times 36 \times 1.0909) \times (22 - 2)\,\text{W}$$
$$= 979.491\,\text{W}, 980\,\text{W to nearest significant watt}$$
$$t_{ei} = t_{ao} + F_1(t_c - t_{ao})\,^\circ\text{C}$$
$$= 2 + 0.9697 \times (22 - 2)\,^\circ\text{C}$$
$$= 21.4\,^\circ\text{C}$$
$$t_{ai} = t_{ao} + F_2(t_c - t_{ao})\,^\circ\text{C}$$
$$= 2 + 1.0909 \times (22 - 2)\,^\circ\text{C}$$
$$= 23.8\,^\circ\text{C}$$

$$t_r = 2t_c - t_{ai}\,^\circ C$$
$$= 2 \times 22 - 23.8\,^\circ C$$
$$= 20.2\,^\circ C$$

Now calculate the room surface temperatures from

$$t_s = t_{ao} + (t_c - t_{ao}) \times F_1 \times (1 - UR_{si})\,^\circ C$$

The R_{si} values are taken from references as

wall and window $R_{si} = 0.12\,m^2\,K/W$
floor $R_{si} = 0.14\,m^2\,K/W$
ceiling $R_{si} = 0.10\,m^2\,K/W$

For this calculation, t_{ao} is the temperature on the other side of the surface, t_{c2} or t_{ao}. The U value is unmodified. The internal surface temperature of the window is found from

$$t_s = 2 + (22 - 2) \times 0.9697 \times (1 - 5.7 \times 0.12)\,^\circ C$$
$$= 8.1\,^\circ C$$

The other surface temperatures are found in the same way, and the data are represented in Table 5.4.

Table 5.4 Area and surface temperature data for Example 5.6

Surface	U	t_{ao}	R_{si}	t_s	A	At_s
Glazing	5.7	2	0.12	8.1	3	24.3
North wall	0.6	2	0.12	20.0	6	120.0
West wall	0.6	2	0.12	20.0	12	240.0
South wall	1.2	18	0.12	21.3	9	191.9
East wall	1.2	18	0.12	21.3	12	255.8
Floor	1.3	22	0.14	22.0	12	264.0
Roof	0.35	2	0.10	20.7	12	248.6

$\Sigma(A) = 66\,m^2$
$\Sigma(At_s) = 1344.6\,m^2\,^\circ C$

$$\text{Area weighted mean } t_s = \frac{\Sigma(At_s)}{\Sigma(A)}\,^\circ C$$
$$= \frac{1344.6}{66}\,^\circ C$$
$$= 20.4\,^\circ C$$

Notice the environmental criteria for this room. A ducted air heating system is used, and there is no source of radiant heat. An air dry-bulb temperature of 23.8 °C is needed to maintain the design dry resultant temperature of 22 °C. The environmental temperature will be 21.4 °C. The mean radiant temperature will be 20.2 °C, and this is practically the same as the mean of the room surface temperatures, 20.4 °C. The air temperature detector that will be used to control the warm air heating system will need to be set at 24 °C. The detector will need to be sampling an average of the room air condition. This could be achieved when it is located in the room air extract duct.

The plant heat output per unit floor area and room volume are found for comparison with the expected values and with those for other rooms:

$$\frac{Q_p}{A} = \frac{Q_p \, W}{\text{floor area } A \text{ m}^2}$$

$$= \frac{980 \, W}{12 \, m^2}$$

$$= 81.7 \, W/m^2$$

$$\frac{Q_p}{V} = \frac{Q_p \, W}{\text{room volume } V \text{ m}^3}$$

$$= \frac{980 \, W}{4 \times 3 \times \times 3 \, m^3}$$

$$= 27.2 \, W/m^3$$

EXAMPLE 5.7

Calculate the heating plant output that is required for the industrial building shown in Fig. 5.3. The factory area is to be maintained at a resultant temperature of 16 °C and the offices at a resultant temperature of 20 °C when the outdoor air is at −5°C. A low-temperature hot-water heating system has panel radiators in the factory and office areas. The rate of infiltration of outdoor air, N, is 0.75 air changes per hour in the factory and one air change per hour in the offices. The thermal transmittances are 6 W/m² for the glazing and external doors, 0.6 W/m² K for the external walls, 0.35 W/m² K for the roof, 0.48 W/m² K for the floor, and 1.7 W/m² K for the interior wall. The designer is to assume that the void above the office flat ceiling is at a temperature of 8 °C. Take the building dimensions from Fig. 5.3. All the windows are 2 m long and 2 m high. Calculate F_u, F_v, F_1, F_2, $\Sigma(AU)/\Sigma(A)$ and $NV/3\Sigma(A)$. Compare the calculated values for F_1 and F_2 with those in Tables A9.1–A9.7 in the *CIBSE Guide A* (CIBSE, 1986a). Calculate the environmental, air, mean surface and mean radiant temperatures and the plant heat output per m² of floor area and per m³ of volume for the room.

Factory area

$$Q_p = [F_1 \Sigma(AU) + 0.33NVF_2] \times (t_c - t_{ao}) \, W$$

The modified U value for the internal wall from the factory to the office is

$$U_2 = U_1 \frac{t_{c1} - t_{c2}}{t_{c1} - t_{ao}} \, W/m^2 \, K$$

$$= 1.7 \times \frac{16 - 20}{16 - -5} \, W/m^2 \, K$$

$$= -0.32 \, W/m^2 \, K$$

This has a minus sign because there will be a negative flow of heat from the factory area. There will be an equal positive flow of heat from the office into the factory. The modified U value for the internal wall to the roof void above the office is

$$U_2 = 1.7 \times \frac{16 - 8}{16 - -5} \, W/m^2 \, K$$

$$= 0.65 \, W/m^2 \, K$$

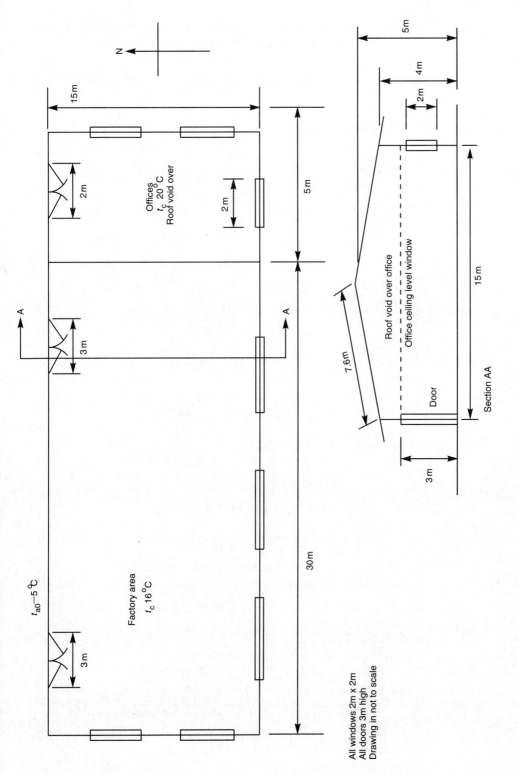

N

15 m

t_{a0} —5 °C

Factory area
t_c 16 °C

3 m

A

3 m

2 m

Offices
t_c 20 °C
Roof void over

2 m

A

5 m

2 m

30 m

All windows 2m x 2m
All doors 3m high
Drawing in not to scale

5 m

4 m

2 m

7.6 m

Roof void over office

Office ceiling level window

Door

3 m

15 m

Section AA

Fig. 5.3 Factory buuilding in Example 5.7.

The data to find $\Sigma(A)$ and $\Sigma(AU)$ for the factory is represented in Table 5.5.

Table 5.5   Area and thermal transmittance data for
Example 5.7

Surface	Area, A (m²)	U (W/m² K)	AU (W/K)
Gross external wall			
$\quad$ $(60 \times 4 + 15 \times 4.5)$	307.5		
Doors $(2 \times 3 \times 3)$	18.0		
Windows $(5 \times 2 \times 2)$	20.0		
$\quad$ openings	38.0	6.0	228.0
$\quad$ net wall	269.5	0.6	161.7
Wall to office (15×3)	45.0	−0.32	−14.4
Wall to void			
$\quad$ $(15 \times 1 + 0.5 \times 15 \times 1)$	22.5	0.65	14.6
Floor (30×15)	450.0	0.48	216.0
Roof $(30 \times 7.6 \times 2)$	456.0	0.35	159.6
$\Sigma(A) = 1281$		$\Sigma(AU) = 765.5$	

$$F_{\mathrm{u}} = \frac{18\Sigma(A)}{\Sigma(AU) + 18\Sigma(A)}$$

$$= \frac{18 \times 1281}{765.5 + 18 \times 1281}$$

$$= 0.968$$

$$F_{\mathrm{v}} = \frac{6\Sigma(A)}{1012 \times 1.1906Q + 6\Sigma(A)}$$

volume of the factory, $V = 30 \times 15 \times 4.5 \, \mathrm{m^3}$

$$= 2025 \, \mathrm{m^3}$$

$$Q = NV/3600 \, \mathrm{m^3/s}$$

$$= 0.75 \times 2025/3600 \, \mathrm{m^3/s}$$

$$= 0.422 \, \mathrm{m^3/s}$$

$$F_{\mathrm{v}} = \frac{6 \times 1281}{1012 \times 1.1906 \times 0.422 + 6 \times 1281}$$

$$= 0.938$$

For a forced warm-air heating system, K is 70%, so $K = 0.7$ and E is 30%, so
$E = 0.3$.

F_1 and F_2 can be found:

$$F_1 = \frac{E \times [F_{\mathrm{v}} \times 1012 \times 1.1906Q + F_{\mathrm{u}}\Sigma(AU)]}{\Sigma(AU) + 18\Sigma(A) \times (KF_{\mathrm{v}} + EF_{\mathrm{u}})} + F_{\mathrm{u}}$$

$$= \frac{0.3 \times (0.938 \times 1012 \times 1.1906 \times 0.422 + 0.968 \times 765.5)}{765.5 + 18 \times 1281 \times (0.7 \times 0.938 + 0.3 \times 0.968)} + 0.968$$

$$F_1 = \frac{365.4}{22601.4} + 0.968$$

$$= 0.01617 + 0.968$$

$$= 0.984$$

$$F_2 = \frac{K \times [F_v \times 1012 \times 1.1906Q + F_u\Sigma(AU)]}{1012 \times 1.1906Q + 6\Sigma(A) \times (KF_v + EF_u)} + F_v$$

$$= \frac{0.7 \times (0.938 \times 1012 \times 1.1906 \times 0.422 + 0.968 \times 765.5)}{1012 \times 1.1906 \times 0.422 + 6 \times 1281 \times (0.7 \times 0.938 + 0.3 \times 0.968)} + 0.938$$

$$= \frac{852.6}{7787.1} + 0.938$$

$$= 1.0475$$

In order to read values for F_1 and F_2 from the *CIBSE Guide A* (CIBSE, 1986a) Table A9.4, it is necessary to calculate the ratios,

$$\frac{\Sigma(AU)}{\Sigma(A)} = \frac{765.5}{1281}$$

$$= 0.6$$

$$\frac{NV}{3\Sigma(A)} = \frac{0.75 \times 2025}{3 \times 1281}$$

$$= 0.4$$

CIBSE Table A9.4 $F_1 = 0.99$, calculated $F_1 = 0.984 = 0.98$.
CIBSE Table A9.4 $F_2 = 1.02$, calculated $F_2 = 1.0475 = 1.05$.
The calculated values are different by 1% and 3%, which are not considered to be significant.

$$Q_p = [F_1\Sigma(AU) + 0.33NVF_2] \times (t_c - t_{ao})\,\text{W}$$
$$= (0.984 \times 765.5 + 0.33 \times 0.75 \times 2025 \times 1.0475) \times (16 - -5)\,\text{W}$$
$$= 26\,843\,\text{W}$$

$$t_{ei} = t_{ao} + F_1 \times (t_c - t_{ao})\,°\text{C}$$
$$= -5 + 0.984 \times (16 - -5)\,°\text{C}$$
$$= 15.7\,°\text{C}$$

$$t_{ai} = t_{ao} + F_2 \times (t_c - t_{ao})\,°\text{C}$$
$$= -5 + 1.0475 \times (16 - -5)\,°\text{C}$$
$$= 17\,°\text{C}$$

$$t_r = 2t_c - t_{ai}\,°\text{C}$$
$$= 2 \times 16 - 17\,°\text{C}$$
$$= 15\,°\text{C}$$

Now calculate the room surface temperatures from

$$t_s = t_{ao} + (t_c - t_{ao}) \times F_1 \times (1 - UR_{si})\,°\text{C}$$

The R_{si} values are taken from references as

wall and window $R_{si} = 0.12\,\text{m}^2\,\text{K/W}$
floor $R_{si} = 0.14\,\text{m}^2\,\text{K/W}$
ceiling $R_{si} = 0.10\,\text{m}^2\,\text{K/W}$

For this calculation, t_{ao} is the temperature on the other side of the surface, t_{c2} or t_{ao}. The U value is unmodified. The internal surface temperature of the window is found from

$$t_s = -5 + (16 - -5) \times 0.984 \times (1 - 6 \times 0.12)\,^{\circ}C$$
$$= 0.8\,^{\circ}C$$

The other surface temperatures are found in the same way, and the data is shown in Table 5.6.

Table 5.6   Surface temperatures for the factory in Example 5.7

Surface	U	t_{ao}	R_{si}	t_s	A	At_s
Glazing/doors	6	−5	0.12	0.8	38	30.4
North wall	0.6	−5	0.12	14.2	102	1448.4
West wall	0.6	−5	0.12	14.2	59.5	844.9
South wall	0.6	−5	0.12	14.2	108	1533.6
Office wall	1.7	20	0.12	16.9	45	760.5
Floor	0.48	−5	0.14	14.3	450	6435
Roof	0.35	−5	0.10	14.9	456	6794.4
Void wall	1.7	8	0.12	14.3	22.5	321.8

$\Sigma(A) = 1281\,m^2$
$\Sigma(At_s) = 18168.95\,m^2\,^{\circ}C$

$$\text{Area-weighted mean } t_s = \frac{\Sigma(At_s)}{\Sigma(A)}\,^{\circ}C$$
$$= \frac{18\,168.95}{1281}\,^{\circ}C$$
$$= 14.2\,^{\circ}C$$

$$\frac{Q_p}{A} = \frac{Q_p\,W}{\text{floor area } A\,m^2}$$
$$= \frac{26\,843\,W}{450\,m^2}$$
$$= 59.7\,W/m^2$$

$$\frac{Q_p}{V} = \frac{Q_p\,W}{\text{room volume } V\,m^3}$$
$$= \frac{26\,843\,W}{2025\,m^3}$$
$$= 13.3\,W/m^3$$

Office

The explanations are omitted for this room. For the internal wall from the office to the factory:

$$U_2 = 1.7 \times \frac{20 - 16}{20 - -5}\,W/m^2\,K$$
$$= 0.27\,W/m^2\,K$$

The data to find $\Sigma(A)$ and $\Sigma(AU)$ for the office is represented in Table 5.7.

Table 5.7　Area and thermal transmittance data for Example 5.7

Surface	Area, A (m^2)	U (W/m^2 K)	AU (W/K)
Gross external wall (25×3)	75		
Doors (2×3)	6		
Windows $(3 \times 2 \times 2)$	12		
openings	18	6.0	108
net wall	57	0.6	34.2
Wall to factory (15×3)	45	0.27	12.2
Floor (15×5)	75	0.48	36
Roof (15×5)	75	0.35	26.3
$\Sigma(A) = 270\,\text{m}^2$		$\Sigma(AU) = 216.7\,\text{W/K}$	

$$F_\text{u} = \frac{18 \times 270}{216.7 + 18 \times 270}$$
$$= 0.957$$

$$F_\text{v} = \frac{6\Sigma(A)}{1012 \times 1.1906Q + 6\Sigma(A)}$$
$$V = 15 \times 5 \times 3\,\text{m}^3$$
$$= 225\,\text{m}^3$$
$$Q = 1 \times 225/3600\,\text{m}^3/\text{s}$$
$$= 0.0625\,\text{m}^3/\text{s}$$

$$F_\text{v} = \frac{6 \times 270}{1012 \times 1.1906 \times 0.0625 + 6 \times 270}$$
$$= 0.956$$

$$F_1 = \frac{0.3 \times (0.956 \times 1012 \times 1.1906 \times 0.0625 + 0.957 \times 216.7)}{216.7 + 18 \times 270(0.7 \times 0.956 + 0.3 \times 0.957)} + 0.957$$

$$F_1 = \frac{83.81}{4864.3} + 0.957$$
$$= 0.0172 + 0.957$$
$$= 0.974$$

$$F_2 = \frac{0.7 \times (0.956 \times 1012 \times 1.1906 \times 0.0625 + 0.957 \times 216.7)}{1012 \times 1.1906 \times 0.0625 + 6 \times 270 \times (0.7 \times 0.956 + 0.3 \times 0.957)} + 0.956$$

$$= \frac{195.6}{1624.5} + 0.956$$
$$= 1.076$$

$$\frac{\Sigma(AU)}{\Sigma(A)} = \frac{216.7}{270}$$
$$= 0.8$$

$$\frac{NV}{3\Sigma(A)} = \frac{1 \times 225}{3 \times 270}$$
$$= 0.28$$

CIBSE Table A9.4 $F_1 = 0.99$, calculated $F_1 = 0.974 = 0.97$.
CIBSE Table A9.4 $F_2 = 1.05$, calculated $F_2 = 1.076 = 1.08$.

$$Q_p = (0.974 \times 216.7 + 0.33 \times 1 \times 225 \times 1.076) \times (20 - -5)\,\text{W}$$
$$= 7274\,\text{W}$$
$$t_{ei} = -5 + 0.974 \times (20 - -5)\,^\circ\text{C}$$
$$= 19.4\,^\circ\text{C}$$
$$t_{ai} = -5 + 1.076 \times (20 - -5)^\circ\text{C}$$
$$= 21.9\,^\circ\text{C}$$
$$t_r = 2 \times 20 - 21.9\,^\circ\text{C}$$
$$= 18.1\,^\circ\text{C}$$

The surface temperatures are shown in Table 5.8.

Table 5.8   Surface temperatures for the office in Example 5.7.

Surface	U	t_{ao}	R_{si}	t_s	A	At_s
Glazing/door	6	−5	0.12	1.8	18	32.4
External walls	0.6	−5	0.12	17.6	57	1003.2
Internal wall	1.7	16	0.12	19.1	45	859.5
Floor	0.48	−5	0.14	17.7	75	1327.5
Roof	0.35	−5	0.10	18.5	75	1387.5

$\Sigma(A) = 270\,\text{m}^2$
$\Sigma(At_s) = 4610.1\,\text{m}^2\,^\circ\text{C}$

$$\text{Area-weighted mean } t_s = \frac{4610.1}{270}\,^\circ\text{C}$$
$$= 17.1\,^\circ\text{C}$$
$$\frac{Q_p}{A} = \frac{7274\,\text{W}}{75\,\text{m}^2}$$
$$= 97\,\text{W/m}^2$$
$$\frac{Q_p}{V} = \frac{7274\,\text{W}}{225\,\text{m}^3}$$
$$= 32.3\,\text{W/m}^3$$

DATA REQUIREMENT

The worksheet has been set to display all numbers in fixed format to two decimal places except where this has been changed in some cells. Zero, one or three decimal places have been used, and these are indicated as F(0), F(1) or F(3) format. The user can enter any number of decimal places into a cell, for example 123.456789, but only the set number of places will be displayed on the screen; 123.4 when the format is F(1). The user can change the display format. The computer uses all the numbers that are typed and entered.

Create new files on drives A and C in subdirectories so that the filename reflects the job title: for example, A:\HOLDEN\HLOSS\LOSS1.WK1. This means that there is a job data disk in drive A. This disk has a directory named HOLDEN. All the files for the job for Holden Motor Company will be located in the HOLDEN directory. A subdirectory of the HOLDEN main directory is reserved for heat loss calculation files; this is the HLOSS area of the disk. A worksheet named LOSS1.WK1 is to be used for the current data and calculations.

Each room can consist of two rectangular spaces. Where the room cannot be simply described in this manner, it may be necessary to divide it into more than one space. The parts of the room can be given names such as LoungeA1, LoungeA2, etc., so that they are correctly added when it is time to find the total heat load. Each room can have four walls. Each wall can have three openings for windows or doors. Where a heated space has more than four walls and three openings in each wall, it may be appropriate to add some of the lengths together so that all the components can be included. Floor, ceiling and roof data can be entered under any of these surface categories provided that the correct data is identified and entered somewhere in the cell range from B95 to F100.

The temperature on the other side of each wall, floor, ceiling and opening is entered by the user. This will be either the outdoor air temperature or the resultant temperature of the room or space on that side. The thermal transmittance of each wall and opening is entered as the unmodified *U* value. These will be taken from the references or the UVAL.WK1 worksheet output by the user. Modifications for intermediate temperature differences are carried out in the cell formulae. Make sure that zero is present, or is typed, in cells for the dimensions of walls and openings, the second temperature and the thermal transmittance where there are no surfaces to be calculated for. This also ensures that previous applications that have used this copy of the file have been overwritten. If cells contain old data that is correct for the new application, there is no need to overwrite it.

Check that the dimensions of the rooms have been accurately entered, because the only formula that will validate the input data is a simple area check. Visual, measurement or typing errors by the user may not be found by this check. Make sure that you have entered only numbers into the cells that require numeric data. If the letter 'o' or 'O' is typed instead of zero '0', an error message will be displayed in the cell where the calculation is produced. Alphabetic characters in cells always have an apostrophe (') at the beginning of the string in the formula line on screen. Spreadsheet programs do not calculate with alphabetic characters, only numbers. You may think that a cell contains numbers because they appear in the cell on the screen, but you may have typed them in as words and then found an error message in the formula cell. This is an easy mistake to make. Review all the data that is on the screen for each room to double check that there have not been any omissions. Look at the dimensions, temperatures and thermal transmittances. Are they correct?

The worksheet is set up for 10 rooms. Each room must have valid data to avoid the formulae generating an error, ERR, message in cells. When all the real rooms have been described, for example seven of them, the remaining three rooms are give fictitious data and they become fictitious rooms. The original data disk contains these non-existent rooms with appropriate data. Each fictitious room must be 1 m long, 1 m wide and 1 m high. It must have a design resultant temperature of 1 °C and an outdoor air temperature of 0 °C. Wall 1 must be 1 m long, 1 m high and have a thermal transmittance of zero W/m^2 K. This is the only data that is needed for each fictitious room. The worksheet calculates the 1 m^2

floor area, 1 m^3 volume and zero heat loss for the room. This is a valid result, and there are no errors. Look at the worksheet for these rooms and notice the values for the output data. All that the user has to do is type the number of these fictitious rooms into cell D887. The fictitious floor areas and room volumes are removed by the user's declaration in this cell. The user is prompted to review how many fictitious rooms have been entered by looking at the output from cell D890. There are no formulae to stop the user from entering incorrect data, such as more than 10 fictitious rooms.

The user will enter new data for:

1. Own name in cell B4.
2. Name of the job or contract in cell B5.
3. Job reference information in cell B6.
4. Current date in cell B7.
5. The name of the new file that holds the working copy of the job data in cell B9, for example A:LOSS1.WK1 or A:\HOLDEN\HLOSS\LOSS1.WK1.
6. Room name or identification letters and numbers in cell C45.
7. Floor level in cell C46.
8. Design resultant temperature in cell C47.
9. Design outdoor air temperature in cell C48.
10. Rate of infiltration of outdoor air in cell C49.
11. Room dimensions for area 1 in cells D51, D52 and D53; room dimensions for area 2 in cells D56, D57 and D58.
12. Select the type of heating system from the range of cells from A20 to A26. Type the appropriate radiant percentage into cell D64 and the convective percentage into cell D65 from the values for the heating system types.
13. The length of the first wall in cell B69 and its height into cell C69.
14. The lengths of the windows and doors in wall 1 into the cell range B70 to C72.
15. Make sure that zero is present, or is entered, in each cell in the range from B69 to C72 where there is no wall to be calculated or the wall has no openings.
16. The temperature on the other side of the wall and opening into the cell range E70 to E73.
17. The thermal transmittance (U value) of the wall and openings in the cell range F70 to F73.
18. Look at the content of the range of input data cells from B69 to F72. Make sure that the correct data have been entered. The gross and net areas in column D are calculated automatically.
19. The repetition of steps 13–18 for the second, third and fourth walls in the cell range B75 to F93.
20. The floor lengths and widths in the cell range B95 to C96.
21. The temperature on the other side of the floor in cells E95 and E96.
22. The thermal transmission of the floor in cells F95 and F96.
23. The ceiling lengths and widths in the cell range B97 to C98.
24. The temperature on the other side of the ceiling in cells E97 and E98.
25. The thermal transmission of the ceiling in cells F97 and F98.
26. The roof lengths and widths in the cell range B99 to C100.
27. The temperature on the other side of the roof in cells E99 and E100.
28. The thermal transmission of the roof in cells F99 and F100.

29. Visually compare the surface areas in cells D102, D62, B103 and the range from D95 to D100 to ensure that they make sense. If there is an error between the expected room surface area, which is automatically calculated from the overall room dimensions, and the sum of the declared areas in cells D69 to D100, an 'error in areas' warning will appear in cell E103.
30. Repeat the procedures for the other rooms.
31. The number of fictitious rooms in cell D887 when there are less than 10 real rooms on the current worksheet.
32. Additional heat loss worksheet files when there are more than 10 rooms.
33. The filename being used for the part of the job on this worksheet in cells A894 and B894.
34. The filenames being used for up to nine more worksheets on this job in cells A895 to B903.
35. The additional plant load in kW to provide the hot-water service in cell C907. Make sure this cell contains zero if there is no load to be added.
36. The surplus plant load in kW for future or additional phases of the job in cell C908.

This concludes the entry of the room data.

OUTPUT DATA

1. Volume of the room for area 1 is given in cell D54.
2. Floor area for area 1 is given in cell D55.
3. Volume of the room for area 2 is given in cell D59.
4. Floor area for area 2 is given in cell D60.
5. Volume of the whole room, V m^3, is given in cell D61.
6. Floor area A m^2 of the whole room is given in cell D62.
7. Area of each surface for wall 1 is given in the cell range D69 to D72.
8. Net wall area, after the openings have been subtracted from the gross surface area, is given in cell D73.
9. Repetition of the information for the other surfaces is given in the cell range D75 to D100.
10. AU W/K for each surface is given in the cell range G70 to G100.
11. Calculated temperature of each surface, t_s °C, is given in the cell range I70 to I100.
12. At_s m^2 °C for each surface is given in the cell range J70 to J100.
13. $\Sigma(At_s)$ m^2 °C for the whole room from the cell range J70 to J100 is given in cell J101.
14. Calculated mean surface temperature t_s °C of the room is given in cell J102.
15. Total area of the room surfaces, A m^2, is given in cell D102.
16. Sum of the room surface areas that were used in the cell range D69 to D100 is given in cell B103.
17. Any difference in room surface areas between the content of cells D102 and B103 is given in cell E103. If cell E103 is non-zero, go back over the dimensions of the room and its surfaces to see if a mistake has been entered.
18. $\Sigma(AU)$ W/K for the whole room is given in cell C104.
19. Infiltration air flow rate Q m^3/s is given in cell C105.

20. Ventilation factor F_v is given in cell B106.
21. Thermal transmittance factor F_u is given in cell B107.
22. Environmental temperature factor F_1 is given in cell B108.
23. Air temperature factor F_2 is given in cell B109.
24. Ratio $\Sigma(AU)/\Sigma(A)$ is given in cell B110.
25. Ratio $NV/3\Sigma(A)$ is given in cell B111.
26. Design resultant temperature t_c °C is given in cell C112.
27. Environmental temperature t_{ei} °C is given in cell C113.
28. Room air temperature t_{ai} °C is given in cell C114.
29. Mean surface temperature t_s °C is given in cell C115.
30. Mean radiant temperature t_r °C is given in cell C116.
31. Outside air temperature t_{ao} °C is given in cell C117.
32. The plant energy output power Q_p W that is needed to maintain the design resultant temperature at the specified outdoor air temperature is shown in cell B120.
33. Plant energy output power Q_p kW is given in cell B121.
34. Plant energy output power in W/m^3 of room volume is given in cell B122.
35. Plant energy output power in W/m^2 of room floor area is given in cell B123.
36. Data is repeated for the other rooms.
37. The number of fictitious rooms on the current worksheet is given in cell B890.
38. The sum of the plant energy outputs for all the rooms on this worksheet is in cell C894.
39. The sum of the plant energy outputs for all the worksheets for this job is in cell C905.
40. The additional heating plant power to balance the heat loss from the distribution pipes or ducts is given in cell C906.
41. The total heating plant power from the ten worksheets is given in cell C910.
42. The floor area of the building on the current worksheet is given in cell D913.
43. The volume of the building on the current worksheet is given in cell D914.
44. The heat loss per m^2 of the building floor area is given in cell D915.
45. The heat loss per m^3 of the building volume is given in cell D916.

The output data includes all the temperatures that have been used as input information and those that have been calculated. This is a summary of the thermal environmental conditions for the room. The performance of a heating system can be validated by measuring the dry-bulb air and globe temperatures in the heated room and then comparing them with the expected values from this worksheet.

The plant energy output in W/m^2 of floor area and W/m^3 of room volume are used for comparison with other similar applications. They may also be used to gain a quick assessment of heat loads for other rooms and buildings before detailed calculations are undertaken. Ensure that zero plant energy output is displayed for the rooms where you know that there should not be any heat load. If numbers appear for rooms that do not exist in this job, then there may be some data left over from an earlier use of this copy of the worksheet. This error needs to be cancelled by the entry of zeros into the room cells. Carry out a final check that the worksheet has the correct data.

FORMULAE

Representative samples of the formulae are given here. Each formula can be read on the spreadsheet by moving the cursor to the cell. The equation is presented in the form in which it would normally be written and in the format that is used by the spreadsheet. This worksheet has been set for two decimal place format for most of the numbers displayed. Zero, one or three decimal places are more appropriate for percentages, temperatures and dimensions. These formats have been specified in the relevant locations.

Cell D54

(F2) +D51*D52*D53

The cell is formatted for two decimal places. It calculates the volume of the room for area 1:

$$\text{volume} = \text{length} \times \text{width} \times \text{height} \, \text{m}^3$$
$$\text{volume D54} = \text{D51} \times \text{D52} \times \text{D53} \, \text{m}^3$$

Cell D55

(F2) +D51*D52

The floor area of the room for area 1 is found:

$$\text{area} = \text{length} \times \text{width} \, \text{m}^2$$
$$\text{area D55} = \text{D51} \times \text{D52} \, \text{m}^2$$

Cell D59

(F2) +D56*D57*D58

It calculates the volume of the room for area 2:

$$\text{volume} = \text{length} \times \text{width} \times \text{height} \, \text{m}^3$$
$$\text{volume D59} = \text{D56} \times \text{D57} \times \text{D58} \, \text{m}^3$$

Cell D60

(F2) +D56*D57

The floor area of the room for area 2 is

$$\text{area} = \text{length} \times \text{width} \, \text{m}^2$$
$$\text{area D60} = \text{D56} \times \text{D57} \, \text{m}^2$$

Cell D61

(F2) +D54+D59

The total volume of the room is

$$\text{volume } V = \text{volume}_1 + \text{volume}_2 \, \text{m}^3$$
$$\text{volume D61} = \text{D54} + \text{D59} \, \text{m}^3$$

Cell D62

(F2) +D55+D60

The total floor area of the room is

$$\text{area } A = \text{area}_1 + \text{area}_2 \text{ m}^2$$
$$\text{area D62} = \text{D55} + \text{D60 m}^2$$

Cell D69

(F2) +B69*C69

The gross surface area of wall 1 includes the openings for windows and doors, and is found from

$$\text{area } A = \text{length} \times \text{height m}^2$$
$$\text{area D69} = \text{B69} \times \text{C69 m}^2$$

Cell D70

(F2) +B70*C70

The surface area of window 1 is

$$\text{area } A = \text{length} \times \text{height m}^2$$
$$\text{area D70} = \text{B70} \times \text{C70 m}^2$$

Cell G70

(F2) +D70*F70*($C$47−E70)/($C$47−$C$48)

The product of the surface area of window 1 and its thermal transmittance is calculated. The standard thermal transmittance is modified for any temperature on the other side of the window by multiplying by the temperature ratio, $(t_{c1} - t_{c2})/(t_{c1} - t_{ao})$. These are in cells ($C$47 − E70)/($C$47 − $C$48). Cells C47 and C48 are absolute cell references and have the $ symbol. The resultant temperature of the space on the other side of the window is taken from cell E70, because this is a variable for each surface. The whole equation is

$$AU \text{ W/K} = A \text{ m}^2 \times U \frac{\text{W}}{\text{m}^2 \text{ K}} \times \frac{t_{c1} - t_{c2}}{t_{c1} - t_{ao}}$$
$$\text{G70 W/K} = \text{D70 m}^2 \times \text{F70} \frac{\text{W}}{\text{m}^2 \text{ K}} \times \frac{\$C\$47 - E70}{\$C\$47 - \$C\$48}$$

Cell I70

(F1) (1−F70*$C$31)*$B$108*($C$47−E70)+E70

The content of this cell commences with a bracket, so it must be a formula, and it does not need a plus sign. The temperature on the room side of the surface is calculated:

$$t_s = (1 - UR_{si}) \times F_1 \times (t_{c1} - t_{c2}) + t_{c2} \,^{\circ}C$$
$$I70 = (1 - F70 \times \$C\$31) \times \$B\$108 \times (\$C\$47 - E70) + E70 \,^{\circ}C$$

The correct cell for the surface thermal resistance R_{si} has been included in the formula. Walls, windows and doors are assumed to have a high emissivity, and the absolute cell reference $\$C\$31$ has been used. The user can change the cell reference for low emissivity if needed. The environmental temperature factor F_1 has been retrieved from cell $\$B\$108$.

Cell J70

(F1) +D70*I70

The product of area and its surface temperature is

$$At_s = A\,m^2 \times t_s \,^{\circ}C$$
$$J70 = D70 \times I70 \, m^2 \,^{\circ}C$$

Cell D71 to J72

These repeat the calculations that were carried out along line 70, but are for window 2 and door 1.

Cell D73

(F2) +D69−D70−D71−D72

The net area of wall 1 is

$$\text{net wall area} \, A = \text{gross wall area} - \text{area of openings} \, m^2$$
$$D73 = D69 - D70 - D71 - D72 \, m^2$$

Cell G73 to J73

These repeat the calculations for AU, t_s and At_s.

Cell D75 to J100

These repeat the previous calculations for the remaining surfaces in the room.

Cell J101

(F1) @SUM(J70..J100)

The @SUM is a mathematical function to sum a range of cells. It is denoted by the @ sign rather than a plus sign. The cells in the range have the product of area and temperature for each surface. There are 22 possible surfaces. The summation for the room is

$$\Sigma(At_\mathrm{s}) = (At_\mathrm{s})_1 + (At_\mathrm{s})_2 + \ldots + (At_\mathrm{s})_{22}\,\mathrm{m^2\,^\circ C}$$
$$\mathrm{J}101 = \mathrm{J}70 + \mathrm{J}71 + \ldots + \mathrm{J}100\,\mathrm{m^2\,^\circ C}$$
$$= @\mathrm{SUM(J70\ldots J100)\,m^2\,^\circ C}$$

The formula @SUM(J70..J100) adds the numerical contents of the cells within the range from J70 to J100. It ignores any blank cells, alphabetic or symbolic characters within the range, because they have no numerical value.

Cell J102

(F1) +J101/D102

The area-weighted mean surface temperature of the room is

$$\mathrm{mean}\ t_\mathrm{s} = \frac{\Sigma(At_\mathrm{s})\,\mathrm{m^2\,^\circ C}}{\Sigma(A)\,\mathrm{m^2}}$$

$$\mathrm{J}102 = \frac{\mathrm{J}101}{\mathrm{D}102}\,^\circ\mathrm{C}$$

Cell D102

(F3) +D69+D75+D81+D89+D95+D96+D97+D98+D99+D100

The gross surface area of the walls, floor and ceiling is summed. There can be up to 10 surfaces.

$$\Sigma(A) = A_1 + A_2 + \ldots A_{10}\,\mathrm{m^2}$$
$$\mathrm{D}102 = \mathrm{D}69 + \mathrm{D}75 + \ldots + \mathrm{D}100\,\mathrm{m^2}$$

Cell B103

(F3) D70+D71+D72+D73+D76+D77+D78+D79+D82+D83+D84+D85+D90+D91+D92+D93+D95+D96+D97+D98+D99+D100

This is to check that the correct data has been entered for the areas of the surfaces. The area of each window, door, net wall, floor, ceiling and roof surface is summed. Notice that the gross wall area is omitted. There are 22 possible surface areas.

$$\Sigma(A) = A_1 + A_2 \ldots + A_{22}\,\mathrm{m^2}$$
$$\mathrm{B}103 = \mathrm{D}70 + \mathrm{D}71 + \ldots + \mathrm{D}100\,\mathrm{m^2}$$

Cell E103

(F3) +D102−B103

A significant difference between the two total areas would show that there is an error somewhere in the data:

$$\mathrm{error\ in\ areas = gross\ area - sum\ of\ net\ areas\ m^2}$$
$$\mathrm{E}103 = \mathrm{D}102 - \mathrm{B}103\,\mathrm{m^2}$$

The difference in these areas should be zero. Three decimal places are displayed, and a minor place error may appear. The user should be satisfied that either there is no error, or that it is not significant: for example, if it is less than $0.1\,\text{m}^2$. Should an error area be found, check that the dimensions of the surfaces and their openings have been entered with sufficient accuracy.

Cell C104

(F2) @SUM(G70..G100)

The summation of AU for the 22 possible surfaces in the room is

$$\Sigma(AU) = (AU)_1 + (AU)_2 + \ldots + (AU)_{22}\,\text{W/K}$$
$$C104 = G70 + G71 + \ldots + G100\,\text{W/K}$$

Cell C105

(F4) +C49*D61/3600

The rate at which outdoor air infiltrates the room is calculated from the room volume, $V\,\text{m}^3$, and the room air change rate, N air changes per hour.

$$Q = N\,\frac{\text{air change}}{\text{h}} \times \frac{V\,\text{m}^3}{1\,\text{air change}} \times \frac{1\,\text{h}}{3600\,\text{s}}$$

$$= \frac{NV}{3600}\,\frac{\text{m}^3}{\text{s}}$$

$$C105 = C49\,\frac{\text{air change}}{\text{h}} \times \frac{D61\,\text{m}^3}{1\,\text{air change}} \times \frac{1\,\text{h}}{3600\,\text{s}}$$

$$= \frac{C49 \times D61}{3600}\,\frac{\text{m}^3}{\text{s}}$$

EXAMPLE 5.8

An office is 4 m long × 4 m wide and 3 m high. It is expected to have a natural infiltration rate of outdoor air of one air change per hour. Calculate the infiltration rate in m^3/s.

$$Q = \frac{1 \times 4 \times 4 \times 3}{3600}\,\frac{\text{m}^3}{3}$$
$$= 0.0133\,\text{m}^3/\text{s}$$

The smallest significant air flow rate is 1 litre/s for comfort heating and air conditioning systems. The third decimal place is 1 litre, and this can be rounded up, rather than down, from a knowledge of the fourth place to $0.014\,\text{m}^3/\text{s}$.

Cell B106

(F4) 6*D102/(1012*1.1906*C105+6*D102)

The formula commences with a number, so it does not need the plus sign. The dimensionless ventilation factor F_v is

$$F_v = \frac{6\Sigma(A)}{1012 \times 1.1906Q + 6\Sigma(A)}$$

$$B106 = \frac{6 \times D102}{1012 \times 1.1906 \times C105 + 6 \times D102}$$

EXAMPLE 5.9

An office has a total room surface area of $80\,m^2$ and an outdoor air infiltration rate of $0.0133\,m^3/s$. Calculate the ventilation factor F_v.

$$F_v = \frac{6 \times 80}{1012 \times 1.1906 \times 0.0133 + 6 \times 80}$$

$$= 0.967\,69$$

The cell for F_v has been formatted to display four decimal places, and it will show 0.9676. For comparison with the tabulated values it will be rounded to 0.97.

Cell B107

(F4) 18*D102/(C104+18*D102)

The dimensionless thermal transmittance factor F_u is

$$F_u = \frac{18\Sigma(A)}{\Sigma(AU) + 18\Sigma(A)}$$

$$B107 = \frac{18 \times D102}{C104 + 18 \times D102}$$

EXAMPLE 5.10

An office has a total room surface area of $80\,m^2$ and a $\Sigma(AU)$ of 56.95 W/K. Calculate the thermal transmittance factor F_u.

$$F_u = \frac{18 \times 80}{56.95 + 18 \times 80}$$

$$= 0.961\,96 \text{ to be rounded to } 0.96$$

Cell B108

(F4) ((D64*(B106*1012*1.1906*C105+B107*C104)/100)/(C104+(18*D102*(B106*(D65/100)+B107*(D64/100)))))+B107

The environmental temperature dimensionless ratio is found from

$$F_1 = \frac{E \times [F_v \times 1012 \times 1.1906Q + F_u\Sigma(AU)]}{\Sigma(AU) + 18\Sigma(A) \times (KF_v + EF_u)} + F_u$$

This corresponds to

$$B108 = \frac{D64 \times (B106 \times 1012 \times 1.1906 \times C105 + B107 \times C104)}{100 \times \left[C104 + 18 \times D102 \times \left(\dfrac{D65}{100} \times B106 + \dfrac{D64}{100} \times B107\right)\right]} + B107$$

EXAMPLE 5.11

Find the environmental temperature factor for an office from the following data: $E = 30\%$, $K = 70\%$, $F_v = 0.9676$, $F_u = 0.962$, $Q = 0.0133\,\text{m}^3/\text{s}$, $\Sigma(AU) = 56.95\,\text{W/K}$, $\Sigma(A) = 80\,\text{m}^2$.

$$F_1 = \frac{30 \times (0.9676 \times 1012 \times 1.1906 \times 0.0133 + 0.9676 \times 56.95)}{100 \times \left[56.95 + 18 \times 80 \times \left(\dfrac{70}{100} \times 0.9676 + \dfrac{30}{100} \times 0.962\right)\right]} + 0.962$$

$$= \frac{2118.318}{144\,787.48} + 0.962$$

$$= 0.014\,63 + 0.962$$

$$= 0.9766$$

Cell B109

(F4) ((D65*(B106*1012*1.1906*C105+B107*C104)/100)/(1012*1.1906*C105+(6*D102*(B106*(D65/100)+B107*(D64/100)))))+B106

The room air temperature dimensionless ratio is found from

$$F_2 = \frac{K \times [F_v \times 1012 \times 1.1906Q + F_u\Sigma(AU)]}{1012 \times 1.1906Q + 6\Sigma(A) \times (KF_v + EF_u)} + F_v$$

This corresponds to

$$B109 = \frac{D65 \times (B106 \times 1012 \times 1.1906 \times C105 + B107 \times C104)}{100 \times \left[1012 \times 1.1906 \times C105 + 6 \times D102 \times \left(\dfrac{D65}{100} \times B106 + \dfrac{D64}{100} \times B107\right)\right]}$$
$$+ B106$$

EXAMPLE 5.12

Find the room air temperature factor for an office from the following data: $E = 30\%$, $K = 70\%$, $F_v = 0.9676$, $F_u = 0.962$, $Q = 0.0133\,\text{m}^3/\text{s}$, $\Sigma(AU) = 56.95\,\text{W/K}$, $\Sigma(A) = 80\,\text{m}^2$.

$$F_2 = \frac{70 \times (0.9676 \times 1012 \times 1.1906 \times 0.0133 + 0.962 \times 56.96)}{100 \times \left[1012 \times 1.1906 \times 0.0133 + 6 \times 80 \times \left(\dfrac{70}{100} \times 0.9676 + \dfrac{30}{100} \times 0.962\right)\right]}$$
$$+ 0.9676$$

$$= \frac{4920.418}{47\,966.66} + 0.9676$$

$$= 0.10258 + 0.9676$$

$$= 1.0702$$

Cell B110

(F2) +C104/D102

The thermal transmittance ratio is

$$\frac{\Sigma(AU)}{\Sigma(A)}$$

This is the mean thermal transmittance of the structure of the room, and it demonstrates the overall level of thermal insulation.

$$\text{mean } U = \frac{\Sigma(AU)}{\Sigma(A)} \ \frac{\text{W}}{\text{m}^2\,\text{K}}$$

$$\text{B110} = \frac{\text{C104}}{\text{D102}} \ \frac{\text{W}}{\text{m}^2\,\text{K}}$$

Cell B111

(F2) +C49*D61/3/D102

The ventilation ratio is

$$\frac{NV}{3\Sigma(A)} \ \frac{\text{m}}{\text{h}}$$

This corresponds to

$$\text{B111} = \frac{\text{C49} \times \text{D61}}{3 \times \text{D102}} \ \frac{\text{m}}{\text{h}}$$

EXAMPLE 5.13

Find the environmental and air temperature factors F_1 and F_2 from the *CIBSE Guide A* (CIBSE, 1986a) Table A9.4 for an office by using the following data: $\Sigma(AU) = 56.95\,\text{W/K}$, $\Sigma(A) = 80\,\text{m}^2$, $N = 1$ air change/h, room volume $V = 48\,\text{m}^3$.

The mean thermal transmittance for the office is

$$\text{mean } U = \frac{\Sigma(AU)}{\Sigma(A)} \ \frac{\text{W}}{\text{m}^2\,\text{K}}$$

$$= \frac{56.95}{80} \ \frac{\text{W}}{\text{m}^2\,\text{K}}$$

$$= 0.712\,\text{W/m}^2\,\text{K}$$

and the ventilation ratio is

$$\frac{NV}{3\Sigma(A)} = \frac{1 \times 48}{3 \times 80} \ \frac{\text{m}}{\text{h}} = 0.2\,\text{m/h}$$

Table A9.4 shows that F_1 is 0.985 and F_2 is 1.05 by interpolation.

Cell C112

(F1)+C47

The design resultant temperature is copied from cell C47 into cell C112 to bring all the relevant data together.

Cell C113

(F1) +C48+B108*(C47−C48)

The environmental temperature that is produced in the room is found from

$$t_{ei} = t_{ao} + F_1 \times (t_c - t_{ao}) \,°C$$

This corresponds to

$$C113 = C48 + B108 \times (C47 - C48) \,°C$$

Cell C114

(F1) +C48+B109*(C47−C48)

The air temperature that is produced in the room is found from

$$t_{ai} = t_{ao} + F_2 \times (t_c - t_{ao}) \,°C$$

This corresponds to

$$C114 = C48 + B109 \times (C47 - C48) \,°C$$

Cell C115

(F1) +J102

The mean surface temperature, $t_s \,°C$, is copied from cell J102 into cell C115 to bring all the relevant data together.

Cell C116

(F1) 2*C112−C114

The mean radiant temperature is found from

$$t_r = 2t_c - t_{ai} \,°C$$

This corresponds to

$$C116 = 2 \times C112 - C114 \,°C$$

Cell C117

(F1) +C48

The outdoor air temperature, $t_{ao} \,°C$, is copied from cell C48 into cell C117 to bring all the relevant data together.

EXAMPLE 5.14

Calculate the environmental, indoor air, and mean radiant temperatures for an office from the following data: $t_{ao} = 0 \,°C$, $t_c = 20 \,°C$, $F_1 = 0.9765$, $F_2 = 1.0702$.

$$t_{ei} = t_{ao} + F_1 \times (t_c - t_{ao}) \,°C$$
$$= 0 + 0.9756 \times (20 - 0) \,°C$$
$$= 19.5 \,°C$$

$$t_{ai} = t_{ao} + F_2 \times (t_c - t_{ao}) \,^\circ C$$
$$= 0 + 1.0702 \times (20 - 0) \,^\circ C$$
$$= 21.4 \,^\circ C$$
$$t_r = 2t_c - t_{ai} \,^\circ C$$
$$= 2 \times 20 - 21.4 \,^\circ C$$
$$= 18.6 \,^\circ C$$

Cell B120

(F0) (B108*C104+B109*C49*D61/3)*(C47−C48)

The plant energy output that is needed to offset the steady-state heat loss from the room is found from

$$Q_p = [F_1 \times \Sigma(AU) + 0.33F_2NV] \times (t_c - t_{ao}) \, W$$

This corresponds to

$$B120 = (B108 \times C104 + B109 \times C49 \times D61 \times 0.33) \times (C47 - C48) \, W$$

Cell B121

(F3) +B120/1000

The plant energy output is converted to k:

$$Q_p \, kW = Q_p \, W \times \frac{1 \, kW}{10^3 \, W}$$
$$= \frac{Q_p \, W}{1000} \, kW$$

This corresponds to

$$B121 = \frac{B120}{1000} \, kW$$

Cell B122

(F1) +B120/D61

The plant energy output per m^3 of room volume is

$$Q_p = \frac{Q_p \, W}{V \, m^3}$$

This corresponds to

$$B122 = \frac{B120}{D61} \frac{W}{m^3}$$

Cell B123

(F1) +B120/D62

The plant energy output per m^2 of room floor area is

$$Q_p = \frac{Q_p \, W}{A_f \, m^2}$$

This corresponds to

$$B123 = \frac{B120}{D62} \frac{W}{m^2}$$

EXAMPLE 5.15

Calculate the plant energy output that is needed to offset the steady state heat loss from an office from the following data: $t_{ao} = 0\,°C$, $t_c = 20\,°C$, $F_1 = 0.9765$, $F_2 = 1.0702$, $\Sigma(AU) = 56.95\,W/K$, $A_f = 16\,m^2$, $V = 48\,m^3$, $N = 1$ air change/h.

$$Q_p = [F_1\Sigma(AU) + 0.33F_2NV] \times (t_c - t_{ao})\,W$$
$$= (0.9765 \times 56.95 + 0.33 \times 1.0702 \times 1 \times 48) \times (20 - 0)\,W$$
$$= 1455\,W$$

$$Q_p\,kW = \frac{Q_p}{1000}\frac{W}{}\,kW$$
$$= \frac{1455}{1000}\,kW$$
$$= 1.455\,kW$$

$$Q_p = \frac{Q_p}{V}\frac{W}{m^3}$$
$$= \frac{1455\,W}{48\,m^3}$$
$$= 30.3\,W/m^3$$

$$Q_p = \frac{Q_p}{A_f}\frac{W}{m^2}$$
$$= \frac{1455\,W}{16\,m^2}$$
$$= 90.9\,W/m^2$$

Cell B890

(F0) +D887

This copies the number of fictitious rooms from cell D887, and a prompt asks whether the correct number has been entered. If it is not correct, the user must go back to cell D887.

Cell C894

(F3) +B121+B205+B289+B373+B457+B541+B625+B709+B793+B877

The total plant heat output power is totalled for the current worksheet. There are 10 possible rooms to be totalled:

$$Q_p = Q_{p1} + Q_{p2} + \ldots + Q_{p10}\,kW$$
$$C894 = B121 + B205 + \ldots + B877\,kW$$

Cell C905

(F3) @SUM(C894..C903)

The total heat loss for the job is found from the totals of the ten worksheets:

$$Q_p = Q_{p1} + Q_{p2} + \ldots + Q_{p10} \text{ kW}$$
$$C905 = C894 + C895 + \ldots + C903 \text{ kW}$$

Cell C906

(F3) 10*C905/100

Ten percent is added to the total heat loss for the job. This is to allow for the heat loss from the heating system distribution network of hot-water pipes or warm-air ducts. This distribution loss does not result in useful heat being supplied to the occupied rooms. The actual percentage is unknown to the designer until the distribution network is correctly sized and the heat emission calculated. Ten percent is an initial assessment that is unlikely to be exceeded. The user may change the 10% by editing the cell. Press the F2 key, move the cursor leftwards, delete the number 10 and replace it with the desired percentage. Press the enter key to activate the edited formula.

$$\text{distribution loss} = 10 \times \frac{\text{heat loss}}{100} \text{ kW}$$
$$C906 = 10 \times \frac{C905}{100} \text{ kW}$$

Cell C910

(F3) @SUM(C905..C908)

The total plant heating power is the sum of the 10 worksheet totals, the distribution system emission, the hot-water service load, and the surplus capacity:

$$Q_p = Q_{p1} + Q_{p2} + \ldots + Q_{p4} \text{ kW}$$
$$C910 = C905 + C906 + C907 + C908 \text{ kW}$$

Cell D913

(F3) +D62+D146+D230+D314+D398+D482+D566+D650+D734+D818−D887

The floor area of the 10 rooms is summed for the worksheet. When fictitious rooms have been declared, each must have a floor area of $1\,\text{m}^2$. The number of such rooms is subtracted from the total of the area for the 10 rooms to produce the correct floor area for the building:

$$A_f = A_{f1} + A_{f2} + \ldots + A_{f10} - A_{f11} \text{ m}^2$$
$$D913 = D62 + D146 + \ldots + D818 - D887 \text{ m}^2$$

Cell D914

(F3) +D61+D145+D229+D313+D397+D481+D565+D649+D733+D817−D887

The volume of the 10 rooms is summed for the worksheet. When fictitious rooms have been declared, each must have a volume of $1\,m^3$. The number of such rooms is subtracted from the total of the volumes of the 10 rooms to produce the correct volume of the building:

$$V = V_1 + V_2 + \ldots + V_{10} - V_{11}\,m^3$$

$$D914 = D61 + D145 + \ldots + D817 - D887\,m^3$$

Cell D915

(F1) +C905*1000/D913

The heat loss per m^2 of floor area for the 10 rooms on the current worksheet is

$$\frac{Q_p}{A_f} = \frac{\text{heat loss}}{\text{floor area}}\,\frac{W}{m^2}$$

$$D915 = \frac{C905}{D913} \times 1000\,W/m^2$$

Cell D916

(F1) +C905*1000/D914

The heat loss per m^3 of volume for the 10 rooms on the current worksheet is

$$\frac{Q_p}{V} = \frac{\text{heat loss}}{\text{building volume}}\,\frac{W}{m^3}$$

$$D916 = \frac{C905}{D914} \times 1000\,W/m^3$$

THE UNHEATED ROOM

The file A:UNHEAT.WK1 is designed to find the expected air temperature that is in an unheated room or space. Garages, store rooms, service voids and pitched roof spaces are typical examples. The designer needs to know what the space temperature is so that the heat flow into it from warmer rooms can be calculated. The format of the worksheet is similar to that for heat loss calculation. The unheated room is specified in terms of surfaces that have heat flow into it from warmer rooms, and surfaces that experience a heat loss. Outdoor air will normally infiltrate the unheated room and cause a loss of heat from the space. It can be possible for warm air to infiltrate the unheated room and produce a heat gain. This possibility would be unreliable to calculate, and would require some guess-work unless a controlled air movement air conditioning system was in operation. Only the gains from thermal transmittance heat flows are calculated.

The input data is arranged into two areas of the worksheet. These are heat gain surfaces and heat loss surfaces. The user enters the dimensions and thermal transmittance of each surface of the unheated room. When the surface is generating a gain of heat, the user specifies the resultant temperature of the warmer room. It is considered appropriate to calculate only the air temperature of the unheated room to the nearest whole degree Celsius below the point of balance between the heat gain and loss. This is of sufficient accuracy, and it is preferable to underestimate the temperature slightly. There is no attempt to provide for the comfort of any occupants of the unheated room, so the environmental and air temperature dimensionless factors F_1 and F_2 are ignored for these calculations, because such accuracy is not needed. The user needs only use columns A to E to enter data and read the output. Column G is used to collect the products of surface area and thermal transmittance for each surface. There is no need to look at this column. When the correct values have been input, the user views the tabulated heat gains and loss output data. These calculations have been prepared for the unheated room air temperature range from $-5\,°C$ to $20\,°C$. There should be some negative values in the gain and loss columns to indicate that the reverse of the expected heat flows would take place.

Column E was formed by subtracting the loss from the gain at each value of the unheated room air temperature. When the difference between gain and loss is zero, the room air temperature is that which balances the gains and losses. Only the nearest whole degree Celsius below the balance point is required. The user selects the lowest positive value of gain minus loss and then reads the corresponding temperature of the unheated room from column A of the same row.

DATA REQUIREMENT

The user will enter new data for:

37. Your name in cell B4.
38. Job name in cell B5.
39. Job reference in cell B6.
40. Current date in cell B7.
41. File name in cell B9.
42. The name of the unheated room or space in cell C27.
43. Floor level of the room in cell C28.
44. Design outdoor air temperature in cell C29.
45. The expected rate of infiltration of outdoor air into the unheated room in cell C30.
46. Length of area 1 in cell D32.
47. Width of area 1 in cell D33.
48. Height of area 1 in cell D34.
49. Dimensions of area 2 in the cell range D35 to D37.
50. Heat gain surfaces: length and height of the wall and its openings, for wall 1, in the cell range B45 to C48.
51. Resultant temperature of the room on the warm side of wall number one in cell E46. The value in cell E46 is automatically copied into the cell range E47 to E48.
52. Thermal transmittances of the wall and its windows and doors for wall 1 in the cell range F46 to F49.

53. Dimensions, temperatures and thermal transmittances for walls 2, 3 and 4 in the cell range B51 to F70.
54. Dimensions, temperatures and thermal transmittances for the floors and ceilings in the cell range B73 to F76.
55. Heat loss surfaces: the length, height and thermal transmittance data for each heat loss surface is entered in the range of cells from B85 to E116 in a similar manner to that for gain surfaces.

OUTPUT DATA

46. The total floor area of the unheated room is shown in cell D38.
47. The total volume of the unheated room is shown in cell D39.
48. Cell range D45 to D116 shows the area of each surface that has been specified. Check that the figures seem sensible, because there is no programmed error trap. Check that zero area is displayed where this is expected.
49. The product of area and thermal transmittance for each surface is shown in the cell range G46 to G118. It is not normally necessary for the user to look at this range unless it is to double check that there is no erroneous data.
50. The heat gain to the unheated room for each value of possible room air temperature is listed in the cell range C130 to C155.
51. The heat loss from the unheated room for each value of possible room air temperature is listed in the cell range D130 to D155.
52. The difference between the heat gain to and the heat loss from the unheated room for each value of possible room air temperature is listed in the cell range E130 to E155.
53. The user visually selects the lowest positive value of gain minus loss by moving the cursor onto that cell: move the cursor leftwards to column B and highlight the temperature of the unheated room.
54. Type the unheated room temperature into cell C158.

FORMULAE

The worksheet is formatted to display two decimal places, with the exception of the output data section below line number 130, where only whole numbers are desired. One example of each type of calculation is explained.

Cell D38

+D32*D33+D35*D36

The total floor area of the unheated room is calculated. The room may have two areas of floor.

$$\text{total area} = \text{area}_1 + \text{area}_2 \, \text{m}^2$$
$$= L_1 W_1 + L_2 W_2 \, \text{m}^2$$
$$D38 = D32 \times D33 + D35 \times D36 \, \text{m}^2$$

Cell D39

+D32*D33*D34+D35*D36*D37

The total volume of the unheated room is calculated. The room may have two volumes.

$$\text{total volume} = \text{volume}_1 + \text{volume}_2 \, \text{m}^3$$
$$= L_1 W_1 H_1 + L_2 W_2 H_2 \, \text{m}^3$$
$$\text{D39} = \text{D32} \times \text{D33} \times \text{D34} + \text{D35} \times \text{D36} \times \text{D37} \, \text{m}^3$$

Cell D45

+B45*C45

The gross surface area of wall 1 is

$$\text{area} = \text{length} \times \text{height} \, \text{m}^2$$
$$\text{D45} = \text{B45} \times \text{C46} \, \text{m}^2$$

Cell E47

+E46

The user types the resultant temperature of the warm room once only for wall 1. This value is common for all the windows and the door in that wall, so it is copied automatically into cells E47 to E49 to save typing. The same procedure is used for the other walls.

Cell D49

+D45−D46−D47−D48

The net surface area of wall 1 is

$$\text{net wall area} = \text{gross area} - \text{area of openings} \, \text{m}^2$$
$$= \text{gross wall} - \text{window}_2 - \text{window}_2 - \text{door} \, \text{m}^2$$
$$\text{D49} = \text{D45} - \text{D46} - \text{D47} - \text{D48} \, \text{m}^2$$

Cell G46

+D46*F46

The product of surface area and thermal transmittance is found for window 1 of wall 1:

$$AU = A \, \text{m}^2 \times U \, \text{W/m}^2 \, \text{K}$$
$$= AU \, \text{W/K}$$
$$\text{G46} = \text{D46} \times \text{F46} \, \text{W/K}$$

Cell G50

@SUM(G46..G49)

The summation of areas and thermal transmittances is found for wall 1. There are four possible surfaces.

$$\Sigma(AU) = A_1 U_1 + \ldots + A_4 U_4 \, \text{W/K}$$
$$G50 = G46 + G47 + G48 + G49 \, \text{W/K}$$

Cell D73

+B73*C73

The gross area of floor area 1 is calculated. Floors and ceilings do not normally have openings in them.

$$\text{area} = \text{length} \times \text{width} \, \text{m}^2$$
$$D73 = B73 \times C73 \, \text{m}^2$$

Cell G118

@SUM (G86..G116)

All the heat losses from the unheated room take place from the unknown room air temperature to the outdoor air temperature. Only one summation of area and thermal transmittance is needed for all the loss surfaces. There are 22 possible surfaces.

$$\Sigma(AU) = (AU)_1 + \ldots + (AU)_{22} \, \text{W/K}$$
$$G118 = G86 + \ldots + G116 \, \text{W/K}$$

Cell C130

(F0) +$G$50*($E$46−B130)+$G$56*($E$52−B130)+$G$62*($E$58−B130)+
$G$71*($E$67−B130)+$G$73*($E$73−B130)+$G$74*($E$74−B130)+$G$75*
($E$75−B130)+$G$76*($E$76−B130)

The heat gained by the unheated room is calcualted for the first possible tempera-ture of $-5\,°C$. The limits of possible room air temperature have been chosen arbitrarily, but they should be wide enough for all practical cases. The summation of the surface area and thermal transmittance for each of the eight surfaces is multiplied by the difference between the resultant temperature on the warm side of the surface and the possible temperature of the unheated room.

$$\text{gain} = [\Sigma(AU) \times (t_{c1} - t_a)]_1 + \ldots + [\Sigma(AU) \times (t_{c1} - t_a)]_8 \, \text{W}$$
$$C130 = G50 \times (E46 - B130) + \ldots + G76 \times (E76 - B130) \, \text{W}$$

Cell D130

(F0) ($G$118+0.33*$C$30*$D$39)*(B130−$C$29)

The heat loss from the unheated room is found by adding the total of the product of the loss surface areas and their thermal transmittances to the ventilation heat loss component. This is then multiplied by the difference between the possible room air temperature and the outdoor air temperature. A negative loss of heat shows that the outdoor air is warmer than the current value of the room air temperature.

$$\text{heat loss} = [\Sigma(AU) + 0.33NV] \times (t_a - t_{ao})\,\text{W}$$
$$D130 = (G118 + 0.33 \times C30 \times D39) \times (B130 - C29)\,\text{W}$$

Cell E130

(F0) +C130−D130

The difference between the heat gain to and the heat loss from the unheated room is found. The user can see which room air temperature satisfies both heat flows; this is also known as the balance temperature.

$$\text{balance heat flow} = \text{gain} - \text{loss}\,\text{W}$$
$$E130 = C130 - D130\,\text{W}$$

EXAMPLE 5.16

Calculate the temperature that is likely to be established in the unheated store of the factory building shown in Fig. 5.4.

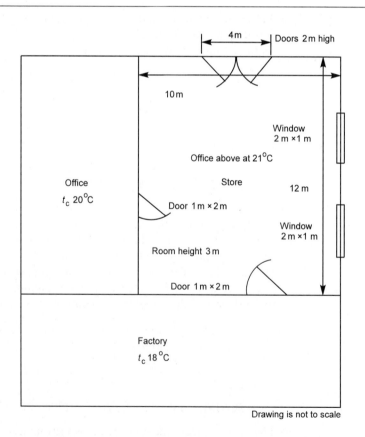

Fig. 5.4   Unheated store room in Example 5.16.

Use the dimensions and data on the figure and the following information: outdoor air temperature for design −2 °C; the store is expected to have 0.75 air changes per hour due to the infiltration of outdoor air; the office above the store is maintained at a resultant temperature of 21°C; the internal doors have a thermal transmittance of $5.3 \, \text{W/m}^2 \, \text{K}$; the store double door and the windows have a thermal transmittance of $5.3 \, \text{W/m}^2 \, \text{K}$; the internal wall thermal transmittance is $1.7 \, \text{W/m}^2 \, \text{K}$; the store ceiling has a thermal transmittance of $1.3 \, \text{W/m}^2 \, \text{K}$; the thermal transmittance of the external wall is $0.7 \, \text{W/m}^2 \, \text{K}$; the thermal transmittance of the floor is $0.8 \, \text{W/m}^2 \, \text{K}$.

The solution is shown on the data disk as file A:UNHEAT.WK1.

Using the worksheet reveals that the store is likely to be at an air temperature of 8 °C. Calculate the heat gains and losses for an assumed temperature of 8 °C for the store. The data is shown in Table 5.9. An example of a heat gain and a heat loss calculation is detailed.

$$\text{factory door heat gain} = 2 \, \text{m}^2 \times 3 \, \text{W/m}^2 \, \text{K} \times (18 - 8)°\text{C}$$
$$= 60 \, \text{W}$$

$$\text{store external wall net area} = [(22 \times 3) - 8 - 4] \, \text{m}^2$$
$$= 54 \, \text{m}^2$$

$$\text{store wall heat loss} = 54 \, \text{m}^2 \times 0.7 \, \text{W/m}^2 \, \text{K} \times (8 - -2) \, \text{K}$$
$$= 378 \, \text{W}$$

Table 5.9   Data for the unheated store in Example 5.16.

Surface	L (m)	H (m)	Area m^2	U (W/m^2 K)	t_{cl} (°C)	Q_p (W)
Heat gains						
Wall	12	3	36			
Door	1	2	2	3	20	72
net			34	1.7	20	693.6
Wall	10	3	30			
Door	1	2	2	3	18	60
net			28	1.7	18	476
Ceiling	12	10	120	1.3	21	2028
				Total heat gain = 3330		
Heat losses						
Wall	22	3	66			
Door	4	2	8	5.3	8	424
Windows	4	1	4	5.3	8	212
net			54	0.7	8	378
Floor	12	10	120	0.8	8	960
				Total conduction heat loss = 1974		

$$\text{ventilation} \; Q_p = 0.33 N V \times (t_a - t_{ao}) \, \text{W}$$
$$= 0.33 \times 0.75 \times 12 \times 10 \times 3 \times (8 - -2) \, \text{W}$$
$$= 891 \, \text{W}$$

$$\text{total heat loss} \; Q_p = 1974 + 891 \, \text{W}$$
$$= 2865 \, \text{W}$$

The difference between the heat gain and heat loss is

$$\text{heat gain} - \text{heat loss} = 3330 - 2865\,\text{W}$$
$$= 465\,\text{W}$$

It can be seen from the worksheet that this is the lowest positive difference. The store is likely to stabilize at around 8 °C provided that there are no other sources of heat, such as working people, solar radiation, lighting or electrically powered equipment. It is also assumed that the doors and windows remain shut.

Questions

Questions 1–16 require the use of a written response and some manual calculations, not the use of the worksheet. When the user understands each part of the calculation procedure and has manually evaluated all the data at least once, further cases should be entered onto the worksheet.

1. Explain the basis for the calculation of the heating plant power needed to provide for the thermal comfort of the occupants of a building. Include in your explanation the factors that are taken into consideration, the physical conditions of the building, the type of heating system, the climate design data used and the method of calculation.
2. Discuss the difference between steady-state and transient heat loss calculation data and methods. Give examples of building types that may require these different approaches. Explain how the heating system is selected in relation to the steady-state and transient heat flows from a building.
3. Explain why resultant temperature is used to specify the thermal comfort condition for an occupied room. Include in your explanation the heat transfer methods that are involved and how they are included in the calculation of heat loss.
4. Calculate the volumetric specific heat capacity for air when it is at a density of $1.185\,\text{kg/m}^3$ and its specific heat capacity is $1008\,\text{J/kg K}$.
5. Explain the reason for, and the methodology of, the use of a modified thermal transmittance for a wall, floor or ceiling that separates two areas of the building that are at different temperatures from that of the outdoor air. Compare this approach with an alternative method of finding the heat flow through such a surface. State which method is preferable and why this is the case.
6. Calculate the thermal transmittance U_1, the modified thermal transmittance U_2 and the heat flow Q_p through an internal wall of $13\,\text{m}^2$ that separates a lounge from an unheated garage in a house. The lounge is maintained at a resultant temperature of 21 °C. The garage air temperature is expected to stabilize at 6 °C when the outdoor air temperature is -1 °C. The wall is constructed from 125 mm thick medium-density concrete blockwork with 15 mm plaster on the lounge side only. The thermal conductivity of the blockwork is $0.9\,\text{W/m K}$ and that of the plaster is $0.5\,\text{W/m K}$. The surface film thermal resistance is $0.12\,\text{m}^2\,\text{K/W}$. The air temperature factor, F_1, is 0.94.
7. Calculate the thermal transmittance, the modified thermal transmittance and the heat flow through a wall between two heated rooms. The wall is 10 m long and 3.5 m high, and is constructed from 150 mm thick lightweight concrete blockwork. The wall is plastered on both sides with 12 mm thickness of lightweight plaster. The rooms are maintained at resultant temperatures of 20 °C and 14 °C when the outdoor air is -6 °C. The thermal conductivity of the blockwork is $0.19\,\text{W/m K}$, and that of the plaster is $0.48\,\text{W/m K}$. The surface film thermal resistance is $0.12\,\text{m}^2\,\text{K/W}$, and the air temperature factor F_1 is 0.89.
8. Calculate the thermal transmittance and the modified thermal transmittance for an intermediate floor of a bedroom that has a lounge directly below it. The floor consists of a 20 mm thickness of carpet and underlay, 25 mm floorboards, 200 mm air space and 25 mm of plasterboard and plaster. Use the data provided to find the heat loss from the lounge and the heat gained by the bedroom. Conduct these two heat flow calculations separately. Data: resultant temperatures, lounge 21 °C, bedroom 16 °C; R_{si} ceiling, heat flow upwards $0.1\,\text{m}^2\,\text{K/W}$; R_{si} floor, heat flow upwards $0.14\,\text{m}^2\,\text{K/W}$; R_a floor air space $0.18\,\text{m}^2\,\text{K/W}$; outdoor air temperature 3°C; thermal conductivity of carpet $0.14\,\text{W/m K}$; thermal conductivity of timber $0.5\,\text{W/m K}$; thermal conductivity of plasterboard $0.5\,\text{W/m K}$; lounge ceiling is $7\,\text{m} \times 4\,\text{m}$; air temperature factor F_1 is 0.97.
9. An unheated store room is expected to be at a resultant temperature of 8 °C when the outdoor air temperature is at 1 °C. One wall of the store is adjacent to an office that is maintained at a resultant temperature of 20 °C. The wall is 8 m long, 2.8 m high and is constructed of 100 mm thick medium-density concrete blockwork with 12 mm dense plaster on both sides. The air temperature factor F_1 is 0.9. The surface film thermal resistance

is $0.12\,m^2\,K/W$. The thermal conductivity of the blockwork is $0.7\,W/m\,K$, and that of the plaster is $0.8\,W/m\,K$. Calculate the heat loss that is expected to flow from the office into the store room.

10. Calculate the surface temperature of an external wall that has a thermal transmittance of $0.72\,W/m^2K$ when the indoor resultant temperature is $19\,°C$ and the outdoor air is at $-3\,°C$. The indoor surface film thermal resistance is $0.12\,m^2\,K/W$ and the air temperature factor F_1 is 0.93.

11. Calculate the minimum value that must be created for the thermal transmittance of an external perimeter construction if the indoor surface temperature is not to fall below $5\,°C$ when the outdoor air temperature remains constant at $-4\,°C$. The surface film thermal resistance of the walling is $0.12\,m^2\,K/W$, the air temperature factor F_1 is 0.95, and the room will be maintained at a resultant temperature of $16\,°C$. State the type of external perimeter construction that would satisfy this requirement.

12. Find the thermal transmittance necessary to avoid surface condensation on a wall that separates a room that is maintained at a resultant temperature of $18\,°C$ from an unheated room which is expected to fall to $4\,°C$ during winter. The room dew point is $11\,°C$ and the air temperature factor F_1 is 0.88. The internal surface film thermal resistance of the wall is $0.12\,m^2\,K/W$.

13. Calculate the area-weighted mean surface temperature of a factory that has an overhead radiant strip heating system. A survey revealed the

following areas and surface temperatures: floor $200\,m^2$, $11\,°C$; roof $230\,m^2$, $13\,°C$; walls $240\,m^2$, $12\,°C$; radiant strip $20\,m^2$, $110\,°C$.

14. A room has an F_1 of 0.96 and an F_2 of 1.03 when the resultant temperature is designed to be $20\,°C$ at an outdoor air temperature of $-2\,°C$. Calculate the environmental, air and mean radiant temperatures that will be produced in the room if the heating plant is correctly sized. Deduce what type of heating system is being used.

15. A factory has a 100% convective heating system, where F_1 is 0.92 and F_2 is 1.23. The resultant temperature is designed to be $16\,°C$ at an outdoor air temperature of $5\,°C$. Calculate the environmental, air and mean radiant temperatures that will be produced in the room if the heating plant is correctly sized.

16. An industrial building has a 90% radiant heating system, where F_1 is 1.05 and F_2 is 0.86. The resultant temperature is designed to be $14\,°C$ at an outdoor air temperature of $1\,°C$. Calculate the environmental, air and mean radiant temperatures that will be produced in the room if the heating plant is correctly sized.

17. Calculate the heating plant output that is required for a hotel dining room from the data provided: room dimensions $20\,m \times 10\,m$ and $3\,m$ high; design resultant temperature $21\,°C$; outdoor air temperature $-3\,°C$; infiltration of outdoor air 1.5 changes per hour; ducted air heating system; one long and one short wall adjoin another room that is at a resultant temperature of $18\,°C$; three exter-

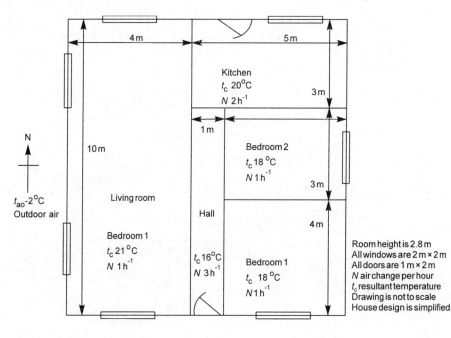

Fig. 5.5 Plan of house in question 18.

nal windows, each 2 m × 2 m; the dining room is a single-storey construction; the thermal transmittances are windows 5.7 W/m²K, external walls 0.55 W/m²K, internal walls 1.3 W/m²K, floor 0.6 W/m²K, flat roof 0.32 W/m²K.

18. Calculate the plant output power needed for the single-storey house shown in Fig. 5.5. A steel panel hot-water radiator system is to be designed for the house. The hot-water service load is 3 kW. Each window is 2 m × 2 m. The house design is simplified. The room height is 2.8 m and the outdoor air temperature for the design is -2 °C. The thermal transmittances are: external walls 0.55 W/m²K, internal walls 1.7 W/m²K, glazing 2.9 W/m²K, external doors 5.7 W/m²K, floor 0.6 W/m²K, roof 0.2 W/m²K. Treat the internal doors as having the same thermal transmittance as the internal walls.

19. Calculate the total heating plant power that will be needed for the three-storey Commodore

House shown in Fig. 5.6. Use the data and dimensions shown on the figure and the following data: low-pressure hot-water convector radiator heating system, 90% convective; 12 kW plant power is needed for the provision of hot-water services; no surplus plant power is required; each storey has a floor-to-ceiling height of 3 m; the floors are ground, first and second; there are a total of nine rooms; outdoor air temperature for design is −1 °C; all windows are 2 m × 2 m; two of the ground floor windows are also doors; internal doors have the same thermal transmittance as the internal walls; thermal transmittances are ground floor 0.6 W/m²K, roof 0.35 W/m²K, windows 2.9 W/m²K, external walls 0.4 W/m²K, internal walls 1.7 W/m²K, first and second floor 1.3 W/m²K.

20. Calculate the temperature that is likely to be established in the unheated garage of the house shown in Fig. 5.7. Use the dimensions and data on the figure and the following information: outdoor

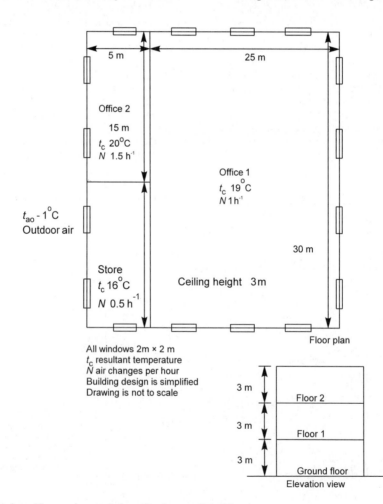

Fig. 5.6 Commodore House plan and elevation in question 19.

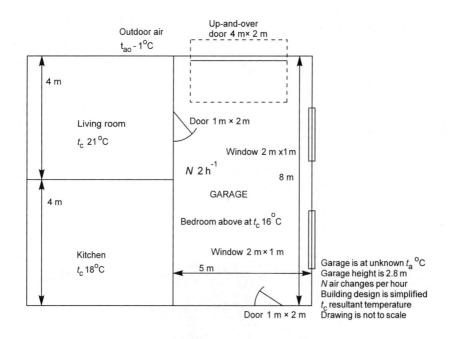

Fig. 5.7 Unheated garage in question 20.

air temperature for design -1 °C; the garage is expected to have 2 air changes per hour due to the infiltration of outdoor air; the bedroom above the garage is maintained at a resultant temperature of 16 °C; the small external door and the internal door have a thermal transmittance of 3 W/m² K; the garage double door and the windows have a thermal transmittance of 5.3 W/m² K; the internal wall thermal transmittance is 1.7 W/m² K; the garage ceiling has a thermal transmittance of 1.3 W/m² K; the thermal transmittance of the gar-

age external wall is 0.6 W/m² K; the thermal transmittance of the garage floor is 0.6 W/m² K.

21. Water storage tanks and water pipes are located in the roof space of the house shown in Fig. 5.8. Use the dimensions and data on the figure and the following information: average house resultant temperature 18 °C; outdoor air for design 0 °C; roof space infiltration 3 air changes per hour of outdoor air; there is 150 mm thick fibreglass wool on the top of the flat plasterboard ceiling; thermal transmittances are flat ceiling between rooms

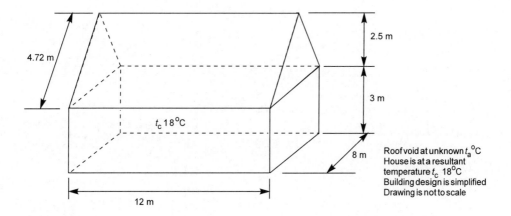

Fig. 5.8 Unheated roof void in question 21.

and roof space 0.25 W/m² K, sloping tiled roof 4.6 W/m² K, vertical walls at ends of roof space 1.2 W/m² K.

(a) Calculate the air temperature that is expected to be found within the roof space.
(b) Experiment with different combinations of outdoor air temperature and infiltration rate, to discover the relationship between the roof space and outdoor air temperatures.

(c) Relocation of the roof space thermal insulation from the flat ceiling to the underside of the tiled roof causes the flat ceiling thermal transmittance to become 3.6 W/m² K and that of the sloping roof to be 0.25 W/m² K. Assess the effects of this change.
(d) Comment on the risk of freezing that is experienced by the water services within the roof space, and the measures that are necessary to reduce the risk.

I was told

6 Fan and system selection

INTRODUCTION

The characteristic curves for the performance of fans are used in graphical form by the design engineer. Students have to learn to recognize the shape of the curve for each type of fan, and be able to manipulate data in graphs of pressure against flow. This chapter makes use of fan graphs, and shows how the pressure rise and flow data from manufacturers' catalogues can be entered into a computer and used by the design engineer.

The selection of a suitable fan is made by plotting the ductwork system design flow rate and its index route pressure drop onto the fan performance curve. This can now be done on a worksheet, and the resulting graph viewed and interrogated. A graphical enlargement of the crossover region between the ductwork system and fan curves allows the designer to make any tolerance adjustments that are desired. Fan speed changes can be easily made and the results displayed numerically and graphically.

Seven independent spreadsheets are presented. They are provided for five specific fan types – forward-curved centrifugal, backward-curved centrifugal, axial flow, propeller and mixed flow, a user-defined fan – and a comparison of fan types. In the user-defined fan spreadsheet, the user types pressure and flow data from the manufacturer's literature into the columns. The spreadsheet has been programmed to plot the fan curves and calculate the performance at other speeds. The seventh worksheet is to demonstrate the common curves for five different types of fan.

LEARNING OBJECTIVES

Study and use of this chapter will enable the user to:

1. recognize the shapes of the performance curves of fans;
2. know the types of fan in use;
3. know representative formulae for fan characteristic curves;
4. understand and use fan performance curves for different fan rotational speeds;
5. know the range of fan speeds used;
6. understand the meaning and use of polynomial equations;
7. use polynomial data in the creation of graphs;
8. use fan performance curves for centrifugal, axial, mixed flow and propeller fans;

9. know how to generate a ductwork system resistance curve;

10. relate the ductwork system resistance curve to fan performance curves;

11. know how to include the commissioning tolerance in the air flow design of ductwork systems;

12. select the intersection point for fan and ductwork system operation;

13. draw an enlargement of the fan and system intersection;

14. enter data for any fan type;

15. understand the gearing effect of fan and motor pulley diameters;

16. know the terms used in fan engineering;

17. be able to convert measurement units;

18. calculate the electrical input power for different fans.

Key terms and concepts

air flow rate	169	gearing	171
axial flow fan	174	index duct route	176
backward-curved centrifugal fan	173	kilovolt-ampere	179
belt drive	171	kilowatt	179
commissioning tolerance	176	manufacturer's test	168
design air flow	176	mathematical software	168
ductwork system	170	mixed-flow fan	175
fan and system intersection	170	motor power	179
fan curves	169	motor speed	170
fan power	179	polynomial equation	172
fan speeds	170	pressure	172
fan static pressure	178	propeller fan	175
fan total pressure	172	pulley	171
fan velocity pressure	178	resistance	172
forward-curved centrifugal fan	171	rotational speed	170

FAN CHARACTERISTIC CURVES

The performance of a fan is specified by the pressure rise that it generates in the air volume flow rate that is passing through it. Typical curves for axial, mixed, propeller, forward-curved centrifugal and backward-curved centrifugal fans are shown in Fig. 6.1.

The data for this figure can be seen in the worksheet file COMPARE.WK1. The shapes can be viewed on graph 1. Fan curves can be reproduced mathematically once the type is known and the test data is published. Manufacturers predict the pressure output of a new size of fan from a knowledge of the performance of geometrically similar models that have been tested. Curves of pressure against volume flow are scaled upwards and downwards depending upon the rotational

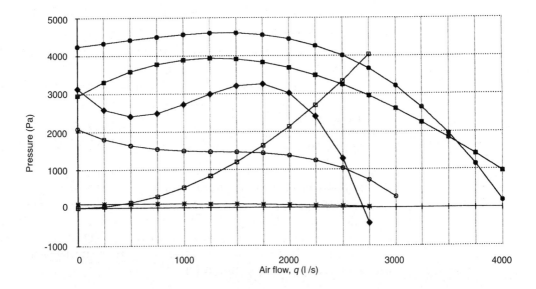

Fig. 6.1 Comparison of fan types. (fans are not directly comparable): ●, backward-curved centrifugal; ■, forward-curved centrifugal; ◆, axial; *, propeller; ○, mixed flow; □, system resistance.

speed at which the new fan is to operate. New fan models are subsequently tested to confirm their predicted performance.

The approach used here has been to read the fan total pressure and volume data from manufacturers' published curves and fit an equation to the curve. The curves of fan static pressure, fan velocity pressure and power consumption are calculated for the range of flow rates employed. All these curves can be automatically scaled up or down when the fan data is required at a different rotational speed.

Manufacturers' data may be fan static pressure against volume flow rate. The minor editing task that is necessary applies only when entering data into the ANYFAN sheet, and is demonstrated in question 3.

The curve of fan total pressure, *FTP* Pa, against air volume flow rate, Q l/s, usually corresponds to a four-term polynomial equation. These are of the form

$$FTP = a \pm bQ \pm cQ^2 \pm dQ^3$$

where a, b, c and d are constants for the particular type and size of fan, *FTP* is the fan total pressure rise in Pa, and Q is the air volume flow rate delivered by the fan in l/s. It may be necessary to add the fifth and higher terms, but four are expected to give a sufficiently good curve fit. The first term includes Q raised to the power of zero. As anything raised to zero is unity, the constant appears on its own. The second term is Q raised to the power of unity, the third is the power 2 and the fourth is the power 3. The second and higher terms can be positive or negative. Notice:

$$FTP = aQ^0 \pm bQ^1 \pm cQ^2 \pm dQ^3$$

as $Q^0 = 1$ and $Q^1 = Q$, so

$$FTP = a \pm bQ \pm cQ^2 \pm dQ^3$$

There are seven worksheets for the calculation and display of fan and system operating conditions:

1. ANYFAN, the user can enter data for any fan;
2. CENTFOR, forward-curved centrifugal fan;
3. CENTBACK, backward-curved centrifugal fan;
4. AXIAL, axial flow fan;
5. MIXED, mixed flow fan.
6. PROPFAN, propeller fan;
7. COMPARE, display any of five fan curves.

The worksheets named ANYFAN, CENTFOR, CENTBACK, AXIAL and MIXED and PROPFAN each have four graphs. The first graph in the CENTFOR worksheet shows the fan and ductwork system characteristic curves for the highest fan speed. A typical chart is shown in Fig. 6.2. A close-up view of the intersection between the fan and system curves is made to facilitate reading the data. A typical chart is shown in Fig. 6.3. The general arrangèment of a forward-curved centrifugal fan is shown in Fig. 6.4.

ANYFAN

This file is for a user-defined fan characteristic. Air flow and pressure data can be entered for axial flow, centrifugal, propeller or mixed flow fans. Four graphs have been prepared using the sample fan data on the worksheet. The graphs are in pairs. Graph 1 is for the first fan, and it shows the whole range of performance at the highest speed. This is expected to be no more than 2900 rev/min, the highest practical motor rotational speed on a 50 Hz electrical power supply. 50 Hz corresponds to 3000 rev/min. A 60 Hz supply corresponds to 3600 rev/min. There is

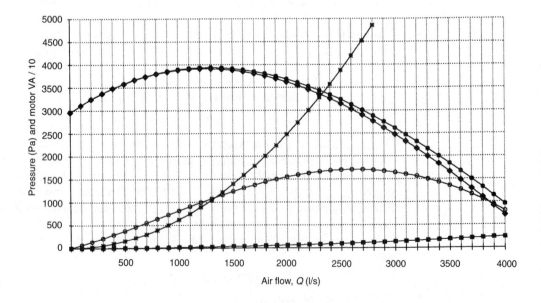

Fig. 6.2   Forward-curved centrifugal fan, highest speed: ●, fan total; ■, fan velocity; ◆, fan static; *, system resistance; ○, motor VA/10.

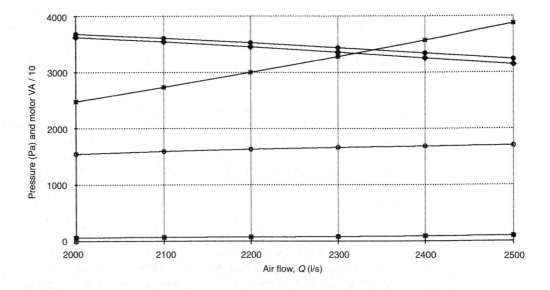

Fig. 6.3 Intersection of fan and system curves for a forward-curved centrifugal fan, highest speed: ●, fan total; ■, fan velocity; ◆, fan static; *, system resistance; ○, motor VA/10.

always a slip between the alternating current frequency in the stator of the motor and the rotor shaft speed. Fans are generally driven with a V-belt drive running on pulleys. There is a larger pulley on the fan shaft than on the motor shaft, so that there is a reduction gearing effect. Fan rotational speeds are usually within the range 350–1450 rev/min in order to minimize the generation of noise. Some manufacturers show curves for fan speeds up to 4000 rev/min. These have a smaller diameter pulley on the fan shaft than on the motor shaft in order to raise the gear ratio. Up to 100 dB may be produced in the outlet duct at such speeds, and attenuators may be required.

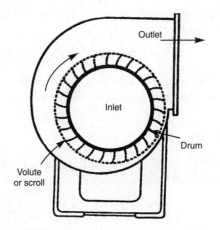

Fig. 6.4 Forward-curved centrifugal fan: general arrangement (reproduced by permission of Woods of Colchester Ltd).

When entering new data into the highest-speed section of the worksheet, acquire the data for the highest possible speed that the manufacturer can provide. The program automatically calculates the performance at lower speeds in the cells further down the sheet. Graph 2 shows a close-up view of the intersection of the fan curves with the system resistance curve to allow improved accuracy of reading from the graph.

Sample data is used on the worksheet, and this can easily be changed by the user. Once a published fan characteristic curve of fan total pressure, *FTP* Pa, against air volume flow rate, Q l/s, has been acquired, it can replace the sample data. The air flow Q l/s axis is divided into equal increments, and at least 10 data points should be used to allow a smooth curve to be generated for the new data. Either use the values of Q that are already in cells A45 to A85, or enter new values from the manufacturer's published graph. Read the value of fan total pressure, *FTP* Pa, from the manufacturer's graph, for each air flow value that has been selected. Type the fan total pressure values into the cell range C45 to C85. Use only whole numbers of pascals and not decimal places, as these would be insignificantly small. There must be the same number of pressures as there are flow rates. If a different number of data points are used, in comparison with the original cell range A45 to A85, then the ranges of data that have been specified in the graph files will have to be edited. If the published fan total pressure is in millimetres H_2O, multiply it by the gravitational acceleration, g m/s^2, to convert it to pascals. That is:

$$\text{pressure, } p = gH$$

where p = pressure (Pa);
  g = gravitational acceleration, 9.907 m/s^2; and
  H = manometer head (mm H_2O).

$$p \text{ Pa} = 9.907 \times H \text{ mm } H_2O$$

CENTFOR

The polynomial equation that was produced from a typical forward-curved blade centrifugal fan is

$$FTP = 2954.4 + 1.606Q - 0.17148 \times 10^{-3}Q^2 + 4.7115 \times 10^{-8}Q^3 \text{ Pa}$$

This equation is located in the range of cells from D36 to D76. The relevant value of Q is read from cells A36 to A76.

CENTBACK

The polynomial equation that was produced from a typical backward-curved blade centrifugal fan is

$$FTP = 4247.5 + 0.315Q + 1.159 \times 10^{-4}Q^2 - 0.1122 \times 10^{-6}Q^3 \text{ Pa}$$

This equation is located in the range of cells from D36 to D76. The relevant value of Q is read from cells A36 to A76. Backward-curved centrifugal fan impellers are shown in Fig. 6.5.

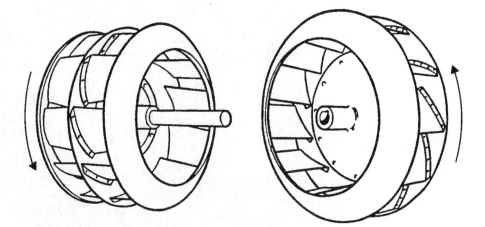

Fig. 6.5 Backward-curved centrifugal fan impellers (reproduced by permission of Woods of Colchester Ltd).

AXIAL

The polynomial equation that was produced from a typical axial flow fan is

$$FTP = 3136 - 3.065Q + 3.791 \times 10^{-3}Q^2 - 0.1144 \times 10^{-5}Q^3 \, \text{Pa}$$

This equation is located in the range of cells from D36 to D76. The relevant value of Q is read from cells A36 to A76. An axial flow fan is shown in Fig. 6.6.

MIXED

The polynomial equation that was produced from a typical mixed flow fan is

$$FTP = 2075 - 1.28Q + 9.33 \times 10^{-4}Q^2 - 0.235 \times 10^{-6}Q^3 \, \text{Pa}$$

This equation is located in the range of cells from D36 to D76. The relevant value of Q is read from cells A36 to A76. A mixed flow fan is shown in Fig. 6.7.

PROPFAN

The polynomial equation that was produced from a typical propeller fan is

$$FTP = 102 + 7.03 \times 10^{-3}Q - 0.23 \times 10^{-5}Q^2 - 0.41 \times 10^{-8}Q^3 \, \text{Pa}$$

This equation is located in the range of cells from D36 to D76. The relevant value of Q is read from cells A36 to A76. A propeller fan is shown in Fig. 6.8.

COMPARE

This file contains comparative performance data and equations for backward-curved centrifugal, forward-curved centrifugal, axial flow, mixed flow and

Fig. 6.6   Axial flow fan (reproduced by permission of Woods of Colchester Ltd).

propeller fans. The file is not meant to be used for calculations. It provides only a visual comparison of the curves for the different types. The fans are not of equal size or performance capacity. They are plotted on the same axes to the same scale against the same system resistance curve. The reader can gain an understanding of how these fans would perform upon the same system.

DATA REQUIREMENT

The general data entry procedure is the same for each of the worksheets. Where there are differences, specific instructions are given with the fan type.
   The user will enter new data for:

1. Your own name in cells B6.
2. The name of the job or the college assignment in cells B9 and C9; further cells along the row can be used.
3. Current date in cell B10.
4. The filename in cell B12.
5. The fan manufacturer's name in cells C16 to E16.
6. The fan model reference numbers in cells C17 to G17.
7. Highest fan speed to be used (rev/min) in cell B18.
8. Fan impeller diameter (mm) in cell B19.

Fig. 6.7   Mixed flow fan (reproduced by permission of Woods of Colchester Ltd).

Fig. 6.8   Propeller fan (reproduced by permission of Woods of Colchester Ltd).

9. Air density in cell B20. This is taken as $1.1906 \, \mathrm{kg/m^3}$ at 20 °C dry bulb, 60% relative humidity and 1013.25 mb atmospheric pressure.
10. Fan air outlet width (mm) in cell D22.
11. Fan air outlet height (mm) in cell F22.
12. Fan motor voltage (V) in cell D24. This will normally be 110, 240 or 415 V depending upon the supply voltage and voltage drop in the distribution cables due to their resistance.
13. Fan motor power factor in cell D25. This is less than 1.0 and is typically between 0.6 and 0.90. Power factor includes the electrical losses in the motor due to the resistance and inductance of the wiring.
14. Fan impeller efficiency (%) in cell D26. Typical values are 85% for a backward-curved centrifugal, 80% for a forward-curved centrifugal, 75% for a mixed flow, 70% for an axial and 60% for a propeller fan impeller. The fan manufacturer and the references can be used as necessary.
15. Fan motor overall efficiency (%) in cell D27. This should include all the mechanical and aerodynamic losses plus those in the drive system. A typical value is 80%. The electrical power losses were included in the power factor.
16. The design air flow, Q l/s, through the air ductwork system is entered into cell C31. This figure may be increased by 0–20% above the calculated design air flow to provide an excess air flow. This will generate the commissioning tolerance. Adding 20% to the design air flow will not necessarily produce a 20% commissioning tolerance because of the shape of the intersecting curves.
17. The total frictional resistance, d_p Pa, of the air ductwork system is entered into cell C32. This includes the pressure drop through the ductwork along the index route, duct fittings, dampers, filters, louvres, heating and cooling coils and the terminal unit. The index route is that series of air flow paths that have the greatest resistance. Other routes are in parallel to the index and they may have surplus pressure available to overcome their resistances.
18. The increments of the air flow, Q l/s, that are delivered by the fan are entered into cells A36 to A76. Commence with 0 l/s and end with the air flow that corresponds to the fan's delivering zero static pressure. Use equal increments of Q, such as 50 l/s or 100 l/s. The values of air flow that are already on the worksheet may be left if they are appropriate.
19. Enter a reduced value for the fan rotational speed, N rev/min, when compared with the previously entered full speed, into cell B92.
20. The design air flow, Q l/s, for the reduced fan speed is entered into cell C105. This figure may be increased by 0–20% above the calculated design air flow to provide an excess air flow. This will produce the commissioning tolerance.
21. The total frictional resistance, d_p Pa, of the air ductwork system that is supplied from the reduced fan speed is entered into cell C106. This includes the pressure drop through the ductwork along the index route, duct fittings, dampers, filters, louvres, heating and cooling coils and the terminal unit.

This concludes the data entry. Most of the other cells are either calculated automatically or are copied from the data that has been entered. Save this file onto the hard disk in drive C for the correct directory of your files, and on your own floppy disk in drive A now. For example, save it as C:\ASEASY\DA-VIDC11\FANS\AXIAL9.WK1 and A:\FANS\AXIAL9.WK1, where ASEASY is the directory on the hard disk containing As-Easy-As, DAVIDC11 is your own subdirectory, FANS is your subdirectory for all work on fans, and AXIAL9.WK1 is the file name of the worksheet.

OUTPUT DATA

1. Fan outlet area is given in cell D23.
2. Air velocity in the fan outlet area is given in cell C36.
3. Fan total pressure rise that is calculated from the polynomial equation for the fan type is in cell D36.
4. Fan velocity pressure is calculated from the fan air discharge velocity and is in cell E36.
5. Fan static pressure is given in cell F36
6. Ductwork system resistance for each value of air flow from the fan is calculated in cell G36.
7. Electrical power input to the motor driving the fan is calculated from the air flow and fan total pressure rise in cell H36.

The calculated data for each air flow is repeated down to line 76.

FORMULAE

Representative samples of the formulae are given. Each formula can be read on the spreadsheet by moving the cursor to the cell. The equation is presented in the form in which it would normally be written and in the format that is used by the spreadsheet.

Cell D23

(F3) +D22*F22/10^6

The fan outlet duct cross-sectional area is calculated from the width and height of the opening. An axial, propeller or mixed flow fan has a circular outlet, and the area is calculated from $\pi d \, \text{mm}^2/4/10^6 \, \text{m}^2$.

$$\text{fan outlet area D23} = \text{width D22 mm} \times \text{height F22 mm} \times \frac{1\text{m}^2}{10^6 \, \text{mm}^2}$$

$$= \frac{\text{D22} \times \text{F22}}{10^6} \, \text{m}^2$$

Cell C36

+A36/1000/$D$23

The air velocity leaving the fan through the discharge duct is found by dividing the air volume flow rate by the cross-sectional area of the outlet. It is defined as the fan air velocity, and is used to calculate the fan velocity pressure.

$$\text{fan air velocity C39} = \frac{\text{air flow A36 l/s}}{\text{outlet area D23 m}^2} \times \frac{1 \, \text{m}^3}{10^3 \, \text{litre}}$$

$$= \frac{\text{A36}}{\text{D23} \times 10^3} \, \text{m/s}$$

Cell D36

+4247.5+0.315*A36+1.159*10^−4*A36^2−0.1122*10^−6*A36^3

Cell D36 contains the fan total pressure rise that is generated from the air flow by the fan type. The appropriate polynomial equation is used in each case. Remember that the polynomials used here were from specific fans, and they are not applicable to every fan of similar construction or rotational speed. Cell A36 contains the air flow, Q l/s, that the fan delivers.

$$D36 = 4247.5 + 0.315 \times A36 + 1.159 \times 10^{-4} \times A36^2 - 0.1122 \times 10^{-6} \times A36^3 \text{ Pa}$$

$$FTP = 4247.5 + 0.315Q + 1.159 \times 10^{-4}Q^2 - 0.1122 \times 10^{-6}Q^3 \text{ Pa}$$

Cell E36

+A36^2*0.5*$B$20/$D$23^2/10^6

$$FVP = 0.5\rho \, \frac{\text{kg}}{\text{m}^3} \times \frac{Q^2 \, \text{litre}^2}{\text{s}^2} \times \frac{1}{\text{area}^2 \, \text{m}^4} \times \frac{1 \, \text{m}^6}{10^6 \, \text{litre}^2}$$

$$= 0.5\rho \, \frac{Q^2 \, \text{kg}}{\text{area}^2 \times 10^6 \, \text{m s}^2} \times \frac{1 \, \text{N s}^2}{1 \, \text{kg m}} \times \frac{1 \, \text{Pa m}^2}{1 \, \text{N}}$$

$$= 0.5\rho \, \frac{Q^2}{\text{area}^2 \times 10^6} \text{ Pa}$$

The air density, $\rho \, \text{kg/m}^3$, is in cell B20, air flow quantity in cell A36, and the fan outlet area is in cell D23.

$$FVP \, E36 = 0.5 \times B20 \times \frac{A36^2}{D23^2 \times 10^6} \text{ Pa}$$

The fan velocity pressure is calculated from the fan air velocity and the density of the air flowing through the discharge duct. The density of the air at this location depends upon the dry-bulb temperature and barometric pressure of the air within the duct. The air is subject to frictional heating by the fan and any heat output from a close-coupled electric driving motor. Belt-driven motors are remote from the air within the duct. The duct air static pressure should be added to the barometric air pressure in order ascertain the air density.

Cell F36

+D36−E36

Fan static pressure F36 = fan total pressure D36 − fan velocity pressure E36.

$$FSP \, E36 = FTP \, D36 - FVP \, E36 \, \text{Pa}$$

Cell G36

+$C$32*A36^2/$C$31^2

The ductwork system resistance varies with the square of the air flow quantity. The general form for this is

$$\frac{dp}{Q^2} = \text{constant for the ductwork system}$$

so

$$\frac{dp_1}{(Q_1)^2} = \frac{dp_2}{(Q_2)^2}$$

where suffixes 1 and 2 refer to different flow rates. Once the design flow rate and pressure drop due to ductwork resistance are known – that is, at condition 1 – the expected system pressure drop at any other rate of flow (condition 2) can be calculated from the ratio:

$$dp_2 = (Q_2)^2 \times \frac{dp_1}{(Q_1)^2}$$

$$= dp_1 \times \frac{(Q_2)^2}{(Q_1)^2}$$

$$\text{system pressure drop G36} = \text{C32} \times \frac{(\text{A36})^2}{(\text{C31})^2}\ \text{Pa}$$

This will calculate the new system resistance for the flow rate found in cell A36.

Cell H36

+A36*D36/$D$25/$D$26/$D$27/100

The input electrical power to be supplied to the motor is found from

$$\text{input power} = Q\,\frac{1}{s} \times FTP\ \text{Pa} \times \frac{1}{PF} \times \frac{1}{\text{imp\%}} \times \frac{1}{\text{motor\%}}$$

where imp% = fan impeller efficiency (%); motor% = overall electrical motor efficiency (%); and PF = motor power factor.

$$\text{power} = Q\frac{1}{s} \times \frac{1\,\text{m}^3}{10^3\,\text{litres}} \times FTP\ \text{Pa} \times \frac{1\,\text{N}}{\text{Pa}\,\text{m}^2} \times \frac{1}{PF} \times \frac{100}{\text{imp\%}} \times \frac{100}{\text{motor\%}}$$

$$= \frac{Q \times FTP \times 100 \times 100}{10^3 \times PF \times \text{imp\%} \times \text{motor\%}}\ \frac{\text{N}\,\text{m}}{\text{s}} \times \frac{1\,\text{W}\,\text{s}}{1\,\text{N}\,\text{m}} \times \frac{1\,\text{kW}}{10^3\,\text{W}} \times \frac{1\,\text{kVA}}{1\,\text{kW}}$$

$$= \frac{Q \times FTP}{100 \times PF \times \text{imp\%} \times \text{motor\%}}\ \text{kVA}$$

$$= \frac{\text{A36}\,\text{l/s} \times \text{D36}\,\text{Pa}}{100 \times \text{D25} \times \text{D26\%} \times \text{D27\%}}\ \text{kVA}$$

The term kVA (kilovolt-ampere) refers to the electrical input power to the motor. When the power factor is 1.0, kVA are equal to kW, but as with most motors, the useful output power of a fan motor in watts is less than the electrical input power in kVA that has to be paid for at the meter. For the convenience of plotting, the input power to the motor is multiplied by 1000 and divided by 10, to be shown as VA/10.

$$\text{power} = \frac{\text{A36}\,\text{l/s} \times \text{D36}\,\text{Pa}}{100 \times \text{D25} \times \text{D26\%} \times \text{D27\%}}\ \text{kVA} \times \frac{1000\,\text{VA}}{1\,\text{kVA}} \times \frac{1}{10}$$

$$= \frac{\text{A36} \times \text{D36}}{\text{D25} \times \text{D26} \times \text{D27}}\ \text{VA (divided by 10)}$$

Read the VA/10 power from the graph, multiply it by 10 to find the input power in VA, divide by 1000 to find the kVA input. The electrical current per phase taken by the motor is

$$\text{current} = \frac{\text{VA}}{\sqrt{(3)} \times \text{V}} = \frac{\text{H}36 \times 10}{\sqrt{(3)} \times \text{D}24 \times 1000} \text{ A}$$

The electrical services designer needs to know the full load running current and starting current.

Rows 36 to 76

The formulae are repeated on these lines.

GRAPHICAL OUTPUT

The graphs can be modified by the user. It may be necessary to alter the range of cells that are displayed in the close-up charts so that the intersection between the fan and system is in view. Instructions are given in Chapter 1. The spreadsheet manual may have to be consulted.

Questions

The solutions to questions that are based on reading graphical data will not be as precise as those from formulae. A tolerance of ±2% should be expected. Two percent of a pressure value of 2000 Pa would be 40 Pa. If the correct answer is 2000 Pa, then a graphical solution of 2000 ± 40 Pa, or between 2040 and 1960 Pa, could be acceptable. Answers can be verified by inserting the flow rate into a row of the worksheet and reading the calculated values of pressure and power. Remember to replace the proper flow value before saving the file.

1. Explain how a fan manufacturer is able to produce curves of fan total pressure against air volume flow rate for a new size of fan that is within a range of fan types that have been regularly made.
2. Acquire fan manufacturers' performance curves for any fan type and digitize the fan total pressure and air volume flow rate data. To do this, read the fan total pressure produced for each air volume flow rate. Choose suitable increments along the horizontal volume flow axis to take data from. At least 10 air flow increments should be taken, because each one will produce one point on a graph. Now plot the digitized data on graph paper to a larger scale or size than on the manufacturer's sheet. Draw a smooth curve through the points to complete the graph. If insufficient increments of air flow were chosen to be able to draw a curve that reproduces the original, then take further data points. A flexible drawing curve

instrument can be used to aid the production of a smooth curve. Avoid any points that appear to be far away from the general curve, because these may have been misread from the original graph. Draw a best-fit smooth curve through the points.

3. Manufacturers often present the fan static air pressure versus air volume flow rate on their data sheets. Use the data from question 2, or another source, and retrieve the ANYFAN worksheet. Enter the air flow rates into column A and the other input data. The spreadsheets will calculate the fan velocity pressure for each air flow rate for you. Manually add the manufacturer's stated fan static pressures to the fan velocity pressures and enter the fan total pressures on the worksheet. Compare the manufacturer's fan static pressures with the calculated worksheet values for accuracy. View the graphs produced.

4. Use the Pascal ABC.123 fan data that is in the original copy of the ANYFAN file. The fan is to be operated at a maximum speed of 2600 rev/min and at a normal speed of 960 rev/min in a factory ventilation system. The outlet air duct is 600 mm wide and 800 mm high. The fan impeller has a diameter of 850 mm. The air density is 1.190 kg/m^3. A 415 V alternating electrical supply is to drive the fan. The motor power factor is 0.72, the fan impeller efficiency is 83%, and the overall motor and drive system efficiency is 78%. The fan is to be used on a ductwork system that has a design air flow requirement of 1250 l/s when the frictional resistance is 1500 Pa. State the air flow rate that will be produced in the ductwork system, and the

effect upon the performance of the ventilation system.

5. An axial flow fan has a maximum speed of 2900 rev/min and an operational speed of 1300 rev/min in an air conditioning system. Use the original copy of the AXIAL.WK1 worksheet. The outlet air duct is 470 mm diameter. The fan impeller has a diameter of 450 mm. The air density is 1.21 kg/m^3. A 415 V alternating electrical supply is to drive the fan. The motor power factor is 0.66, the fan impeller efficiency is 62%, and the overall motor and drive system efficiency is 75%. The fan is to be used on a ductwork system that has a design air flow requirement of 2000 l/s when the frictional resistance is 1500 Pa. There is to be a commissioning tolerance of ±10% on the air flow delivered to the air conditioning system.

(a) State the air flow rate that will be produced when the fan is operated at 2900 rev/min, and the effect upon the performance of the air conditioning system.
(b) Find the air flow rate and fan total pressure that correspond to the fan being run at 1300 rev/min.
(c) Find the correct fan speed that will satisfy the commissioning procedure. State the air flow provided, the fan total, static and velocity pressures, the air velocity leaving the fan, the electrical power taken by the fan, and the current that will be taken by the motor.

6. The axial flow fan performance that is shown on the original copy of the AXIAL.WK1 worksheet is to be used for an extract ventilation duct. The fan has a maximum speed of 2900 rev/min and an operational speed of 1400 rev/min. The outlet air duct is 550 mm in diameter. The fan impeller has a diameter of 550 mm. The air density is 1.205 kg/m^3. A 415 V alternating electrical supply is to drive the fan. The motor power factor is 0.82, the fan impeller efficiency is 74%, and the overall motor and drive system efficiency is 82%. The design air flow requirement is 2000 l/s when the ductwork frictional resistance is 1000 Pa. There is to be a commissioning tolerance of ±10% on the air flow delivered to the air conditioning system.

(a) State the result of using the fan at the highest speed on the extract system, and the motor power used.
(b) Find the fan speed that must be used to meet the commissioning criteria. State the air flow provided, the fan total, static and velocity pressures, the air velocity leaving the fan, the electrical power taken by the fan, and the current that will be taken by the motor.
(c) State the excess air pressure that the fan will generate in part (b), and how it can be dealt with.

7. A backward-curved centrifugal fan is to be used as the main supply fan in a variable-volume air conditioning system. Use the original copy of the CENTBACK.WK1 worksheet. The fan has a manufacturer's catalogue speed of 2200 rev/min. The outlet air duct is 500 mm wide and 400 mm high. The fan impeller has a diameter of 480 mm. The air density is 1.205 kg/m^3. A 415 V alternating electrical supply is to drive the fan. The motor power factor is 0.75, the fan impeller efficiency is 82%, and the overall motor and drive system efficiency is 75%. The fan reference code is ZYT46.33, and the fan is manufactured by Pascal Industries plc. The design air flow requirement is 2200 l/s when the ductwork frictional resistance is 3000 Pa.

(a) Find the operational values for the fan and system.
(b) Find the fan speed that must be used to meet a commissioning tolerance of $Q \pm 5\%$ l/s. State the operating data.
(c) State the excess air pressure from the fan.

8. A forward-curved centrifugal fan is to be used as the main extract fan in a variable-volume air conditioning system. The fan has a manufacturer's catalogue speed of 2600 rev/min. Use the original CENTFOR.WK1 worksheet. The outlet air duct is 400 mm wide and 350 mm high. The fan impeller has a diameter of 400 mm. The air density is 1.205 kg/m^3. A 415 V alternating electrical supply is to drive the fan. The motor power factor is 0.65, the fan impeller efficiency is 72%, and the overall motor and drive system efficiency is 70%. The fan reference code is X4AT.77, and the fan is manufactured by Pascal Industries plc. The design air flow requirement is 3000 l/s when the ductwork frictional resistance is 1500 Pa.

(a) Find the operational values for the fan and system.
(b) Find the fan speed that must be used to meet a commissioning tolerance of design $Q \pm 10\%$ l/s. State the operating data.
(c) State the excess air pressure from the fan.

9. A mixed flow fan is used in an industrial air extract duct. The fan is tested at a speed of 1500 rev/min by the manufacturer, and has the following performance:

Q (l/s)	*Fan total pressure, FTP* (Pa)
0	900
250	950
500	900
750	900
1000	850
1250	780
1500	600
1750	350

Enter this data into the ANYFAN.WKl worksheet. The outlet air duct is 450 mm diameter. The fan impeller has a diameter of 430 mm. The air density is 1.205 kg/m³. A 240 V alternating electrical supply is to drive the fan. The motor power factor is 0.6, the fan impeller efficiency is 70%, and the overall motor and drive system efficiency is 60%. The fan reference code is MXFA.56, and the fan is manufactured by Joule Fan Co. Ltd. The extract duct design air flow requirement is 1100 l/s when the ductwork frictional resistance is 750 Pa.

(a) Find the operational values for the fan and system.
(b) Find the fan speed that must be used to meet a commissioning tolerance of $Q \pm 15\%$ l/s. State the operating data.
(c) State the excess air pressure from the fan.

10. A propeller fan is connected to a short extract duct from toilet accommodation. The fan reference code is WGP800.4, and the fan is manufactured by Lyndhurst Fans Ltd. The fan is tested at a speed of 920 rev/min by the manufacturer and has the following performance:

Q (l/s)	Fan static pressure, FSP (Pa)
0	200
500	197
1000	193
1500	190

2000	180
2500	168
3000	160
3500	153
4000	144
4500	125
5000	90
5500	20

The outlet air duct is 750 mm diameter. The fan impeller has a diameter of 720 mm. The air density is 1.22 kg/m³. Enter this data into the ANYFAN.WKl worksheet. The air flow quantities, density and outlet diameter are entered first. Note the fan velocity pressure that the worksheet calculates. Manually add the fan velocity pressure to the fan static pressure in the question, and enter these fan total pressure values into the *FTP* column of the worksheet. Then progress to the remainder of the question. A 240 V alternating electrical supply is to drive the fan. The motor power factor is 0.7, the fan impeller efficiency is 60%, and the overall motor and drive system efficiency is 68%. The extract duct design air flow requirement is 3500 l/s when the ductwork frictional resistance is 120 Pa.

(a) Find the operational values for the fan and system.
(b) State the commissioning percentage tolerance that is produced.

The prison architect.

7 Air duct design

INTRODUCTION

The design procedure for air duct systems is explained. The principles are shown from the use of Bernoulli's equation for straight ducts, changes in duct cross-section, flows through branches, grilles and fans, through to a spreadsheet that can be used for duct routes. Both supply and extract air duct systems are used. The spreadsheet can be used for individual air ducts and systems having up to five branches. It can be expanded by the user to accommodate larger systems once the methods are understood. Node numbers are used throughout the duct systems, and these can easily be changed by the user for applications. The air temperature can be changed for each section of duct. Air pressure terms and methods of measurement are explained. Detailed pressure changes at duct fittings and fans are calculated. The worksheet calculates all the total and static air pressures throughout the duct system and in the ventilated room or space. The index duct route and the fan pressure rise are automatically selected. There is no need to refer to reference data, tables or charts to be able to calculate the correct sizes of air ducts, because sufficient data for many applications is on the worksheet. Data on fittings' velocity pressure loss factors can be entered from other sources.

The method that is used to calculate the duct carrying capacity is the Colebrook and White equation for the design friction factor in the D'Arcy formula including the fluid properties and pipe material factors. The data is for clean galvanized sheet metal ducts having joints made in accordance with good practice (CIBSE, 1986c). The overall formula for duct air flow is in a reliable and usable form, which reproduces published data within acceptable accuracy. It can easily be verified with manual calculations and by reference to the source equations and duct chart.

LEARNING OBJECTIVES

Study and use of this chapter will enable the user to:

1. understand the terms total, static and velocity air pressures;
2. calculate the density of air;
3. calculate total, static and velocity air pressures;
4. understand and use frictional resistance air pressure drops in duct systems and changes in duct size;
5. understand and use static regain air pressure;

6. understand the use of velocity pressure loss factors;
7. calculate the changes in air pressure when air flows through ductwork and branches;
8. calculate the carrying capacities of air ducts;
9. calculate air flow and pressure data for air at different temperatures and pressures;
10. calculate the carrying capacity of air ducts for values of pressure drop rate and duct diameter;
11. understand how air pressure changes take place though duct systems;
12. draw air pressure gradient graphs;
13. understand and use the changes in air pressure that occur at a fan;
14. understand and use the terms fan total, static and velocity pressures;
15. calculate the equivalent diameter of rectangular air ducts;
16. calculate the static air pressure in a mechanically ventilated room;
17. use numbered nodes in the design of air duct systems;
18. find the correct dimensions for air duct systems;
19. make adjustments to air flows in duct systems for variations in air temperature;
20. use limiting air velocity and pressure drop rate data;
21. know the limiting air velocities for low-velocity duct systems;
22. analyse supply and extract air ductwork systems;
23. know the equations for air duct sizing and their method of use;
24. find the index route that offers the highest frictional resistance to the fan;
25. know how a spreadsheet uses conditional 'IF' statements to select a highest value;
26. know how to specify the performance criteria for a fan;
27. know how to calculate the balancing pressure in branch ducts from the index route.

Key terms and concepts

AIR PRESSURES IN A DUCT

Air flowing through a duct exerts two types of pressure on its surroundings. The first is dynamic pressure due to its motion, or kinetic energy. This is velocity pressure, p_v; it acts only in the direction of the flow of air.

$$p_v = 0.5\rho v^2 \text{ Pa}$$

where ρ = air density (kg/m^3) and v = air velocity (m/s).

The second is the bursting pressure of the air trying to escape from the enclosure or of the surrounding air trying to enter the enclosing duct. This is static pressure, p_s Pa, and it acts in all directions.

The sum of these two pressures is the total pressure p_t:

$$\text{total pressure} = \text{static pressure} + \text{velocity pressure}$$

$$p_t = p_s + p_v \text{ Pa}$$

Losses of pressure due to the frictional resistance of the duct, its bends, branches, changes in cross-section, dampers, filters, air heating or cooling coils, the entry of air into a duct system and the exit of air from a duct, cause loss of total pressure.

Bernoulli's theorem means that, for a fluid flowing from position 1 to position 2, the total pressure energy at point 1 equals the total pressure energy at point 2 plus the loss of energy due to friction between the two locations. All energy lost in friction is dissipated as heat, and in some cases this can cause a noticeable rise in the air temperature. A plot of the pressures that occur when air flows through an enlargement in a duct is shown in Fig. 7.1. The data in Fig. 7.1 relates to Example 7.1. It can be seen that the air static pressure increases through the enlargement because the drop of velocity pressure is greater than the loss of pressure due to friction.

The balance of air pressures between 1 and 2 is:

$$p_{t1} = p_{t2} + dp_{1-2}$$

where p_{t1} = total pressure at 1; p_{t2} = total pressure at 2; and dp_{1-2} = pressure drop due to friction between nodes 1 and 2.

The balance of pressures between nodes 2 and 3 is

$$p_{t2} = p_{s2} + p_{v2}$$
$$p_{t3} = p_{s3} + p_{v3}$$

but

$$p_{t2} = p_{t3} + dp_{2-3}$$

and

$$dp_{2-3} = p_{t2} - p_{t3}$$
$$= p_{s2} + p_{v2} - (p_{s3} + p_{v3})$$
$$= p_{v2} - p_{v3} + p_{s2} - p_{s3}$$
$$p_{s3} - p_{s2} = (p_{v2} - p_{v3}) - dp_{2-3}$$

This shows that a regain of static pressure occurs when a reduction in the velocity pressure is greater than the pressure drop due to the frictional resistance of the

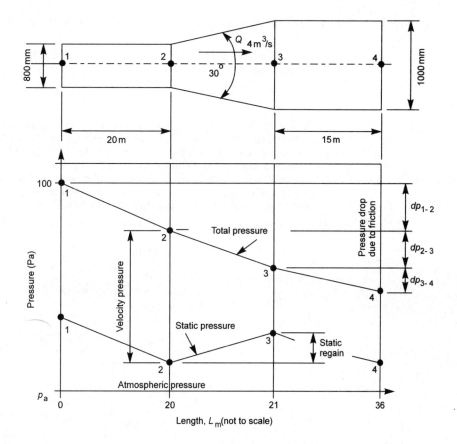

Fig. 7.1 Changes of pressure as air flows through an enlargement.

section of duct. This can take place at a branch fitting, at a change of duct dimensions, or when air discharges into a room or plenum space such as a ceiling void.

The worked examples can be entered onto an appropriate worksheet. The worksheet uses more decimal places than the user would write, so minor differences between calculated and computed answers will arise. These differences will normally be within 1 Pa, 1 mm or 1 litre/s. Two worksheet files are provided on the disk. The file DUCTFIT.WK1 is to calculate the pressure changes through air duct enlargements, contractions and branch fittings. The file DUCT1.WK1 is for the most common application, where there is a supply air fan and ductwork which has up to five branches. These two files have been used for all the worksheet calculations in this chapter. When the user has gained experience from using the information and methods that have been provided, worksheets may be adapted for other applications. It is suggested that the user may produce adaptations to the file DUCT1.WK1 for the following applications:

1. DUCT2.WK1 extract fan and ductwork with five branches;
2. DUCT3.WK1 supply and extract fans for one room and no recirculation of room air;

3. DUCT4.WK1 supply and extract fans for one room with room air recirculation;
4. DUCT5.WK1 supply and extract fans for multiple rooms and with room air recirculation.

EXAMPLE 7.1

Calculate the static regain and pressure changes that occur when $4\,\text{m}^3/\text{s}$ flow through a 20 m long, 800 mm diameter duct that enlarges at 30° to 1000 mm diameter and remains at 1000 mm for 15 m, as shown in Fig. 7.1. The air total pressure at the commencement of the 800 mm duct is 100 Pa above the atmospheric air pressure. The air density is $1.1906\,\text{kg/m}^3$. The frictional pressure loss rates are 0.7 Pa/m in the 800 mm diameter and 0.25 Pa/m in the 1000 mm diameter ducts. The solution is shown on the original file named DUCTFIT.WK1.

$$p_{t1} = 100\,\text{Pa}$$

$$v_1 = 4\,\frac{\text{m}^3}{\text{s}} \times \frac{4}{\pi \times 0.8^2\,\text{m}^2}$$

$$= 7.96\,\text{m/s}$$

$$p_{v1} = 0.5 \times 1.1906\,\frac{\text{kg}}{\text{m}^3} \times 7.96^2\,\frac{\text{m}^2}{\text{s}^2}$$

As

$$1\,\text{N} = 1\,\text{kg} \times 1\,\frac{\text{m}}{\text{s}^2}$$

$$p_{v1} = 0.5 \times 1.1906\,\frac{\text{kg}}{\text{m}^3} \times 7.96^2\,\frac{\text{m}^2}{\text{s}^2} \times \frac{1\,\text{N}\,\text{s}^2}{1\,\text{kg}\,\text{m}}$$

$$= 38\,\frac{\text{N}}{\text{m}^2}$$

and

$$1\,\text{Pa} = 1\,\text{N/m}^2$$

$$p_{v1} = 38\,\text{Pa}$$

$$p_{s1} = p_{t1} - p_{v1}$$

$$= 100 - 38\,\text{Pa}$$

$$= 62\,\text{Pa}$$

$$dp_{1-2} = 0.7\,\frac{\text{Pa}}{\text{m}} \times 20\,\text{m}$$

$$= 14\,\text{Pa}$$

$$p_{t2} = p_{t1} - dp_{1-2}$$

$$= 100 - 14\,\text{Pa}$$

$$= 86\,\text{Pa}$$

$$p_{s2} = p_{t2} - p_{v2}$$

$$p_{v2} = p_{v1}$$

$$p_{s2} = 86 - 38\,\text{Pa}$$

$$p_{s2} = 48\,\text{Pa}$$

$$v_3 = 4\frac{m^3}{s} \times \frac{4}{\pi \times 1^2 m^2}$$

$$= 5.09 \, m/s$$

$$p_{v3} = 0.5 \times 1.1906 \times 5.09^2 \, Pa$$

$$= 15 \, Pa$$

Figures 7.14–7.16 show the velocity pressure loss factors for air duct fittings (CIBSE, 1986c).

For an abrupt enlargement of a circular duct:

$$\frac{A_2}{A_3} = \frac{\pi d_2^2}{4} \times \frac{4}{\pi d_3^2}$$

$$= \frac{d_2^2}{d_3^2}$$

$$= \frac{0.8^2}{1^2}$$

$$= 0.64$$

so k_1 can be taken as 0.2 and k_2 as 0.8:

$$k = k_1 k_2$$

$$= 0.2 \times 0.8$$

$$= 0.16$$

$$dp_{2-3} = k p_{v2}$$

$$= 0.16 \times 38 \, Pa$$

$$= 6 \, Pa$$

$$p_{t3} = p_{t2} - dp_{2-3}$$

$$= 86 - 6 \, Pa$$

$$= 80 \, Pa$$

$$p_{s3} = p_{t3} - p_{v3}$$

$$= 80 - 15$$

$$= 65 \, Pa$$

$$dp_{3-4} = 15 \, m \times 0.25 \frac{Pa}{m}$$

$$= 4 \, Pa$$

$$p_{t4} = p_{t3} - dp_{3-4}$$

$$= 80 - 4 \, Pa$$

$$= 76 \, Pa$$

$$p_{s4} = p_{t4} - p_{v4}$$

$$p_{v3} = p_{v4}$$

$$p_{s4} = 76 - 15 \, Pa$$

$$= 61 \, Pa$$

The regain of static pressure is

$$SR = p_{s3} - p_{s2} \, \text{Pa}$$
$$= 65 - 48 \, \text{Pa}$$
$$= 17 \, \text{Pa}$$

EXAMPLE 7.2

Calculate the total pressure of air flowing at $0.75 \, \text{m}^3/\text{s}$ in a 400 mm internal diameter duct when the air temperatures is $18 \, ^\circ\text{C}$ d.b. and the static pressure in the duct is 40 mm water gauge above the atmospheric pressure of 101 560 Pa. Also calculate the total pressure when the air is at $-5 \, ^\circ\text{C}$ and $30 \, ^\circ\text{C}$ during winter and summer operation of the air conditioning system. A worksheet is not used for this example.

The atmospheric pressure within the duct:

$$p_a = 101\,560 + p_s \, \text{Pa}$$
$$p_s = 9.807 \times 40 \, \text{mm H}_2\text{O Pa}$$
$$= 392.3 \, \text{Pa}$$
$$p_a = 101\,560 + 392.3$$
$$= 101\,952.3 \, \text{Pa}$$

It is reasonable to ignore decimal parts of pascal pressures, so $p_s = 392 \, \text{Pa}$ and $p_a = 101\,952 \, \text{Pa}$.

$$\rho = 1.1906 \times \frac{p_a}{101\,325} \times \frac{273 + 20}{273 + t} \, \frac{\text{kg}}{\text{m}^3}$$
$$= 1.1906 \times \frac{101\,952}{101\,325} \times \frac{273 + 20}{273 + 18} \, \frac{\text{kg}}{\text{m}^3}$$
$$= 1.2062 \, \text{kg/m}^3$$
$$v = \frac{4 \times 0.75}{\pi \times 0.4^2} \, \frac{\text{m}}{\text{s}}$$
$$= 5.97 \, \text{m/s}$$
$$p_v = 0.5 \times 1.2062 \times 5.97^2 \, \text{Pa}$$
$$= 21.5 \, \text{Pa}$$
$$= 22 \, \text{Pa}$$

total pressure $p_t = p_s + p_v$

$$p_t = 392 + 22 \, \text{Pa}$$
$$= 414 \, \text{Pa}$$

When the air is at $-5 \, ^\circ\text{C}$ but Q remains at $0.75 \, \text{m}^3/\text{s}$:

$$\rho = 1.1906 \times \frac{101\,952}{101\,325} \times \frac{273 + 20}{273 + -5} \, \frac{\text{kg}}{\text{m}^3}$$
$$= 1.3097 \, \text{kg/m}^3$$
$$v = 5.97 \, \text{m/s}$$
$$p_v = 0.5 \times 1.3097 \times 5.97^2 \, \text{Pa}$$
$$= 23 \, \text{Pa}$$

$$\text{total pressure } p_t = p_s + p_v$$

$$p_t = 392 + 23 \,\text{Pa}$$

$$= 415 \,\text{Pa}$$

When the air is at 30 °C but Q remains at 0.75 m³/s:

$$\rho = 1.158 \,\text{kg/m}^3$$

$$v = 5.97 \,\text{m/s}$$

$$p_v = 21 \,\text{Pa}$$

$$p_t = 413 \,\text{Pa}$$

AIR FLOW IN DUCTS

The flow of air at 20 °C d.b., 43% percentage saturation and a barometric pressure of 101 325 Pa through clean galvanized sheet metal ducts having joints made in accordance with good practice is given by

$$Q = -2.0278 \times dp^{0.5} \times d^{2.5} \times \log\left(\frac{4.05 \times 10^{-5}}{d} + \frac{2.933 \times 10^{-5}}{dp^{0.5} \times d^{1.5}}\right)$$

(CIBSE 1986c, adapted for fluid properties stated; Chadderton, 1997)
where Q = air flow rate (m³/s); dp = pressure loss rate (Pa/m); and d = duct internal diameter (m).

EXAMPLE 7.3

A 700 mm internal diameter galvanized sheet steel duct is carrying air at 20 °C d.b. and has a design pressure loss rate of 0.8 Pa/m. Calculate the maximum air volume flow rate that can be passed and the air velocity. A worksheet is not used for this example.

$$d = 0.70 \,\text{m}$$

$$dp = 0.8 \,\text{Pa/m}$$

$$Q = -2.0278 \times 0.8^{0.5} \times 0.7^{2.5} \times \log\left(\frac{4.05 \times 10^{-5}}{0.7} + \frac{2.933 \times 10^{-5}}{0.8^{0.5} \times 0.7^{1.5}}\right)$$

$$= 2.932 \,\text{m}^3/\text{s}$$

$$v = 2.932 \frac{\text{m}^3}{\text{s}} \times \frac{4}{\pi \times 0.7^2} \,\text{m}^2$$

$$= 7.62 \,\text{m/s}$$

EXAMPLE 7.4

Find a suitable diameter for a galvanized sheet steel duct that is to carry 500 l/s of air at 20 °C d.b. at a maximum velocity of 5.0 m/s when the pressure loss rate is not to exceed 0.6 Pa/m. Duct diameters are in increments of 50 mm. A worksheet is not used for this example.

An iterative procedure is needed; try $d = 350\,\text{mm}$.

$$\text{air flow } Q = 0.5\,\text{m}^3/\text{s}$$

$$v = 0.5\,\frac{\text{m}}{\text{s}} \times \frac{4}{\pi \times 0.35^2}\,\text{m}^2 = 5.2\,\text{m/s}$$

This is over the limit. Try $d = 0.4\,\text{m}$ at $Q = 0.5\,\text{m}^3/\text{s}$:

$$v = 3.98\,\text{m/s}$$

A 400 mm diameter duct will be used, provided that the frictional pressure loss rate does not exceed $dp = 0.6\,\text{Pa/m}$. Insert $dp = 0.6\,\text{Pa/m}$ and $d = 0.4\,\text{m}$ into the duct formula and calculate the maximum carrying capacity, irrespective of limiting velocity:

$$Q = -2.0278 \times 0.6^{0.5} \times 0.4^{2.5} \times \log\left(\frac{4.05 \times 10^{-5}}{0.4} + \frac{2.933 \times 10^{-5}}{0.6^{0.5} \times 0.4^{1.5}}\right)$$

$$= 0.572\,\text{m}^3/\text{s}$$

This is greater than required, so the actual pressure loss rate will be less than 0.6 Pa/m, around 0.47 Pa/m.

VARIATIONS OF PRESSURE ALONG A DUCT

Pressure gradients are produced along the length of a duct, as shown in Fig. 7.1. The gradient caused along the constant-diameter duct is calculated from the pressure drop rate due to friction. The gradients (slope) of the total and static pressure lines are equal. Friction losses can be calculated as total or static pressure drops. Pressure drops will be treated as reductions in the available total pressure.

Duct fittings cause loss of pressure through surface friction and additional turbulence because of the shape of the enclosure. Such losses are found from experiment, and require the use of unique factors that are multiplied by the velocity pressure. Data for commonly used duct fittings are given in Figs 7.14–7.16 (CIBSE, 1986c).

To produce a pressure gradient diagram similar to Fig. 7.1, draw the ductwork to be analysed directly above a graph as shown. Clearly mark each point where a change of section takes place, and identify each with a number. These points are the nodes that will be used for the worksheet pressure analysis and for finding the sizes of the ducts. This method is common to all duct-sizing programs. The sequence is:

1. Find the total pressure at 1, p_{t1}, and plot it to an appropriate scale above, or below, atmospheric pressure p_a Pa.
2. Calculate the pressure drop between 1 and 2 in the straight duct, dp_{1-2}. Note that dp_{1-2} means the pressure drop due to friction between nodes 1 and 2.
3. Subtract dp_{1-2} from p_{t1} to find the total pressure at point 2, p_{t2} Pa.
4. Draw the total pressure straight line from p_{t1} to p_{t2}.
5. Find the velocity pressure loss factor for the duct fitting, k, from Figs 7.14–7.16 or the reference.
6. Find the velocity pressure to be used for the frictional resistance calculation for the fitting. In this case, it is the velocity pressure in the smaller duct, p_{v3}.

7. Multiply the fitting k factor by the velocity pressure p_{v3} to calculate the pressure drop through the fitting, dp_{2-3}.
8. Subtract pressure drop dp_{2-3} from the total pressure p_{t2} to find the total pressure at 3, p_{t3}.
9. Plot p_{t3} and connect the total pressure straight line from p_{t2}.
10. Calculate the pressure drop between 3 and 4 in the straight duct, dp_{3-4}.
11. Subtract dp_{3-4} from p_{t3} to find the total pressure at point 4, p_{t4} Pa.
12. Draw the total pressure straight line from p_{t3} to p_{t4}.
13. Calculate the velocity pressure at 1, p_{v1} Pa. Note that p_{v1} is equal to p_{v2}, because the velocity remains constant along a uniform-diameter duct.
14. Subtract p_{v1} from both p_{t1} and p_{t2}. The differences are the static pressures at 1 and 2, p_{s1} and p_{s2}.
15. Plot p_{s1} and p_{s2}, and connect them with a straight line.
16. Subtract p_{v3} from p_{t3} and find static pressure at 3, p_{s3}.
17. Draw a straight line from p_{s2} to p_{s3}.
18. Subtract p_{v3} from p_{t4} and find p_{s4}. Note that p_{v3} equals p_{v4}.
19. Draw a straight line from p_{s3} to p_{s4}.
20. Mark the upper gradient as the total pressure line.
21. Mark the lower line as the gradient of static pressure.
22. Mark the differences between the two gradients as velocity pressures.
23. Identify each of the corresponding points of the duct diagram and the graph.
24. Mark on the graph all the calculated pressures.

This procedure applies to other air duct cases and to the changes of pressure across a fan.

Notice that for this increase in diameter of the duct, the static pressure rises. This is because the velocity pressure has reduced. The available total pressure has fallen because of friction losses. Such a regain of static pressure depends on the amount of frictional pressure drop that occurs. Static regain can also take place at branches because of a reduction in the air flow quantity and velocity. Excess static pressure in branch ducts can be absorbed by reducing the duct diameter, provided that the velocity limit is not exceeded, or by partial closure of a balancing damper. The worksheet shows a warning when there is excessive static pressure at a branch duct outlet. The user then enters new data until a satisfactory system balance is produced. This forces the user to understand the procedure of designing a correctly balanced air duct network. High-cost software will make such iterations automatically, leaving the designer unaware of how the design was balanced.

EXAMPLE 7.5

Air is blown at 2.5 m³/s into a 35 m long 700 mm diameter duct, which reduces to 500 mm diameter and remains at 500 mm for 22 m. The air is discharged into a room that is at atmospheric pressure. The air total pressure at the commencement of the 700 mm duct is 192 Pa above the atmospheric pressure. The reducer is the 30° concentric type. The air density is 1.23 kg/m³. The frictional pressure loss rates are 0.6 Pa/m in the 700 mm diameter duct and 3.2 Pa/m in the 500 mm diameter duct. Calculate the air pressures in the duct system and draw them on a graph. Enter the data into a copy of file DUCTFIT.WK1 and save it as file DUCTFIT2.WK1.

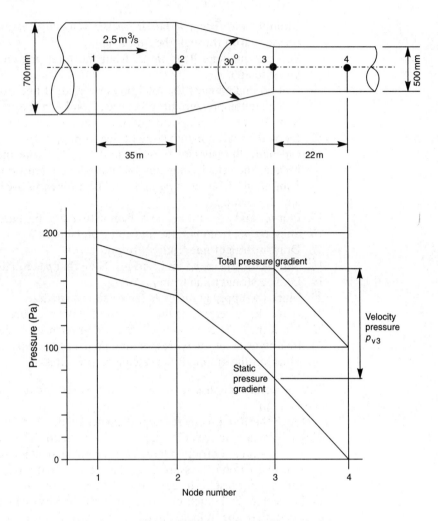

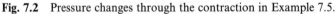

Fig. 7.2   Pressure changes through the contraction in Example 7.5.

Using the node numbers from Fig. 7.2 and the sequence numbers listed:

1. $p_{t1} = 192\,\text{Pa}$; plot graph axes.
2. $dp_{1-2} = 0.6\,\text{Pa/m} \times 35\,\text{m}$
   $= 21\,\text{Pa}$
3. $p_{t2} = p_{t1} - dp_{1-2}$
   $= 192 - 21\,\text{Pa}$
   $= 171\,\text{Pa}$
4. Draw the total pressure line 1 to 2.
5. The reducer has a velocity pressure loss factor k of 0.02, and this is multiplied by the smaller duct velocity pressure, p_{v3}.
6. $v_3 = 2.5\,\dfrac{\text{m}^3}{\text{s}} \times \dfrac{4}{\pi \times 0.5^2\,\text{m}^2}$
   $= 12.7\,\text{m/s}$
   $p_{v3} = 0.5 \times 1.23 \times 12.7^2\,\text{Pa}$
   $= 99\,\text{Pa}$

7. Pressure drop through reducer:

$$dp_{2-3} = \text{factor} \times p_{v3}$$
$$= 0.02 \times 99 \, \text{Pa}$$
$$= 2 \, \text{Pa}$$

8. $p_{t3} = p_{t2} - dp_{2-3}$
$$= 171 - 2 \, \text{Pa}$$
$$= 169 \, \text{Pa}$$

9. Plot total pressure line 2 to 3.

10. Pressure drop through 500 mm duct:

$$dp_{3-4} = 22 \, \text{m} \times 3.2 \, \text{Pa/m}$$
$$= 70 \, \text{Pa}$$

11. $p_{t4} = p_{t3} - dp_{3-4}$
$$= 169 - 70 \, \text{Pa}$$
$$= 99 \, \text{Pa}$$

12. Draw total pressure line 3 to 4.

13. $v_1 = 2.5 \dfrac{\text{m}^3}{\text{s}} \times \dfrac{4}{\pi \times 0.7^2} \, \text{m}^2$
$$= 6.5 \, \text{m/s}$$
$$p_{v1} = 0.5 \times 1.23 \times 6.5^2 \, \text{Pa}$$
$$= 26 \, \text{Pa}$$
$$p_{v2} = 26 \, \text{Pa}$$

14. $p_{s1} = p_{t1} - p_{v1}$
$$= 192 - 26 \, \text{Pa}$$
$$= 166 \, \text{Pa}$$
$$p_{s2} = p_{t2} - p_{v2}$$
$$p_{v2} = p_{v1}$$
$$p_{s2} = 171 - 26 \, \text{Pa}$$
$$= 145 \, \text{Pa}$$

15. Plot static pressures 1 and 2, and join them with a line.

16. $p_{s3} = p_{t3} - p_{v3}$
$$= 169 - 99 \, \text{Pa}$$
$$= 70 \, \text{Pa}$$

17. Draw static pressure line from 2 to 3.

18. $p_{s4} = p_{t4} - p_{v4}$
$$p_{v3} = p_{v4}$$
$$p_{s4} = 99 - 99 \, \text{Pa}$$
$$= 0 \, \text{Pa}$$

This is atmospheric pressure.

19. Draw static pressure line from 3 to 4.

20–24. Figure 7.2 shows these answers plotted to scale.

PRESSURE CHANGES AT A FAN

The fan produces a rise of total pressure in the whole system. This rise is the fan total pressure, *FTP* Pa. The fan total pressure rise is equal to the drop of total pressure due to friction in the ductwork system plus the increase of air velocity pressure that has been generated by the fan. The fan total pressure is calculated from the total pressure in the fan outlet duct, node 2, minus the total pressure in the fan inlet duct, node 1:

$$FTP = p_{t2} - p_{t1} \, \text{Pa}$$

Fan velocity pressure, *FVP*, is defined as the velocity pressure in the discharge area from the fan, p_{v2}:

$$FVP = p_{v2} \, \text{Pa}$$

Fan static pressure, *FSP*, is defined as fan total pressure minus fan velocity pressure. Note that this is not necessarily the same as the change in static pressure across the fan connections:

$$FSP = FTP - FVP \, \text{Pa}$$

EXAMPLE 7.6

The 1200 mm × 1200 mm air-handling plant shown in Fig. 7.3 supplies 3 m³/s of air at 30 °C into a room. The axial flow fan has a diameter of 550 mm. Outdoor air at 5 °C is drawn through a louvre and filter. The room is at atmospheric pressure. The fixed pressure drops are: filter 450 Pa, heater coil 125 Pa, and discharge grille 45 Pa. The pressure drop rate in the plant ductwork is estimated to be 0.1 Pa/m. Calculate the pressures at each node. Enter the data into a copy of the file DUCT1.WK1 and save it as file DUCT7%6.WK1. The node numbers to be used are shown in Fig. 7.3.

From Figs 7.14–7.16 the velocity pressure loss factors are as follows.
Air intake louvre for entry, louvre and wire mesh:

$$k = 0.5 + 4.5 + 1.7$$
$$= 6.7$$

45° gradual contraction:

$$k = 0.04$$

30° gradual enlargement:

$$\frac{A_1}{A_2} = \frac{\pi \times 0.55^2 \, \text{m}^2}{4} \times \frac{1}{1.2 \, \text{m} \times 1.2 \, \text{m}}$$
$$= 0.17$$

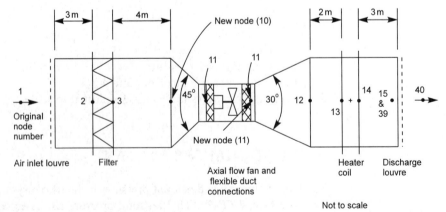

Fig. 7.3 Air-handling plant in Example 7.6.

$$k_1 = 0.7 \text{ estimated}$$
$$k_2 = 0.8$$
$$k = 0.7 \times 0.8$$
$$= 0.56$$

Duct exit:

$$k = 1$$

The fan air volume flow at 5°C is

$$Q_1 = 3\frac{\text{m}^3}{\text{s}} \times \frac{273 + 5}{273 + 30}$$
$$= 2.753 \,\text{m}^3/\text{s}$$

After the heater coil:

$$Q_{14} = 3 \,\text{m}^3/\text{s}$$

The circular equivalent of the rectangular plant duct is found from (CIBSE, 1986c)

$$d_2 = 1.265 \times \left[\frac{(WD)^3}{W + D}\right]^{0.2} \text{mm}$$

$$= 1.265 \times \left[\frac{(1200 \times 1200)^3}{1200 + 1200}\right]^{0.2} \text{mm}$$

$$= 1322 \,\text{mm}$$

$$v_2 = 2.753\frac{\text{m}^3}{\text{s}} \times \frac{4}{\pi \times 1.322^2 \,\text{m}^2}$$
$$= 2 \,\text{m/s}$$

$$v_{11} = 2.753\frac{\text{m}^3}{\text{s}} \times \frac{4}{\pi \times 0.55^2 \,\text{m}^2}$$
$$= 11.59 \,\text{m/s}$$

$$v_{14} = 3\frac{\text{m}^3}{\text{s}} \times \frac{4}{\pi \times 1.322^2 \,\text{m}^2}$$
$$= 2.19 \,\text{m/s}$$

$$\rho_2 = 1.1906\frac{\text{kg}}{\text{m}^3} \times \frac{273 + 20}{273 + 5}$$
$$= 1.255 \,\text{kg/m}^3$$

$$\rho_{14} = 1.1906\frac{\text{kg}}{\text{m}^3} \times \frac{273 + 20}{273 + 30}$$
$$= 1.151 \,\text{kg/m}^3$$

$$p_{v2} = 0.5 \times \rho_2 \times v_2^2 \,\text{Pa}$$
$$= 0.5 \times 1.255 \times 2^2 \,\text{Pa}$$
$$= 2.5 \,\text{Pa}$$

$$p_{v11} = 0.5 \times \rho_2 \times v_{11}^2 \,\text{Pa}$$
$$= 0.5 \times 1.255 \times 11.59^2 \,\text{Pa}$$
$$= 84.3 \,\text{Pa}$$

$$p_{v14} = 0.5 \times \rho_{14} \times v_{14}^2 \, \text{Pa}$$
$$= 0.5 \times 1.151 \times 2.19^2 \, \text{Pa}$$
$$= 2.8 \, \text{Pa}$$

$$dp_{1-2} = k p_{v2} + L \, \text{m} \times dp \frac{\text{Pa}}{\text{m}}$$
$$= 6.7 \times 2.5 + 3 \times 0.1 \, \text{Pa}$$
$$= 17 \, \text{Pa}$$

$$dp_{2-3} = 450 \, \text{Pa}$$

$$dp_{3-10} = 4 \times 0.1 \, \text{Pa}$$
$$= 0.4 \, \text{Pa}$$

$$dp_{10-11} = k p_{v11} \, \text{Pa}$$
$$= 0.04 \times 84.3 \, \text{Pa}$$
$$= 3.4 \, \text{Pa}$$

$$dp_{11-12} = k p_{v11} \, \text{Pa}$$
$$= 0.56 \times 84.3 \, \text{Pa}$$
$$= 47.2 \, \text{Pa}$$

$$dp_{12-40} = \text{duct} + \text{heater coil} + \text{grille} + \text{discharge} \, k$$
$$= 5 \times 0.1 + 125 + 45 + k p_{v39} \, \text{Pa}$$
$$= 170.5 + 1 \times 2.5 \, \text{Pa}$$
$$= 173 \, \text{Pa}$$

drop of total pressure, $dp_{1-40} = 17 + 450 + 0.4 + 3.4 + 47.2 + 173 \, \text{Pa}$
$$= 691 \, \text{Pa}$$

fan total pressure rise, $FTP = dp_{1-10} + p_{v40} - p_{v1} \, \text{Pa}$

$$p_{v40} = p_{v2} = 2.5 \, \text{Pa}$$
$$p_{v1} = p_{v2} = 2.5 \, \text{Pa}$$

fan total pressure rise, $FTP = 691 + 2.5 - 2.5 \, \text{Pa}$
$$= 691 \, \text{Pa}$$

fan velocity pressure, $FVP = p_{v11}$
$$= 84 \, \text{Pa}$$

fan static pressure, $FSP = 691 - 84 \, \text{Pa}$
$$= 607 \, \text{Pa}$$

$$p_{t1} = 0 \, \text{Pa, atmospheric pressure}$$
$$p_{s1} = 0 \, \text{Pa, atmospheric pressure}$$
$$p_{t2} = p_{t1} - dp_{1-2} \, \text{Pa}$$
$$= 0 - 17 \, \text{Pa}$$
$$= -17 \, \text{Pa}$$
$$p_{s2} = p_{t2} - p_{v2} \, \text{Pa}$$
$$= -17 - 2.5 \, \text{Pa}$$
$$= -20 \, \text{Pa}$$

$$p_{t3} = -17 - 450\,\text{Pa}$$
$$= -467\,\text{Pa}$$
$$p_{s3} = -467 - 2.5\,\text{Pa}$$
$$= -470\,\text{Pa}$$
$$p_{t10} = -467 - 0.4\,\text{Pa}$$
$$= -468\,\text{Pa}$$
$$p_{s10} = -468 - 2.5\,\text{Pa}$$
$$= -471\,\text{Pa}$$
$$p_{t11} = -468 - 3.4\,\text{Pa}$$
$$= -471\,\text{Pa}$$
$$p_{s11} = -471 - 84.3\,\text{Pa}$$
$$= -555\,\text{Pa}$$
$$p_{t11} = p_{t11} + \text{fan total pressure rise Pa}$$
$$= -471 + 691\,\text{Pa}$$
$$= 220\,\text{Pa}$$
$$p_{s11} = 220 - 84.3\ \text{Pa}$$
$$= 136\,\text{Pa}$$
$$p_{t12} = 220 - 47.2\,\text{Pa}$$
$$= 173\,\text{Pa}$$
$$p_{s12} = 173 - 2.5\,\text{Pa}$$
$$= 171\,\text{Pa}$$
$$p_{t13} = 173 - 2 \times 0.1\,\text{Pa}$$
$$= 173\,\text{Pa}$$
$$p_{s13} = 173 - 2.5\,\text{Pa}$$
$$= 171\,\text{Pa}$$
$$p_{t14} = 173 - 125\,\text{Pa}$$
$$= 48\,\text{Pa}$$
$$p_{s14} = 48 - 2.5\,\text{Pa}$$
$$= 45\,\text{Pa}$$
$$p_{t40} = p_{t14} - \text{duct } dp_{14-15} - \text{grille } dp - \text{discharge } k \times p_{v15}$$
$$= 48 - 3 \times 0.1 - 45 - 2.8\,\text{Pa}$$
$$= 0\,\text{Pa}$$

This is the pressure in the room due to the fan.

Also, as the air velocity leaving the discharge grille has reduced to zero within the room:

$$p_{s40} = p_{t40}$$
$$= 0\,\text{Pa}$$

This example is typical of other cases where there are fewer nodes in the duct-work system than are available on the worksheet. The user is required to edit the problem to fit the worksheet and ensure that the worksheet has the relevant data. This will involve renumbering some of the nodes in the example and ignoring some of the facilities of the worksheet. The fan is located between node numbers that need to remain as they are displayed on the worksheet. This is because the formula for calculating the total pressure at the fan outlet is not the same as those for other sections of duct. The user can change the fan location provided that the formulae are also changed. It may be simpler to leave the formulae where they are and change the node numbers on the drawing of the ductwork to suit the worksheet. This is the recommended method. The procedure for entering Example 7.6 into the worksheet file DUCT7%6.WK1 is as follows.

1. The node numbers in Fig. 7.3 have been arranged to suit the worksheet numbers.
2. Describe each duct section in column B.
3. Enter 5 °C into cells C50, C51, C52, C59, C60, C68 and C69.
4. Enter 30 °C into cells C70, C71 and C47. The other temperature cells are not needed in this example, and their values need not be changed.
5. Enter the duct lengths into column E. Make sure that zero length is in the cells where there should not be any duct.
6. The air flow quantity at 5 °C is found from the known value at 30 °C by multiplying by the ratio of the absolute temperatures:

$$Q_1 = Q_{10} \times \frac{273 + 5}{273 + 30} \frac{m^3}{s}$$
$$= 3 \times \frac{273 + 5}{273 + 30} \frac{m^3}{s}$$
$$= 2.753 \, m^3/s$$

7. Enter 2.753 m³/s (only the number and not the units) into cells F50, F51, F52, F59, F60, F68 and F69.
8. Enter 3 m³/s into cells F70, F71 and F110.
9. Rows that have zero duct length and zero fittings velocity pressure loss factor k will always generate zero pressure drop in that duct section.
10. Enter 2.5 m/s for the maximum air velocity in the sections of plant duct and 15 m/s for the fan. Other values may be tried after the first complete calculation.
11. Enter 0.1 Pa/m for the plant duct pressure drop rate.
12. Ignore the Q error in column K, likely duct diameter in column L and velocity in column M, as the duct sizes and pressure drop rate are known.
13. Enter 1200 mm for both the width and depth of the plant ducts in columns N and O.
14. Enter 500 mm for both the width and depth of the fan duct connections, because this produces the correct circular equivalent diameter of 551 mm.
15. Enter the fittings velocity pressure loss factors in the cell range S50 to S110.
16. Enter the plant pressure drops in cell range V50 to V110.
17. This completes the data entry, and the results can be compared with the calculated solution.

DUCTWORK SYSTEM DESIGN

The worksheet can be used for various applications. Each application, worked example and question requires that the ductwork system is identified with node numbers, known air flow rates, and a limiting air velocity for each duct section. Typical velocity limits are given in Table 7.1.

Table 7.1 Limiting air velocities in ducts for low-velocity system design

Application	Main duct v (m/s)	Branch duct v (m/s)
Hospital, concert hall, library, sound studio	5.0	3.5
Cinema, restaurant, hall	7.5	5.0
General office, dance hall, shop, exhibition hall	9.0	6.0
Factory, workshop, canteen	12.5	7.5

The user is required to enter a duct pressure drop rate that produces the least error between the maximum flow capacity of the duct and the design air flow. This is done iteratively, and can produce a satisfactory result quickly. The user selects the appropriate dimensions for circular and rectangular ducts. These are normally in increments of 50 mm.

EXAMPLE 7.7

Calculate the fan total pressure for the supply air duct system shown in Fig. 7.4. The design air flow is 1 m³/s with the air at 20 °C d.b. and 101 325 Pa. The heating and cooling coils are in place, but they are not operational. The fresh air intake louvre has a 50% free area, and it is backed with wire mesh. The plant pressure drops are: filter 250 Pa, heater 100 Pa, cooler 100 Pa, and discharge diffuser 25 Pa. The limiting air velocities are: intake and discharge grilles 2.5 m/s, plant 2.5 m/s, and distribution ductwork 5 m/s. Duct and plant dimensions are available in increments of 50 mm. The limiting air velocities are not to be exceeded. Find the smallest sizes of duct that can be achieved within the limits. Use only the dimensions and duct construction shown; do not add contractions or enlargements on the connections to the fan or other items in the system. Investigate the effects of realistic winter and summer operating conditions on the fan total pressure.

The original file DUCT1.WK1 shows the data and solution for this example. The air temperature remains at 20 °C throughout the system. Winter and summer air intake temperatures can range from −5 °C to 40 °C. Operation of the heating and cooling coils will produce supply air temperatures that will range from 30 °C in winter to 15 °C during summer. The nodes in Fig. 7.4 coincide with those in the original file. This leaves a gap between the end of the ductwork at node 12 and the next duct point at node 40. The user may alter the node numbering to suit the application or leave them as shown. The duct fitting velocity pressure loss factors are:

fresh air intake louvre 0.5 free area, $k = 4.5$;
wire mesh, $k = 1.7$;
fresh air intake duct entry, $k = 0.5$;
flat blade damper, $k = 0.5$;

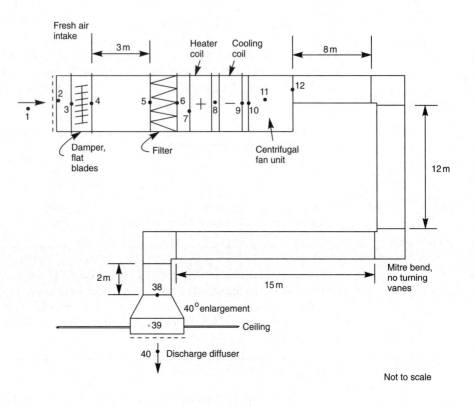

Fig. 7.4 Supply fan ductwork for Example 7.7.

mitre bend with no turning vanes, $k = 1.25$;
$40°$ enlargement from the duct to the discharge diffuser; this will not be known until the duct sizes have been determined because the area ratio between the two ducts is needed.

The density of air at $20°C$ d.b. and $101\,325\,Pa$ atmospheric pressure is 1.1906 kg/m^3.

An air-handling plant of $650\,mm \times 650\,mm$ cross-section will have an equivalent circular diameter of

$$d_2 = 1.265 \times \left[\frac{(WD)^3}{W + D}\right]^{0.2} mm$$

$$= 1.265 \times \left[\frac{(650 \times 650)^3}{650 + 650}\right]^{0.2} mm$$

$$= 716\,mm$$

$$= 0.716\,m$$

The air velocity produced is from

$$Q\frac{m^3}{s} = \frac{\pi(d\,m)^2}{4} \times v\,\frac{m}{s}$$

$$v_2 = \frac{4Q}{\pi d_2^2} \frac{\text{m}}{\text{s}}$$

$$= \frac{4 \times 1}{\pi \times 0.716^2} \frac{\text{m}}{\text{s}}$$

$$= 2.5 \, \text{m/s}$$

The air velocity pressure is

$$p_{v2} = 0.5 \rho v_2^2 \, \text{Pa}$$

$$= 0.5 \times 1.1906 \times 2.5^2 \, \text{Pa}$$

$$= 3.7 \, \text{Pa}$$

The overall velocity pressure loss factor for the outside air inlet louvre is

$$k = 0.5 + 4.5 + 1.7$$

$$= 6.7$$

The pressure drop through the outside air intake louvre is

$$dp_{1-2} = k p_{v2} \, \text{Pa}$$

$$= 6.7 \times 3.7 \, \text{Pa}$$

$$= 24.6 \, \text{Pa}$$

Pressure drop through the damper is

$$dp_{3-4} = k p_{v3} \, \text{Pa}$$

$$= 0.5 \times 3.7 \, \text{Pa}$$

$$= 1.8 \, \text{Pa}$$

A pressure drop rate of 0.1 Pa/m of straight duct is appropriate, because it produces an air flow rate of

$$Q = -2.0278 \times dp^{0.5} \times d^{2.5} \times \log\left(\frac{4.05 \times 10^{-5}}{d} + \frac{2.933 \times 10^{-5}}{dp^{0.5} \times d^{1.5}}\right)$$

$$= -2.0278 \times 0.1^{0.5} \times 0.716^{2.5} \times \log\left(\frac{4.05 \times 10^{-5}}{0.716} + \frac{2.933 \times 10^{-5}}{0.1^{0.5} \times 0.716^{1.5}}\right)$$

$$= 1.023 \, \text{m}^3/\text{s}$$

The duct pressure drop is

$$dp_{4-5} = \frac{dp}{L} \frac{\text{Pa}}{\text{m}} \times L \, \text{m}$$

$$= 0.1 \frac{\text{Pa}}{\text{m}} \times 3 \, \text{m}$$

$$= 0.3 \, \text{Pa}$$

The total length of duct from the fan outlet to the discharge diffuser is

$$L = 8 + 12 + 15 + 2 \, \text{m}$$

$$= 37 \, \text{m}$$

The three mitre bends have an overall velocity pressure loss factor of

$$k = 3 \times 1.25$$
$$= 3.75$$

The expected diameter of the distribution duct is

$$d = 1000 \times \left(\frac{Q\,\mathrm{m}^3}{\mathrm{s}} \times \frac{4}{\pi} \times \frac{1}{v} \frac{\mathrm{s}}{\mathrm{m}} \right)^{0.5} \mathrm{mm}$$

$$= 1000 \times \left(\frac{1\,\mathrm{m}^3}{\mathrm{s}} \times \frac{4}{\pi} \times \frac{1}{5} \frac{\mathrm{s}}{\mathrm{m}} \right)^{0.5} \mathrm{mm}$$

$$= 505\,\mathrm{mm}$$

A pressure drop rate of 0.48 Pa/m in the duct between nodes 12 and 38 produces an air flow rate of

$$Q = -2.027 \times 0.48^{0.5} \times 0.505^{2.5} \times \log\left(\frac{4.5 \times 10^{-5}}{0.505} + \frac{2.933 \times 10^{-5}}{0.48^{0.5} \times 0.505^{1.5}} \right)$$

$$= 0.943\,\mathrm{m}^3/\mathrm{s}$$

A 500 mm $\times$ 450 mm rectangular duct has an equivalent diameter of

$$d = 1.265 \times \left[\frac{(500 \times 450)^3}{500 + 450} \right]^{0.2} \mathrm{mm}$$

$$= 522\,\mathrm{m}$$

$$v_{12} = 1\,\frac{\mathrm{m}^3}{\mathrm{s}} \times \frac{4}{\pi \times 0.522^2}\,\mathrm{m}^2$$

$$= 4.7\,\mathrm{m/s}$$

$$p_{v12} = 0.5 \times 1.1906 \times 4.7^2\,\mathrm{Pa}$$

$$= 13.2\,\mathrm{Pa}$$

The distribution duct pressure drop is

$$\text{duct } dp_{12-38} = 0.48 \frac{\mathrm{Pa}}{\mathrm{m}} \times 37\,\mathrm{m}$$

$$= 17.8\,\mathrm{Pa}$$

$$\text{fittings } dp_{12-13} = 3.75 \times 13\,\mathrm{Pa}$$

$$= 48.8\,\mathrm{Pa}$$

The duct enlarges to the same dimensions as the plant, 650 mm $\times$ 650 mm, to reduce the air velocity to 2.5 m/s prior to connection to the diffuser.

$$p_v = 3.7\,\mathrm{Pa} \text{ as before.}$$

For an abrupt enlargement:

$$\frac{A_1}{A_2} = \frac{450 \times 450}{650 \times 650}$$

$$= 0.48$$

k_1 may be taken as 0.36, and this is multiplied by the factor for a 40° gradual enlargement, k_2, of 1.

$$\text{Enlargement } k = k_1 k_2$$
$$= 0.36 \times 1$$
$$= 0.36$$

Add the velocity pressure loss factor for the exit from the duct, and the overall becomes

$$k = 1 + 0.36$$
$$= 1.36$$

Pressure drop through the discharge diffuser at the design air flow rate is 25 Pa.

$$\text{fitting } dp_{38-40} = 1.36 \times 3.7 + 25\,\text{Pa}$$
$$= 5 + 25\,\text{Pa}$$
$$= 30\,\text{Pa}$$

The overall drop of pressure through the duct and air-handling plant system is

$$dp_{1-40} = \Sigma dp_{1-40}\,\text{Pa}$$
$$= 24.6 + 1.8 + 0.3 + 250 + 100 + 100 + 17.8 + 48.8 + 5 + 25\,\text{Pa}$$
$$= 573\,\text{Pa}$$

The fan total pressure rise is

$$FTP = dp_{1-40} + p_{v40} - p_{v1}$$
$$= 573 + 3.7 - 3.7\,\text{Pa}$$
$$= 573\,\text{Pa}$$

The fan velocity pressure FVP is, as far as we know in this case, $p_{v12}\,\text{Pa}$:

$$FVP = 13\,\text{Pa}$$
$$FSP = FTP - FVP\,\text{Pa}$$
$$= 573 - 13\,\text{Pa}$$
$$= 560\,\text{Pa}$$

Check the data on a working copy of file DUCT1.WK1, and save it as file name DUCT7%7.WK1 to represent Example 7.7. Make whatever changes seem appropriate to validate the answers given. There will be other possible solutions. It may be possible to reduce the fan total pressure by selecting other duct sizes. Find out what happens when the winter and summer air temperatures are used for the relevant sections of ductwork. (In summer the fan total pressure becomes 572 Pa, and in winter it is 574 Pa; the differences are not significant.)

The worksheet is arranged for a supply air duct system that has up to five branch ducts. The user may decide to alter the worksheet to suit an extract duct system with five branches. The contents of line numbers 60 and 67, which relate to the fan data, need to be moved down the worksheet until the correct number of suction and discharge branches are either side of the fan.

When a supply and extract duct system is to be analysed, enter the supply air duct system onto the worksheet and make a paper record of the results. This will include the room air pressure and the grille from the outside air. Then enter the data for the extract air duct system and fan from the room to the exhaust air grille. This will overwrite the previous data, which will be lost unless a separate file copy was made. The starting air pressure for the extract system will be the room static air pressure that was created by the supply system. The results for the extract system are then found.

Duct dimensions are quoted as width × depth. Note that the equivalent diameter equation produces a different result when these dimensions become reversed.

If the original spreadsheet does not have enough sections, copy rows down the screen to make more. Edit the new lines to ensure that the correct cell reference are maintained, particularly for the air density. New users should beware when making such changes until they fully understand the worksheet.

DATA REQUIREMENT

The user will enter new data for:

1. Your name in cell B4.
2. Name of the job or assignment in cell B5.
3. Reference number or job details in cell B6.
4. Name of the file in cell B9.
5. Duct air temperature in cell C14.
6. Duct air pressure in cell C15.
7. Number all the system nodes on a schematic layout of the ductwork using the numbers that are on the worksheet.
8. Ensure that the section node numbers in column A of the worksheet agree with those on the drawing of the ductwork. There will be occasions when you have to adapt the worksheet node numbers to suit the drawing.
9. Enter the air temperature t °C d.b. in cell C50, duct length L m in cell E50, air flow Q m^3/s in cell F50, maximum air velocity v m/s in cell G50, and the maximum allowable pressure loss rate dp Pa/m in cell H50, and subsequent rows, for each section of duct.
10. The estimated duct diameter and its maximum carrying capacity are displayed in cells I50 and J50.
11. The difference between the maximum carrying capacity and the required Q m^3/s is calculated and expressed as Q error % in cell K50. If this is more than, say, ±5%, enter values for maximum dp Pa/m in cell H50 until close agreement is reached. Use only the first two decimal places for pressure drop rate, because this is sufficiently accurate.
12. Enter the likely duct diameter using increments of 50 mm in cell L50.
13. Enter the dimensions of the rectangular duct to be used in millimetres in cells N50 and O50. The circular equivalent diameter and actual air velocity are displayed in cells P50 and Q50. Change the duct dimensions until a satisfactory result is obtained. Check that the duct dimensions being selected will fit into the building. This is a crucial part of the design.
14. The velocity pressure p_v Pa is displayed in cell R50.
15. Enter the sum of the velocity pressure loss factors k for all the fittings in the duct section in cell S50.
16. Enter the pressure drop through the air-handling plant, such as filters, heating and cooling coils, in the plant dp Pa column in cell V50.

17. Enter the starting total pressure p_t Pa for node 1 in cell X50. This will normally be atmospheric pressure, 0 Pa gauge pressure.
18. Each line is self-contained, and calculations take place whenever new data is entered.
19. The total pressure at each node is displayed in cells X50 and Y50.
20. The static pressure at each node is displayed in cells AA50 and AB50.
21. Enter a line of data for each section of duct.
22. Check that zero is present in the cells for duct length L, air flow quantity Q and fittings velocity pressure loss factor k, where there is no ductwork in that row. Check that zero pressure drop is displayed for that section of ductwork.
23. Save the spreadsheet into a unique filename such as EX003. This preserves the original worksheet for future use.
24. Make the supply air outlet for the main duct route at node number 40 at line 110.
25. Enter the data for branches A to E as for the main route.
26. If the original worksheet does not have enough sections, copy rows down the screen to make more. Edit the new rows to ensure that the correct cell references and formulae are maintained. New users should be careful when making such changes (to a copy of the original file).

OUTPUT DATA

1. The current date is given in cell B7.
2. The current time is given in cell B8.
3. The air density for the temperature and pressure is stated in cells C14 and C15.
4. The air density given in cell D50 is that to be used in line number 50.
5. The first estimate of duct diameter is given in cell I50.
6. The maximum carrying capacity of the estimated duct diameter is given in cell J50.
7. The difference between the design air flow and the maximum carrying capacity of the duct is expressed as a percentage of the design flow in cell K50. This percentage is to be minimized by changing the maximum pressure drop rate in cell H50 until either zero error has been achieved, or as small an error as possible is obtained while using only two decimal places for the pressure drop. An error of less than $\pm 1\%$ is normally achievable. An error of $\pm 5\%$ should not be exceeded.
8. The air velocity that will be produced in the likely duct diameter from cell L50 is shown in cell M50.
9. The equivalent diameter of the rectangular duct selected in cells N50 and O50 is shown in cell P50.
10. The air velocity of the duct selected is shown in cell Q50.
11. The air velocity pressure is given in cell R50.
12. The air pressure drop through the straight duct is given in cell T50.
13. The air pressure drop through the duct fittings is given in cell U50.
14. The frictional resistance of the section of duct is given in cell W50.
15. The total air pressure at the commencement of the duct is given in cell X50.
16. The total air pressure at the end of the duct is given in cell Y50.
17. The static air pressure at the commencement of the duct is given in cell AA50.

18. The static air pressure at the end of the duct is given in cell AB50.
19. The air velocity pressure is repeated in cell AC50 from cell R50 for information only.
20. The node numbers and the description of the duct section are repeated in cells AE50 and AF50 for information.
21. The same information is produced for each line of duct section and branch.
22. The pressure drop through the main run of ductwork between node 1 and node 40 is shown in cell W112. It is not necessary to read this cell.
23. The air pressure drop through each section of duct is displayed in the cell range C228 to C238.
24. The pressure drop through the duct route from nodes 1 to 40 is displayed in cell D233.
25. The pressure drop from node 1 to the node at the end of each branch duct is shown in the cell range D234 to D238.
26. When the pressure drop through a branch duct is less than that which is available at the commencement of the branch, the excess pressure is displayed in cells E233 to E38. This is the pressure drop that is needed across the volume control damper in order to remove the excess air pressure.
27. Cells F233 to F238 display the duct route that has the highest resistance to air flow. This is the index route. The word 'index' is displayed.
28. The air pressure drop through the index duct route is given in cell D240.
29. The index route of duct nodes is given in cell F240.
30. The fan total pressure is given in cell E245.
31. The fan velocity pressure is given in cell E246.
32. The fan static pressure is given in cell E247.
33. The air volume flow rate through the fan is given in cell E248.
34. The air temperature at the fan is given in cell E249.
35. The air density at the fan is given in cell E250.

FORMULAE

Cell B7

@TODAY

The cell is formatted for the @TODAY function to display the current date.

Cell B8

@TODAY

The cell is formatted for the @TODAY function to display the current time.

Cell C17

1.1906*293*C15/101325/(273+C14)

The density of the air is calculated. This density is for reference only.

$$\text{density, } \rho = 1.1906 \, \frac{\text{kg}}{\text{m}^3} \times \frac{(273 + 20)\text{K}}{(273 + t)\text{K}} \times \frac{b \, \text{Pa}}{101 \, 325 \, \text{Pa}}$$

where b = barometric air pressure (Pa), and t = air temperature (°C d.b.).

$$\text{C17} = 1.1906 \, \frac{293}{273 + \text{C14}} \times \frac{\text{C15}}{101325} \, \frac{\text{kg}}{\text{m}^3}$$

Cell D50

1.1906*293*$C$15/101325/(273+C50)

The air density to be used in that section of duct is calculated for each line of the worksheet. The standard atmospheric pressure is taken from cell C15, because this would normally be common to all the duct system.

$$\text{D50} = 1.1906 \, \frac{293}{273 + \text{C50}} \times \frac{\$\text{C}\$15}{101325} \, \frac{\text{kg}}{\text{m}^3}$$

Cell I50

1000*@SQRT(F50*4/G50/@PI)

Finds the duct diameter to carry the design air flow.

$$d = 1000 \times \sqrt{\left(\frac{4Q}{\pi v}\right)} \, \text{mm}$$

$$\text{I50} = 1000 \times \sqrt{\left(\frac{4 \times \text{F50}}{\pi \times \text{G50}}\right)} \, \text{mm}$$

EXAMPLE 7.8

Calculate the diameter of an air duct that will pass 2.25 m³/s of air at a velocity that does not exceed 5 m/s.

$$d = 1000 \times \sqrt{\left(\frac{4Q}{\pi v}\right)} \, \text{mm}$$

$$Q = 2.25 \, \text{m}^3/\text{s}$$

$$v = 5 \, \text{m/s}$$

$$d = 1000 \times \sqrt{\left(\frac{4 \times 2.25}{\pi \times 5}\right)} \, \text{mm}$$

$$= 757 \, \text{mm}$$

Cell J50

−2.027*SQRT(H50)*(I50)^2.5*LOG(((4.05*10^−5)/I50)+(2.933*10^−5)/SQRT(H50)/I50^1.5)

This equation calculates the maximum carrying capacity of an air duct from the duct diameter d m, and the pressure drop rate dp Pa/m. The flow of air at 20 °C d.b., 43% percentage saturation and a barometric pressure of 1013.25 mbar through clean galvanized sheet metal ducts having joints made in accordance with good practice is given by

$$Q = -2.0278 \times dp^{0.5} \times d^{2.5} \times \log\left(\frac{4.05 \times 10^{-5}}{d} + \frac{2.933 \times 10^{-5}}{dp^{0.5} \times d^{1.5}}\right)$$

where Q = air flow rate (m³/s); dp = pressure loss rate (Pa/m) (cell H50); and d = duct internal diameter (m) (cell I50).

$$J50 = -2.0278 \times (H50)^{0.5} \times (I50)^{2.5} \times \log\left[\frac{4.05 \times 10^{-5}}{I50} + \frac{2.933 \times 10^{-5}}{(H50)^{0.5} \times (I50)^{1.5}}\right]$$

EXAMPLE 7.9

Find the maximum air-carrying capacity and air velocity in a 350 mm diameter galvanized sheet steel duct at 20 °C d.b. when the pressure loss rate is 0.85 Pa/m.

$d = 0.35\,\text{m}$

$dp = 0.85\,\text{Pa/m}$

$Q = \text{air flow (m}^3/\text{s})$

$$Q = -2.0278 \times dp^{0.5} \times d^{2.5} \times \log\left(\frac{4.05 \times 10^{-5}}{d} + \frac{2.933 \times 10^{-5}}{dp^{0.5} \times d^{1.5}}\right)$$

$$= -2.0278 \times 0.85^{0.5} \times 0.35^{2.5} \times \log\left(\frac{4.05 \times 10^{-5}}{0.35} + \frac{2.933 \times 10^{-5}}{0.85^{0.5} \times 0.35^{1.5}}\right)$$

$$= -0.1355 \times \log(2.6935 \times 10^{-4})$$

$$= -0.1355 \times -3.57$$

$$= 0.484\,\text{m}^3/\text{s}$$

EXAMPLE 7.10

A 500 mm internal diameter galvanized sheet steel duct is carrying air at 20 °C d.b. at a design pressure loss rate of 1.0 Pa/m. Calculate the air volume flow rate being passed and the air velocity.

$d = 0.50\,\text{m}, \quad dp = 1.0\,\text{Pa/m}$

$$Q = -2.0278 \times 1.0^{0.5} \times 0.50^{2.5} \times \log\left(\frac{4.05 \times 10^{-5}}{0.50} + \frac{2.933 \times 10^{-5}}{1.0^{0.5} \times 0.50^{1.5}}\right)$$

$$= 1.357\,\text{m}^3/\text{s}$$

$$\text{Air velocity } v = 1.357\frac{\text{m}^3}{\text{s}} \times \frac{4}{\pi \times 0.5^2\,\text{m}^2}$$

$$= 6.91\,\text{m/s}$$

Cell K50

(F50−J50)*100/F50

Calculates the percentage error between the design air flow Q m³/s and the maximum carrying capacity of the duct Q_1 m³/s at the current air pressure drop rate dp/l Pa/m. The user enters new values for the duct air pressure drop rate to minimize the error. When the user is satisfied with the error, the current value of air pressure drop rate is used to calculate the air pressure drop along the straight ducts.

$$\text{error} = \frac{Q - Q_1}{Q} \times 100\%$$

$$K50 = \frac{F50 - J50}{F50} \times 100\%$$

Cell M50

+F50*4/@PI/(L50/1000)^2

Calculates the air velocity that will be produced for the duct diameter shown in cell L50.

$$\text{air velocity } v = Q \frac{\text{m}^3}{\text{s}} \times \frac{4}{\pi d^2 \, \text{m}^2}$$

$$M50 = F50 \frac{\text{m}^3}{\text{s}} \times \frac{4}{\pi} \times \left(\frac{1000}{L50}\right)^2 \frac{1}{\text{m}^2}$$

Cell P50

1.265*((N50*O50)^3/(N50+O50))^0.2

Calculates the equivalent circular diameter of the rectangular duct selected in cells N50 and O50 for equal volume, pressure drop rate and surface roughness (CIBSE, 1986c).

$$d = 1.265 \times \left[\frac{(ab)^3}{a + b}\right]^{0.2} \text{mm}$$

where a = duct width (mm); b = duct depth (mm); and d = equivalent duct diameter (mm).

$$P50 = 1.265 \times \left[\frac{(N50 \times O50)^3}{N50 + O50}\right]^{0.2} \text{mm}$$

Cell Q50

+F50*4/@PI/(P50/1000)^2

Calculates the air velocity in the circular equivalent of the rectangular duct selected.

$$\text{air velocity } v = Q\frac{\text{m}^3}{\text{s}} \times \frac{4}{\pi d^2\,\text{m}^2}$$

$$M50 = F50\frac{\text{m}^3}{\text{s}} \times \frac{4}{\pi} \times \left(\frac{1000}{P50}\right)^2\frac{1}{\text{m}^2}$$

EXAMPLE 7.11

A 700 mm wide and 600 mm deep galvanized steel air duct is to carry 2200 l/s of air at 20 °C d.b. at a maximum velocity of 6 m/s and at a pressure loss rate not exceeding 0.72 Pa/m. Calculate the equivalent diameter of the duct, the air velocity, and the error between the design air flow and the maximum air-carrying capacity of the duct.

$$\text{design air flow } Q = 2.2\,\text{m}^3/\text{s}$$

The equivalent diameter of a 700 mm × 600 mm duct is

$$d = 1.265 \times \left[\frac{(ab)^3}{a+b}\right]^{0.2}\text{mm}$$

$$= 1.265 \times \left[\frac{(700 \times 600)^3}{700 + 600}\right]^{0.2}\text{mm}$$

$$= 713\,\text{mm}$$

The air velocity in the duct is

$$v = Q\frac{\text{m}^3}{\text{s}} \times \frac{4}{\pi d^2\,\text{m}^2}$$

$$= 2.2\frac{\text{m}^3}{\text{s}} \times \frac{4}{\pi\,0.713^2\,\text{m}^2}$$

$$= 5.51\,\text{m/s}$$

The maximum air-carrying capacity of a 713 mm diameter duct at a pressure drop rate of 0.72 Pa/m is

$$Q_1 = -2.0278 \times dp^{0.5} \times d^{2.5} \times \log\left(\frac{4.05 \times 10^{-5}}{d} + \frac{2.933 \times 10^{-5}}{dp^{0.5} \times d^{1.5}}\right)$$

$$= -2.0278 \times 0.72^{0.5} \times 0.713^{2.5} \times \log\left(\frac{4.05 \times 10^{-5}}{0.713} + \frac{2.933 \times 10^{-5}}{0.72^{0.5} \times 0.713^{1.5}}\right)$$

$$= -0.7386 \times \log(1.1422 \times 10^{-4})$$

$$= 2.912\,\text{m}^3/\text{s}$$

$$\text{error} = \frac{Q - Q_1}{Q} \times 100\%$$

$$= \frac{2.2 - 2.91}{2.2} \times 100\%$$

$$= -32.3\%$$

It can be seen that the duct is carrying 32.3% less than its maximum capacity; this is indicated by the negative result. A pressure drop rate of 0.43 Pa/m produces a maximum carrying capacity of $2.22\,\text{m}^3/\text{s}$ and an error of -0.9%; this is as close as can be found from an accuracy of two decimal places.

Cell R50

0.5*D50*Q50*Q50

The air velocity pressure p_v Pa is given by

$$p_v = 0.5\rho v^2 \,\text{Pa}$$
$$R50 = 0.5 \times D50 \times Q50 \times Q50 \,\text{Pa}$$
$$D50 \text{ density kg/m}^3$$
$$Q50 \text{ velocity m/s}$$

Cell T50

+E50*H50

Air pressure drop dp_1 Pa in the straight duct length is

$$dp_1 = dp\frac{\text{Pa}}{\text{m}} \times L\,\text{m}$$
$$T50 = H50\frac{\text{Pa}}{\text{m}} \times E50\text{m}$$

Cell U50

+S50*R50

Air pressure drop through the duct fittings dp_2 Pa is

$$dp_2 = kp_v \,\text{Pa}$$
$$R59 = S50 \times R50 \,\text{Pa}$$

Cell W50

+T50+U50+V50

The sum of the duct pressure drop dp_1, duct fittings resistance dp_2 and the plant pressure drop dp_3 for that section of duct is

$$dp = dp_1 + dp_2 + dp_3 \,\text{Pa}$$
$$W50 = T50 + U50 + V50 \,\text{Pa}$$

Cell X51

+Y50

The total air pressure at the commencement of the duct section. This is normally zero for node 1. Node 2 has a total pressure that is equal to the total air pressure at the end of section 1. The content of the previous duct end total pressure is copied into cell X51.

Cell Y50

+X50−W50

The air total pressure at the end of this section of duct is equal to the total pressure at the beginning of the duct less the frictional resistance of the section:

$$p_{t2} = p_{t1} - dp_{1-2} \, \text{Pa}$$
$$Y50 = X50 - W50 \, \text{Pa}$$

Cell AA51

+AB50

The static air pressure at the start of the duct section, p_s Pa, is either:

1. zero, if this is the first node in the duct system and is at atmospheric air pressure; or
2. the room air static pressure, for example up to ±50 Pa depending upon how the space is pressurized; or
3. equal to the static air pressure at the end of the preceding section of duct.

Cell AB50

+Y50−R50

The static air pressure p_s Pa at the end of the duct is

$$p_{s2} = p_{t2} - p_{v2} \, \text{Pa}$$
$$AB50 = Y50 - R50 \, \text{Pa}$$

Cell AC50

+R50

A copy of the velocity pressure in the section for information only.

Cell AE50

+A50

Copies the node numbers that are used for the duct section.

Cell AF50

+B50

Copies the section description for reference only.

Cell Y60

+X60+D240

The total air pressure at the air discharge from the fan p_{t2} is the total pressure at the fan inlet node p_{t1} plus the fan total pressure rise *FTP*.

$$p_{t1} = \text{total pressure at fan inlet Pa}$$
$$FTP = \text{fan total pressure rise Pa}$$
$$= \text{loss of total pressure through the ductwork } d_p \text{ Pa}$$
$$p_{t2} = p_{t1} + FTP \text{ Pa}$$
$$Y60 = X60 + D240 \text{ Pa}$$

Cell Y110

+X110−W110+R110

The total pressure of the air that is discharged from the air outlet grille or diffuser at the end of the duct route into the air conditioned or ventilated space is equal to the total pressure at the final duct node minus the pressure drop through the grille and minus one velocity pressure. There is one velocity pressure loss factor for the discharge of air from a duct. This produces a final total pressure of zero in the ventilated space; this is not correct, because the discharge air velocity pressure is produced by the total pressure. It is the final static air pressure that is zero, or whatever is held within the ventilated space. The final total air pressure must contain the discharge air velocity and zero, or other value, static pressure. The final velocity pressure is added back into the total pressure because it was taken out by the velocity pressure loss factor.

$$p_{t2} = p_{t1} - d_{p1-2} + p_{v2} \text{ Pa}$$
$$Y110 = X110 - W110 + R110 \text{ Pa}$$

EXAMPLE 7.12

The total pressure at node 1 in a duct system is 235 Pa above the atmospheric air pressure. The duct is 25 m long, 600 mm in diameter, and carries 1400 l/s of air at a density of 1.2 kg/m^3. The air pressure loss rate through the straight duct is 0.5 Pa/m, and the duct fittings have a combined velocity pressure loss factor of 3.6. The duct has an air filter and a heating coil, which have a combined resistance to air flow of 125 Pa. Calculate the total and static air pressures at the end of the duct.

$$p_{t1} = 235 \text{ Pa}$$

The air velocity in the duct is

$$v_1 = Q \frac{\text{m}^3}{\text{s}} \times \frac{4}{\pi d^2 \, \text{m}^2}$$
$$= 1.4 \frac{\text{m}^3}{\text{s}} \times \frac{4}{\pi \, 0.6^2 \, \text{m}^2}$$
$$= 4.95 \, \text{m/s}$$
$$p_{v1} = 0.5 \rho v_1^2 \text{ Pa}$$
$$= 0.5 \times 1.2 \times 4.95^2 \text{ Pa}$$
$$= 15 \text{ Pa}$$

$$dp_{1-2} = dp\frac{\text{Pa}}{\text{m}} \times L\,\text{m}$$

$$= 0.5\frac{\text{Pa}}{\text{m}} \times 25\,\text{m}$$

$$= 13\,\text{Pa}$$

Air pressure drop through the duct fittings:

$$dp = kp_{v1}\,\text{Pa}$$

$$= 3.6 \times 15\,\text{Pa}$$

$$= 54\,\text{Pa}$$

Loss of total pressure through the duct:

$$dp_{1-2} = (\text{duct} + \text{fittings} + \text{plant})\,\text{Pa}$$

$$= 13 + 54 + 125\,\text{Pa}$$

$$= 192\,\text{Pa}$$

Total pressure at node 2:

$$p_{t2} = p_{t1} - dp_{1-2}\,\text{Pa}$$

$$= 235 - 192\,\text{Pa}$$

$$= 43\,\text{Pa}$$

$$p_{s1} = p_{t1} - p_{v1}\,\text{Pa}$$

$$= 235 - 15\,\text{Pa}$$

$$= 220\,\text{Pa}$$

$$p_{s2} = p_{t2} - p_{v2}\,\text{Pa}$$

$$= 43 - 15\,\text{Pa}$$

$$= 28\,\text{Pa}$$

EXAMPLE 7.13

Part of a ventilation duct system has three nodes. Node 1 is within the duct system, some distance before node 2, which is at the entry into a discharge air diffuser. The ventilated room is maintained at atmospheric air pressure. Node 3 is in the ventilated room after the diffuser. The total pressure at node 1 in the duct system is 115 Pa above the atmospheric air pressure. The duct has pressure losses that amount to 80 Pa between nodes 1 and 2. The air velocity at node 1 is 6 m/s and that at node 2 is 3 m/s. The velocity pressure loss factor for the discharge of air from the end of the duct, k, is 1, and the diffuser has a resistance of 30 Pa. The air velocity in the room is zero. The air density is 1.2 kg/m³. Calculate the total and static air pressures at the end of the duct.

Sketch the duct system, identify the nodes and sketch the pressure gradients.

$$p_{t1} = 115\,\text{Pa}$$

$$v_1 = 6\,\text{m/s}$$

$$p_{v1} = 0.5 \times 1.2 \times 6^2\,\text{Pa}$$

$$= 22\,\text{Pa}$$

$$dp_{1-2} = 80\,\text{Pa}$$
$$p_{t2} = p_{t1} - dp_{1-2}\,\text{Pa}$$
$$= 115 - 80\,\text{Pa}$$
$$= 35\,\text{Pa}$$
$$p_{s1} = p_{t1} - p_{v1}\,\text{Pa}$$
$$= 115 - 22\,\text{Pa}$$
$$= 93\,\text{Pa}$$
$$v_2 = 3\,\text{m/s}$$
$$v_3 = 3\,\text{m/s}$$
$$p_{v2} = 0.5 \times 1.2 \times 3^2\,\text{Pa}$$
$$= 5\,\text{Pa}$$
$$p_{v3} = 5\,\text{Pa}$$
$$p_{s2} = p_{t2} - p_{v2}\,\text{Pa}$$
$$= 35 - 5\,\text{Pa}$$
$$= 30\,\text{Pa}$$

Air pressure drop through the diffuser and due to the discharge of air from the end of the duct is

$$dp_{2-3} = 30 + kp_{v2}\,\text{Pa}$$
$$= 30 + 1 \times 5\,\text{Pa}$$
$$= 35\,\text{Pa}$$

The final air velocity on the discharge side, ignoring the diffusing air stream velocity within the room, is that through the diffuser:

$$v_3 = 3\,\text{m/s}$$
$$p_{v3} = 5\,\text{Pa}$$

The air pressure loss from node 2 to node 3 includes the discharge air velocity pressure. The total pressure of the discharge air is the velocity pressure p_{v3}, and it is not zero, so:

$$p_{t3} = p_{t2} - dp_{2-3} + p_{v3}\,\text{Pa}$$
$$= 35 - 35 + 5\,\text{Pa}$$
$$= 5\,\text{Pa}$$
$$p_{s3} = p_{t3} - p_{v3}\,\text{Pa}$$
$$= 5 - 5\,\text{Pa}$$
$$= 0\,\text{Pa}$$

Cell C228

@SUM(W50.W79)

The air pressure drop from node 1 to node 23 is found by the addition of the drop of total pressure in each section. This calculation is repeated for the other sections of duct in the main route from node 23 to the final discharge of air into the ventilated space at node 40, in cells C229 to C233.

$$dp_{1-23} = dp_{1-2} + \ldots + dp_{22-23}\,\mathrm{Pa}$$
$$\mathrm{C228} = \mathrm{W50} + \ldots + \mathrm{W79}\,\mathrm{Pa}$$

Cell C234

@SUM(W133..W137)

The air pressure drop from the beginning of the first branch duct at node 23 to the final discharge of air into the ventilated space at node 54 is found by the addition of the drop of total pressure in each section:

$$dp_{23-54} = dp_{23-50} + \ldots + dp_{53-54}\,\mathrm{Pa}$$
$$\mathrm{C234} = \mathrm{W133} + \ldots + \mathrm{W137}\,\mathrm{Pa}$$

This calculation is repeated for the other duct branches at nodes 26, 29, 32 and 35 in cells C235 to C238.

Cell D233

@SUM(C228..C233)

This cell accumulates the drop of total pressure along the main through-route of ductwork from node 1 to node 40:

$$dp_{1-40} = dp_{1-23} + \ldots + dp_{35-40}\,\mathrm{Pa}$$
$$\mathrm{D233} = \mathrm{C228} + \mathrm{C229} + \mathrm{C230} + \mathrm{C231} + \mathrm{C232} + \mathrm{C233}\,\mathrm{Pa}$$

Cell D234

+C228+C234

This adds the air pressure drop from node 1 to node 23, the main duct route to the first branch duct, to the pressure drop through the first branch duct:

$$dp_{1-54} = dp_{1-23} + dp_{23-54}\,\mathrm{Pa}$$
$$\mathrm{D234} = \mathrm{D228} + \mathrm{C234}\,\mathrm{Pa}$$

Cell D235

+ C228 + C229 + C235

This adds the air pressure drop from node 1 to node 26, the main duct route to the second branch duct, to the pressure drop through the second branch duct:

$$dp_{1-64} = dp_{1-23} + dp_{23-26} + dp_{26-64}\,\mathrm{Pa}$$
$$\mathrm{D235} = \mathrm{C228} + \mathrm{C229} + \mathrm{C235}\,\mathrm{Pa}$$

Cell D236

+ C228 + C229 + C230 + C236

This adds the air pressure drop from node 1 to node 29, the main duct route to the third branch duct, to the pressure drop through the third branch duct:

$$dp_{1-74} = dp_{1-23} + dp_{23-26} + dp_{26-29} + dp_{29-74}\,\mathrm{Pa}$$
$$\mathrm{D236} = \mathrm{C228} + \mathrm{C229} + \mathrm{C230} + \mathrm{C236}\,\mathrm{Pa}$$

Cell D237

+ C228 + C229 + C230 + C231 + C237

This adds the air pressure drop from node 1 to node 32, the main duct route to the fourth branch duct, to the pressure drop through the fourth branch duct:

$$dp_{1-84} = dp_{1-23} + dp_{23-26} + dp_{26-29} + dp_{29-32} + dp_{32-84}\,\mathrm{Pa}$$
$$D237 = C228 + C229 + C230 + C231 + C237\,\mathrm{Pa}$$

Cell D238

+ C228 + C229 + C230 + C231 + C232 + C238

This adds the air pressure drop from node 1 to node 35, the main duct route to the fifth branch duct, to the pressure drop through the fifth branch duct:

$$dp_{1-94} = dp_{1-23} + dp_{23-26} + dp_{26-29} + dp_{29-32} + dp_{32-35} + dp_{35-94}\,\mathrm{Pa}$$
$$D238 = C228 + C229 + C230 + C231 + C232 + C238\,\mathrm{Pa}$$

Cells G221 to J239

There is no need for the user to read these cells once their function is understood. A series of comparisons are made in order to discover which of the combinations of main and branch ducts forms the highest-resistance path. This path is known as the index circuit. The index circuit pressure drop is the one that creates the highest pressure drop for the fan to overcome. The fan total pressure rise is equal to the drop of pressure through the index route.

Cell H233

@IF($D$233>$D$234,D233,D234)

This formula is a conditional 'IF' statement, or logical equation. The contents of cell D233 are compared with the contents of cell D234. Cell D233 contains the air pressure drop from node 1 to node 40; this is the main duct route, and it may be the index circuit. Cell D234 contains the air pressure drop from node 1 to node 54. This is the route from the first node to the discharge of air from the first branch duct; it may be the index circuit. If the route from node 1 to node 40 has a greater pressure drop than that from node 1 to node 54 at the end of the first branch, the conditional 'IF' statement is proved to be true: that is, the content of cell D233 has a greater value than that of cell D234, in which case the next cell reference in the formula is used, D233. If the statement is not true – that is, D234 is more than D233 – then the result is negative and the next cell reference is used, D234. The explanation of this formula can be confirmed by reference to the spreadsheet instruction manual or on-screen help menu. This formula selects the higher of the two pressure drops and holds that value in cell H233. This process is repeated with the other possible index circuit combinations until the final value selected is the index pressure loss.

Cell H235

@IF($H$233›$D$235,H233,D235)

The contents of cell H233 are compared with the contents of cell D235. The higher value is held in cell H235.

Cell H236

@IF($H$235›$D$236,H235,D236)

The contents of cell H235 are compared with the contents of cell D236. The higher value is held in cell H236.

Cell H237

@IF($H$236›$D$237,H236,D237)

The contents of cell H236 are compared with the contents of cell D237. The higher value is held in cell H237.

Cell H238

@IF($H$237›$D$238,H237,D238)

The contents of cell H237 are compared with the contents of cell D238. The higher value is held in cell H238. Cell H238 now contains the highest value of air pressure drop between entry node 1 and the air discharge node from the system. This is the index circuit air flow resistance.

Cell I233

@IF ($D$233›$D$234,"1 −40","branch A")

The air pressure drops in cells D233 and D234 are compared. If D233 contains a higher value, the text contained within the first set of quotation marks is written in cell I233. If cell D234 contains a larger number than that in cell D233, the text within the second set of quotation marks is put into cell I233. Cell I233 now holds the description of the duct route that has the greater air pressure drop. This is part of the index circuit. The choice is between the duct route from node 1 to node 40, and that from node 1 to branch A.

Cell I235

@IF ($H$233›$D$235,I233,"branch B")

The air pressure drops in cells H233 and D235 are compared. Cell I235 now holds the description of the duct route that has the greater air pressure drop. The choice is between the duct route from node 1 to node 40, and that from node 1 to branch B.

Cell I236

@IF ($H$235>$D$236,I235,"branch C")

The air pressure drops in cells H235 and D236 are compared. Cell I236 now holds the description of the duct route that has the greater air pressure drop. The choice is between the duct route from node 1 to node 40, and that from node 1 to branch C.

Cell I237

@IF ($H$236>$D$237,I235,"branch D")

The air pressure drops in cells H236 and D237 are compared. Cell I237 now holds the description of the duct route that has the greater air pressure drop. The choice is between the duct route from node 1 to node 40, and that from node 1 to branch D.

Cell I238

@IF ($H$237>$D$238,I237,"branch E")

The air pressure drops in cells H237 and D238 are compared. Cell I238 now holds the description of the duct route that has the greater air pressure drop. The choice is between the duct route from node 1 to node 40, and that from node 1 to branch E.

Cell D240

+H238−X50

Cell H238 contains the index circuit air pressure drop. Cell X50 contains the total air pressure at node 1; this will normally be zero, because it is at atmospheric air pressure. Cell D240 now contains the index circuit air pressure drop that is copied from cell H238.

Cell F240

+I238

The contents of cell I238 are copied from cell I238 into cell F240. Cell F240 contains the description of the index circuit.

Cell E245

+D240

The index circuit air pressure drop is copied from cell D240 into cell E245 and is listed with the fan performance specification as the fan total pressure.

Cell E246

+R60

The velocity pressure of the air discharging from the fan is copied from cell R60 into cell E246, and is listed with the fan performance specification as fan velocity pressure.

Cell E247

+E245−E246

The fan static pressure rise is the fan total pressure rise in cell E245 minus the fan velocity pressure in cell E246:

$$FSP = FTP - FVP \, \text{Pa}$$
$$E247 = E245 - E246 \, \text{Pa}$$

Cell E248

+F60

The air volume flow rate through the fan is copied from cell F60 into the fan performance specification list.

Cell E249

+C60

The air temperature at the fan is copied from cell C60 into the fan performance specification list.

Cell E250

+D60

The air density at the fan is copied from cell D60 into the fan performance specification list.

Cell E233

+D240−D233

The difference between the air pressure drop through the index circuit and that through the main duct route from node 1 to node 40 is found. This difference is the excess pressure at the commencement of the duct or branch. When this difference is 0 Pa, this is the index route. Any surplus air pressure that is available at a branch can be used to reduce the duct sizes, provided that the air velocity limits are not exceeded, or it must be lost by being absorbed through a balancing volume control damper:

$$\text{damper } dp = \text{index } dp - dp_{1-40} \, \text{Pa}$$
$$E240 = D240 - D233 \, \text{Pa}$$

Cell E234

+D240−D234

The excess air pressure at node 50 for branch A is calculated:

$$\text{damper } dp = \text{index } dp - dp_{1-54} \text{ Pa}$$
$$E234 = D240 - D234 \text{ Pa}$$

Cell E235

+D240−D235

The excess air pressure at node 60 for branch B is calculated:

$$\text{damper } dp = \text{index } dp - dp_{1-64} \text{ Pa}$$
$$E235 = D240 - D235 \text{ Pa}$$

Cell E236

+D240−D236

The excess air pressure at node 70 for branch C is calculated:

$$\text{damper } dp = \text{index } dp - dp_{1-74} \text{ Pa}$$
$$E236 = D240 - D236 \text{ Pa}$$

Cell E237

+D240−D237

The excess air pressure at node 80 for branch D is calculated:

$$\text{damper } dp = \text{index } dp = dp_{1-84} \text{ Pa}$$
$$E237 = D240 - D237 \text{ Pa}$$

Cell E238

+D240−D238

The excess air pressure at node 90 for branch E is calculated:

$$\text{damper } dp = \text{index } dp - dp_{1-94} \text{ Pa}$$
$$E238 = D240 - D238 \text{ Pa}$$

Questions

The worksheet is needed from question 12 onwards.

1. State the measurements of air pressure that are needed in the calculation of air duct sizes. Explain how these air pressures are used in the design calculations.

2. Explain, with the aid of sketches, how the three airway pressures are measured. List all the equipment that would be needed. State how each item would be used.

3. Explain how the volume flow rate and mass flow rate of air are found from site measurements on an operational ventilation ductwork system. List the measurements that are necessary. State the

formulae that would be needed, their method of use, and their units of measurement.

4. Sketch air pressure gradient graphs of the airway pressures changing along a duct of length L m and constant diameter d mm, for the following cases.

   (a) Total and static pressures are above atmospheric.
   (b) Total pressure is above atmospheric, but static pressure is below atmospheric.
   (c) The duct is on the suction side to a fan, and air total pressure is below atmospheric pressure.
   (d) The duct tapers from 1 m diameter to 350 mm diameter along length L m. Pressures remain above atmospheric pressure.
   (e) The duct enlarges from 400 mm diameter to 950 mm diameter while the total pressure remains below atmospheric pressure.
   (f) A 300 mm diameter duct is above atmospheric pressure. The commissioning engineer omitted to seal a test hole halfway along the length of the duct.
   (g) Room air returns to the air-handling plant through ducts that have inadequate joint sealing along their entire length, causing significant leakage. Total pressure at the commencement of the duct is above atmospheric. Static pressure within the duct starts at below atmospheric pressure.

5. Calculate the density of air for a temperature of 5 °C d.b. when the atmospheric pressure is 101 210 Pa.

6. Calculate the temperature of air that corresponds to a density of 1.05 kg/m³ at standard atmospheric pressure.

7. Calculate the total pressure of air flowing at 2250 l/s in a 950 mm internal diameter air duct when the air temperature is 38 °C d.b. and the static pressure of the air is at −125 Pa relative to the atmospheric air pressure of 101 450 Pa.

8. 5.5 m³/s flows through a 1300 mm diameter duct. Calculate the air velocity.

9. Calculate the carrying capacities of air ducts of 250 mm, 400 mm, 950 mm and 1200 mm diameter when the maximum allowable air velocity is 6 m/s.

10. Air flowing in a 750 mm diameter duct was 12 °C d.b. on a day when the atmospheric pressure was 101 280 Pa. The static pressure of the air in the duct was 120 Pa above atmospheric. The average air velocity was measured as 5.5 m/s. Calculate the density, velocity pressure and total pressure of the air.

11. An air flow of 1750 l/s at a temperature of 35 °C d.b. passes through a 650 mm diameter duct at a static pressure of 60 Pa below the atmospheric pressure of 101 300 Pa. Calculate the air density, velocity and total pressures.

12. Calculate the pressures that occur when 3500 l/s flow through a 30 m long, 900 mm diameter duct that then reduces to 750 mm diameter and remains at 750 mm for 15 m. The air total pressure at the commencement of the 900 mm duct is 320 Pa above atmospheric. The reducer is the 60° concentric type. The air density is 1.18 kg/m³. The frictional pressure loss rates are 0.3 Pa/m in the 900 mm diameter duct and 0.85 Pa/m in the 750 mm diameter duct.

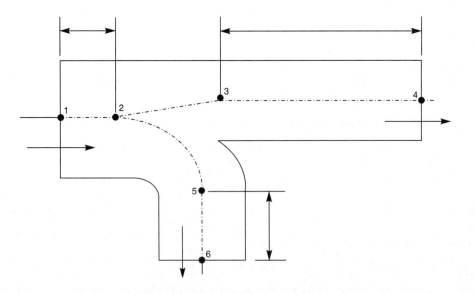

Fig. 7.5   Notation for the pressure analysis of a duct branch in question 16.

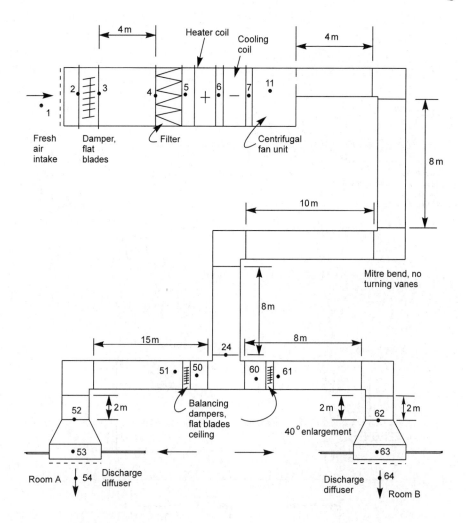

Fig. 7.6 Supply air ductwork for question 19.

13. Calculate the pressures that occur when 6 m³/s flow through an 8 m long, 1000 mm diameter duct that then reduces to 700 mm diameter and remains at 700 mm for 12 m. The air total pressure at the commencement of the 1000 mm duct is 800 Pa above atmospheric. The reducer is the 45° concentric type. The air density is 1.21 kg/m³. The frictional pressure loss rates are 0.5 Pa/m in the 1000 mm diameter duct and 3.2 Pa/m in the 800 mm diameter duct.

14. Calculate the static regain and pressure changes that occur when 1.2 m³/s flow through a 25 m long, 400 mm diameter duct that then enlarges to 600 mm diameter and remains at 600 mm for 35 m. The air total pressure at the commencement of the 400 mm duct is 200 Pa above atmospheric. The enlarger is the 30° concentric type. The air density is 1.2 kg/m³. The frictional pressure loss

rates are 2.6 Pa/m in the 400 mm diameter duct and 0.3 Pa/m in the 600 mm diameter duct.

15. A 550 mm diameter duct supply air duct enlarges at 40° into a 1200 mm × 1000 mm plenum chamber containing filters. Calculate the static regain and pressure changes that occur when 2.75 m³/s flows through the 10 m long, 550 mm diameter duct, enlarges, and then flows through the plenum for 2 m. The air total pressure at the commencement of the 550 mm duct is 25 Pa below atmospheric. The air density is 1.19 kg/m³. The frictional pressure loss rate is 2.3 Pa/m in the 550 mm diameter duct and 0.06 Pa/m in the plenum.

16. Calculate the duct pressures at the nodes of the air duct branch shown in Fig. 7.5, from the following data and velocity pressure loss factors in Tables 7.15 and 7.16:

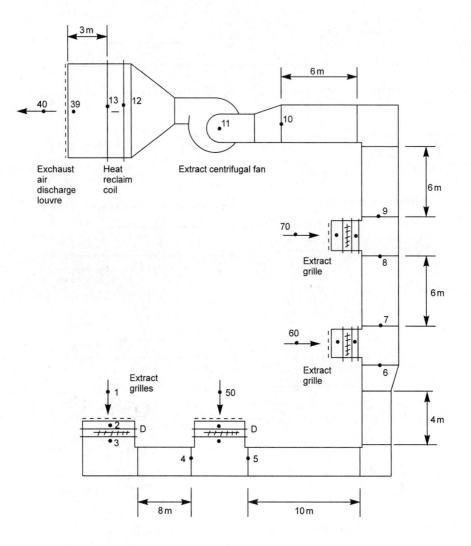

Fig. 7.7   Exhaust air duct system for question 20.

900 mm diameter duct 1–2 is 16 m long and carries 4 m³/s.

500 mm diameter duct 3–4 is 25 m long and carries 2 m³/s.

700 mm diameter duct 5–6 is 8 m long and carries 2 m³/s.

The branch offtake is a short-radius bend.

The straight-through contraction has a velocity pressure loss factor of 0.05. Air density is 1.2 kg/m³.

Pressure drop rates are:

duct 1–2	0.4 Pa/m
duct 3–4	2.1 Pa/m
duct 5–6	0.4 Pa/m

The total pressure at node 1 is 250 Pa.

17. Calculate the duct pressures at the nodes of the air duct branch from the following data, and velocity pressure loss factors in Tables 7.15 and 7.16:

1000 mm × 900 mm duct 1–2 is 15 m long and carries 4 m³/s.

800 mm × 700 mm duct 3–4 is 10 m long and carries 2.5 m³/s.

600 mm × 500 mm duct 5–6 is 6 m long and carries 1.5 m³/s.

The branch offtake is a right-angled bend having several turning vanes. The vanes are in the branch duct. Air density is 1.2 kg/m³.

Pressure drop rates are:

duct 1–2	0.23 Pa/m

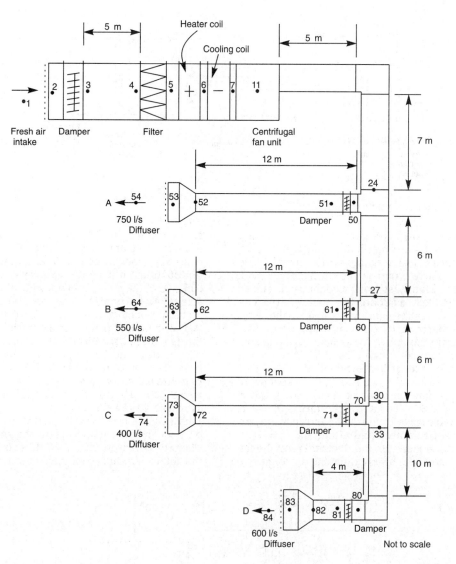

Fig. 7.8 Multi-room supply air duct system for question 21.

duct 3–4   0.3 Pa/m
duct 5–6   0.5 Pa/m
The total pressure at node 1 is 75 Pa.

18. A centrifugal fan delivers 5.5 m³/s into an outlet duct of 700 mm × 600 mm. The inlet duct to the fan is 900 mm diameter. The static pressure at the fan inlet is measured as −275 Pa. The ductwork system has a calculated resistance of 560 Pa. Air density at the fan is 1.19 kg/m. Calculate the pressures on either side of the fan.

19. The duct system shown in Fig. 7.6 is to be installed in a false ceiling over offices and corridors. The air-handling plant comprises the fresh air inlet, filter, hot and chilled water coils and a fan, and is located in a plant room. Room A is supplied with 1.5 m³/s and room B has 2.5 m³/s. The limiting air velocities are 2.5 m/s through the plant and grilles, 12 m/s through the fan, and 5 m/s in the ducts. The outdoor air inlet is constructed from 45° louvres having a free area of 60% and backed with a wire mesh screen. The filter, supply grille, heating and cooling coils have air pressure drops of 65 Pa, 30 Pa, 50 Pa and 40 Pa respectively. The enlargements are at 40°. The air temperature is 20 °C d.b. Find suitable sizes for the ducts, and state the performance specification for the fan.

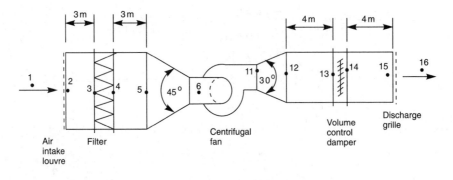

Fig. 7.9   Fan and duct system for question 22.

20. An air extract duct system is shown in Fig. 7.7. Air flow quantities are marked on the figure. Each extract grille has a multi-leaf flat metal blade volume control damper. The air-handling plant comprises a fan, heat reclaim cooling coil, and an exhaust grille to outdoors. The limiting air velocities are 2.5 m/s through the exhaust grille, 3 m/s through the cooling coil, 10 m/s through the fan, and 6 m/s in the ducts. The exhaust grille is constructed from 45° louvres having a free area of 70%. The extract grilles and cooling coil have air pressure drops of 30 Pa and 65 Pa respectively. The contraction is at an angle of 45°, and the enlargement is at 30°. The air temperature is 20°C d.b. Find suitable sizes for the ducts, and state the performance specification for the fan.

21. Find suitable sizes for the ductwork and the fan duty for the supply of conditioning air to the five rooms shown in Fig. 7.8. The pressure drops through the plant items are filter 100 Pa, heating coil 45 Pa, cooling coil 65 Pa and diffuser 25 Pa. The 45° fresh air intake louvres have a free area of 60%, and they are backed with a wire mesh screen. All the dampers have aerofoil blades. The limiting velocities are louvres and diffusers 2.5 m/s, air-handling plant from the fresh air intake to the fan discharge 2.5 m/s, fan 10 m/s, ductwork from nodes 11 to 30, 5 m/s, and 4 m/s for all the branch ducts to the diffusers. The duct air temperature has an average value of 20 °C d.b. for the whole year. The air flows at each diffuser are shown in Fig. 7.8 in l/s.

22. Find suitable sizes of the ductwork and the fan duty for the air-handling plant shown in Fig. 7.9. The pressure drop through the filter is 125 Pa, and the diffuser requires an inlet pressure of 35 Pa.

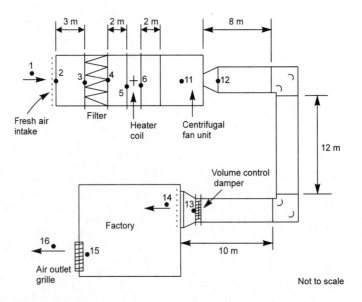

Fig. 7.10   Supply and air duct system for the factory in question 23.

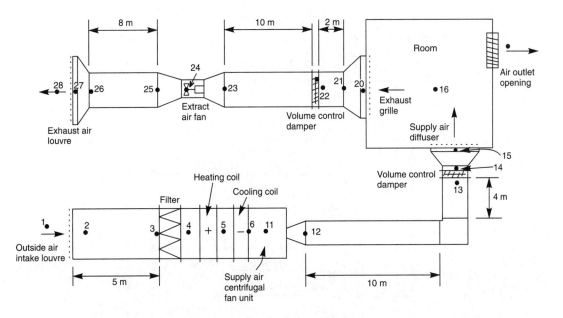

Fig. 7.11 Single-duct 100% outside air conditioning system in question 24.

The 45° outdoor air intake louvres have a free area of 50%, and are backed with a wire mesh screen. The volume control damper has aerofoil blades. The limiting velocities are louvres and diffusers 2.5 m/s, air-handling plant from the fresh air intake to the fan discharge 2.5 m/s, fan 11 m/s, and discharge ductwork from the fan 5 m/s. The duct air temperature has an average value of 20 °C d.b. The discharge air flow rate is to be 3500 l/s.

23. The single-duct heating and ventilation system shown in Fig. 7.10 serves a production area where recirculated air is not permitted. The factory is 24 m long, 12 m wide and 6 m high, and has 5 air changes per hour. The pressure drop through the

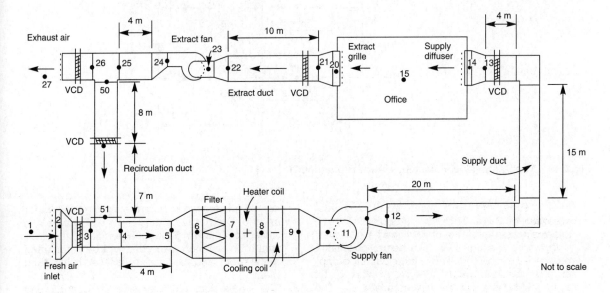

Fig. 7.12 Single-duct air conditioning system with recirculation in question 25. VCD – volume control damper; VAVR – variable air volume regulator.

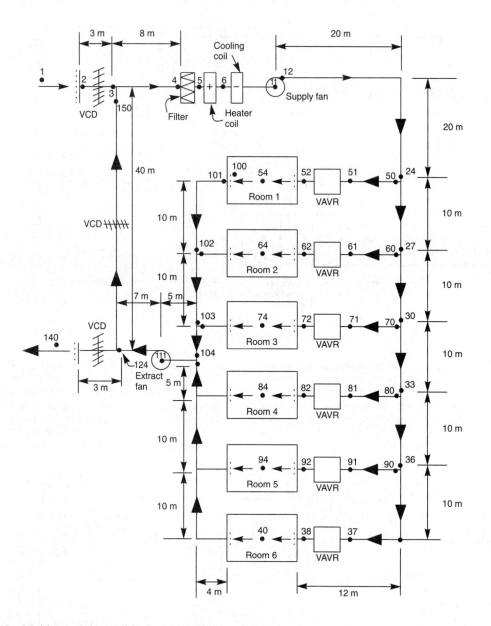

Fig. 7.13 Multi-room air conditioning system in question 26.

filter is 90 Pa, and the diffuser requires an inlet pressure of 25 Pa. The 45° outdoor air intake louvres have a free area of 60%, and are backed with a wire mesh screen. The heater coil has a pressure drop of 60 Pa. The volume control damper has aerofoil blades. The air outlet grille from the factory has a resistance of 10 Pa. The limiting velocities are louvres, grilles and diffusers 2.5 m/s, air-handling plant from the fresh air intake to the fan discharge 2.5 m/s, fan 10 m/s, and

supply air ductwork from the fan 6 m/s. Outdoor air at 0 °C d.b. enters the heater coil, where it is raised to 30 °C d.b. Find suitable sizes of the ductwork, the static pressure that will be maintained in the room, and the fan duty for the air-handling plant.

24. The single-duct 100% outside air conditioning system shown in Fig. 7.11 serves a commercial kitchen. All the air supplied is exhausted directly to the atmosphere. The room air temperature is to

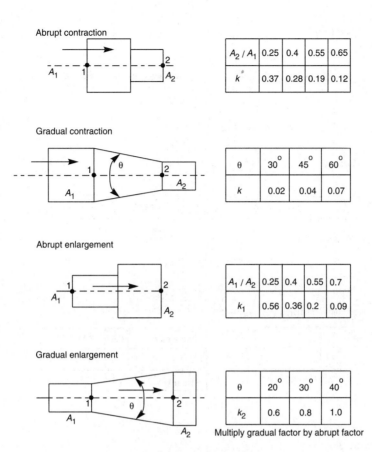

Abrupt contraction

A_2 / A_1	0.25	0.4	0.55	0.65
k		0.37	0.28	0.19	0.12

Gradual contraction

θ		30°	45°	60°
k		0.02	0.04	0.07

Abrupt enlargement

A_1 / A_2	0.25	0.4	0.55	0.7
k_1	0.56	0.36	0.2	0.09

Gradual enlargement

θ	20°	30°	40°
k_2	0.6	0.8	1.0

Multiply gradual factor by abrupt factor

Fig. 7.14 Velocity pressure loss factors for air duct fittings (CIBSE, 1986c). Factors are multiplied by the velocity pressure in the smaller area: $dp_{12} = k_1 k_2 p_{v1}$.

be maintained at 24 °C d.b. by 1.2 m³/s of air that is supplied at 14 °C d.b. when the outside air temperature is 35 °C d.b. The combined effect of window and door openings is expected to allow 120 l/s of air at room temperature to flow directly to outdoors. Outside air is not to be allowed to infiltrate the room through these openings. The velocity pressure of the air flow within the room is effectively zero. The outside air intake and the exhaust air louvres have a free area of 60%, and they are backed with wire mesh. The air pressure drops through the plant are filter 75 Pa, heater 40 Pa, cooling coil 55 Pa, supply air diffuser 25 Pa, room air openings to outdoors 10 Pa, and the exhaust grille 20 Pa. The duct enlargements are at 40°, contractions are at 45°, there is a right-angle mitre bend in the supply duct, and the volume control dampers have aerofoil blades. The limiting air velocities are louvres, diffuser, grilles, filter, heater and cooler 2.5 m/s, fans 10 m/s and duct-

work 5 m/s. Find the duct sizes and the supply and extract air fan performance specifications.

25. A single-duct air conditioning system for an open plan office is shown in Fig. 7.12. The supply air has a mixture of outside and recirculated room air that is filtered, heated and cooled prior to being supplied at 35 °C d.b. in winter and 13 °C d.b. in summer. The room air temperature is to be maintained at 22 °C d.b. by 2000 l/s of air that is supplied at 35 °C d.b. when the outside air temperature is −1 °C d.b. The mixed air that enters the air-handling plant is at 19 °C d.b. A balanced supply and extract air system is to be provided so that the office remains at atmospheric pressure. The outside air intake and the exhaust air louvres have a free area of 50%, and they are backed with wire mesh. The air pressure drops through the plant are filter 95 Pa, heater 35 Pa, cooling coil 60 Pa, supply air diffuser 25 Pa, and the extract air grille 20 Pa. The duct enlargements are at 30°;

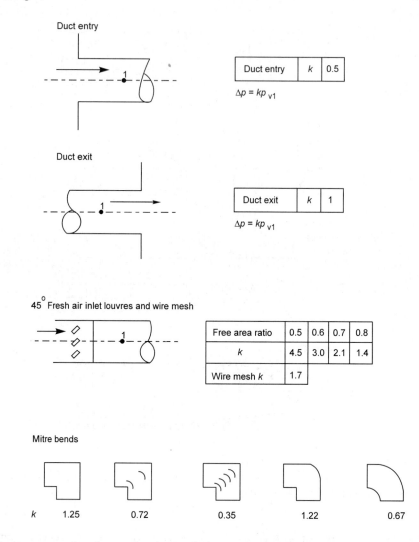

Duct entry

Duct entry	k	0.5

$\Delta p = kp_{v1}$

Duct exit

Duct exit	k	1

$\Delta p = kp_{v1}$

45° Fresh air inlet louvres and wire mesh

Free area ratio	0.5	0.6	0.7	0.8
k	4.5	3.0	2.1	1.4
Wire mesh k	1.7			

Mitre bends

k	1.25	0.72	0.35	1.22	0.67

Fig. 7.15   Velocity pressure loss factors for air duct fittings (CIBSE, 1986c).

contractions are at 45°. Right-angle mitre bends with turning vanes are used. The volume control dampers have flat metal blades. The limiting air velocities are louvres, diffuser, grilles, filter, heater and cooler 2.5 m/s, fans 10 m/s, and ductwork 5 m/s. Find the duct sizes and the supply and extract air fan performance specifications.

26. A single-duct air conditioning system that serves six rooms is shown in Fig. 7.13. The supply air has a mixture of outside and recirculated room air that is filtered, heated and cooled prior to being supplied to the rooms through a variable-volume controller. Summer design conditions are for an outside air temperature of 32 °C d.b. when the room air is maintained at 24 °C d.b. Mixed return and outside air enters the filter at 26 °C d.b. A balanced supply and extract air system is to be provided so that the rooms are at atmospheric pressure. The outside air intake and the exhaust air louvres have a free area of 50% and they are backed with a wire mesh screen. The air pressure drops through the plant are filter 125 Pa, heater 40 Pa, cooling coil 70 Pa, supply air diffuser 25 Pa, and the extract air grille 20 Pa. The duct enlargements are at 30° and contractions are at 45°. All dampers have aerofoil blades. Radiused bends are used. The limiting air velocities are louvres, diffuser, grilles, filter, heater and cooler 2.5 m/s, fans 10 m/s, and ductwork 5 m/s. The variable-volume control box has a resistance of 40 Pa when its damper is fully open at the design flow of 500 l/s. Find the duct sizes and the supply and extract air fan performance specifications for summer operation.

Rectangular branch

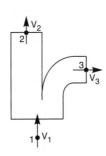

$\dfrac{V_2}{V_1}$	k_2	$\dfrac{V_3}{V_1}$	$k_3 \times$ bend k
0.6	0.34	0.6	3.5 "
0.8	0.18	0.8	1.8 "
1.0	0.09	1.0	1.0 "
1.2	0.05	1.2	1.0 "

Open dampers, single or multi-blade

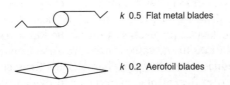

k 0.5 Flat metal blades

k 0.2 Aerofoil blades

Fig. 7.16   Velocity pressure loss factors for air duct fittings (CIBSE, 1986c).

You should have used ...

8 Water pipe sizing

INTRODUCTION

This chapter uses the worksheet file PIPE.WKS to find the sizes of pipes and the pump specification for two-pipe heating hot water and chilled water circulation systems. The user assesses which is the index circuit by inspection of the pipework layout on the scale drawing of the building. The heat load and pipe length are entered onto the worksheet for each section of the index route and then for the branches. A few iterations of the value for the pressure drop rate, as is done during manual design, produce a suitable pipe diameter for the section.

Data for a range of different pipe materials is provided on the worksheet. The user copies the required data into a staging area on the worksheet. Additional pipe data may be entered by the user for any material or range of diameters. An initial estimate is entered for the pipe heat emission. The worksheet lists the lengths of all the specified pipe diameters, and calculates the actual heat emission and the total installed cost of the pipe system. The designer can then change the input data to try different options to minimize the cost of the pipework. Such redesign would often be avoided during manual design because of the time involved in recalculation. The worksheet is a powerful design tool, which finds pipe sizes and the pump duty with the minimum of data entry, and also generates a list of pipe lengths and cost for the estimator.

LEARNING OBJECTIVES

Study and use of this chapter will enable the user to:

1. understand how to find the sizes of a two-pipe water heating or cooling system;
2. know how to calculate the pump performance specification for a two-pipe system;
3. know the use of the term index circuit;
4. calculate the water flow rates that are required for a two-pipe system;
5. find the heat emission from a pipe system;
6. use the hydraulic resistance of items of plant;
7. use the equations for water flow in pipes;
8. use water density, dynamic viscosity, pipe surface roughness and pipe internal diameter in pipe sizing calculations;

9. use velocity pressure loss factors in the calculation of pump pressure rise;
10. know what is meant by the equivalent length of a pipeline system;
11. know how to maintain a convenient source of pipe sizing data on disk;
12. know how to calculate pipe systems by manual methods;
13. calculate the installed cost of 1 m of a pipeline system from updated price data;
14. find the total installed cost of a two-pipe system;
15. use the total installed cost of the pipe system as part of the design decision work.

Key tems and concepts

boiler	236	maximum carrying capacity	240
branch	235	pipe heat emission	243
chilled water	236	pipe section	235
cost	243	pipeline fittings	239
density	237	pressure drop	241
diameter	238	pressure drop rate	240
dynamic viscosity	237	pressure loss	241
emitter	236	pressure rise	246
equivalent length	240	pump	234
flow rate	240	pump specification	234
heat exchanger	236	single-pipe system	235
heat load	237	specific heat capacity	236
hot water	237	surface roughness	238
hydraulic resistance	236	total installed cost	244
index	240	two-pipe system	235
installed cost	243	velocity	240
iteration	247	water flow rate	237

TWO-PIPE CIRCULATION SYSTEM

Most heating hot water and chilled water circulation systems are of the two-pipe flow and return design. This worksheet is not used for single-pipe systems. A typical two-pipe reticulation system is shown in Fig. 8.1. A numbering or lettering code is used for each section of pipe. A format block is shown in Fig. 8.1 to label each pipe section and to allow for the recording of the output data on the schematic drawing.

The worksheet has lines of data for 20 pipes along the index circuit and for 20 branch pipes. More pipe sections can be added by the user to enlarge the worksheet when necessary for a particular project. The worksheet should not be enlarged without a good reason. Move the cursor down a screen page at a time using the Page Down, Page Up and Home cursor control keys.

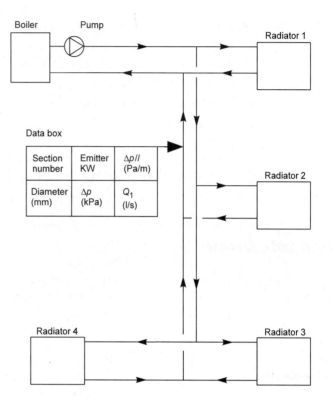

Fig. 8.1 Typical two-pipe system schematic.

The heat load at each terminal heat emitter is found from the heat loss and heat gain worksheets in conjunction with the designer's further calculations and system decisions. The heat loads are summed from the heat emitters, along the pipe routes, back to the heat source. The heat source will be a heat exchanger, boiler or water-chilling machine. The worksheet calculates the water flow rate required for each section of pipe, and adds the specified percentage for the heat lost from the pipework. Designers will usually overestimate the pipe heat emission. The worksheet calculates the percentage heat emission that is produced by the design. The user enters the correct value of percentage heat emission and checks the pipe and pump sizes again. The hydraulic resistance of all the plant and pipeline fittings is entered, as would be done with manual calculation methods. For cooling systems, the term 'heat emission' refers to the chilled water cooling load at the terminal unit and heat gains to the chilled water circulation system.

WATER FLOW EQUATIONS

The water flow rate that is required to provide the heat load is given by

$$Q_1 = H \text{ kW} \times \frac{1}{SHC} \frac{\text{kg K}}{\text{kJ}} \times \frac{1}{T_f - T_r} \times \frac{100 + L\%}{100} \times \frac{1 \text{ kJ}}{1 \text{ kW s}}$$

$$= \frac{H(100 + L)}{SHC(T_f - T_r) \times 100} \frac{\text{kg}}{\text{s}}$$

$$= \frac{H(100+L)}{SHC(T_f - T_r) \times 100} \frac{kg}{s} \times \frac{1}{\rho} \frac{m^3}{kg} \times \frac{10^3 \, litre}{1 \, m^3}$$

$$= \frac{H(100+L) \times 10}{SHC(T_f - T_r) \times \rho} \frac{1}{s}$$

The pipe diameter that is needed for a specified water velocity is found from

$$Q_1 = \frac{\pi d^2}{4} \, m^2 \times v \, \frac{m}{s} \times \frac{10^3 \, litre}{1 \, m^3}$$

$$d^2 = \frac{4Q_1}{\pi v \times 1000} \, m^2$$

$$d = \left(\frac{4Q_1}{\pi v \times 1000}\right)^{0.5} m \times \frac{10^3 \, mm}{1 \, m}$$

where H = heat emitter load (kW); Q_1 = water flow rate required (l/s); SHC = specific heat capacity of water, 4.186 kJ/kg K; T_f = water flow temperature at heat emitter (°C); T_r = water return temperature leaving heat emitter (°C); L = estimated pipe heat emission (%); ρ = density of water (kg/m³); and d = pipe diameter (m).

The turbulent flow of water in a pipe is given by (CIBSE, 1986c)

$$M = -\pi\rho \left(\frac{dp_1 \times d^5}{2\rho}\right)^{0.5} \times \log_{10}\left[\frac{k_s}{3.7d} + \frac{1.255\mu}{\rho} \times \left(\frac{2\rho}{dp_1 \times d^3}\right)^{0.5}\right]$$

where M = water flow rate (kg/s); dp_1 = water pressure drop rate (Pa/m); d = pipe internal diameter (m); μ = dynamic viscosity of water (Pa s); and k_s = absolute surface roughness of pipe (m).

This equation is the basis of the published data tables. It would not normally be used by designers, but is easily handled on a spreadsheet. An example is given here, and also a question that requires its use by manual calculation. These are the only occasions on which the user sees the equation. The water density and viscosity data used on the worksheet are given in Table 8.1. The absolute roughness for pipes is given in Table 8.2. Table 8.3 shows the range of nominal pipe diameters and actual internal diameters for copper pipe Table X, BS 2871 : Part 1.

Table 8.1 Density and viscosity for water (Rogers and Mayhew, 1987)

Temperature (°C)	Density (kg/m³)	Dynamic viscosity (mPa s)
10	999.7	1.300
60	983.3	0.463
70	977.5	0.400
75	974.7	0.374
80	971.8	0.351
85	969.0	0.330
90	965.3	0.311
100	957.9	0.279
120	943.4	0.230
130	934.6	0.211
140	925.9	0.195
150	916.6	0.181

Table 8.4 shows the similar data for medium black mild steel pipes to BS 1387. Table 8.5 shows typical values for the velocity pressure loss factors for pipe fittings in the range of diameters from 32 mm to 50 mm. Further data may be added to this table as required. This, and other, reference data is repeated in the data bank section in the worksheet.

Table 8.2   Absolute roughness for pipes

Material	Roughness, k_s (mm)
Copper	0.0015
Plastic	0.0030
New black mild steel	0.0460
Rusted black mild steel	2.5000
Galvanized mild steel	0.1500

Table 8.3   Copper pipe Table X, BS 2871 : Part 1

Nominal diameter, d (mm)	Internal diameter, d (mm)
15	13.60
22	20.22
28	26.22
35	32.63
42	39.63
54	51.63
67	64.27
76	73.22
108	105.12
133	130.38
159	155.38

Table 8.4   Medium black mild steel BS 1387

Nominal diameter, d (mm)	Internal diameter, d (mm)
15	16.10
20	21.60
25	27.30
32	36.00
40	41.90
50	53.00
65	68.70
80	80.70
90	93.15
100	105.10
125	129.95
150	155.40

Table 8.5 Velocity pressure loss factors for pipe fittings

Pipe fitting	*k factor*
Copper pipe tee branch	1.35
25 mm branch equal sweep tee	1
Mild steel tee branch	1.05
25 mm screwed mild steel bend	0.7
25 mm copper pipe elbow	1
Angle radiator valve	5
Steel panel radiator	2.5
Cast iron boiler	2.5
Fire tube horizontal boiler	8
25 mm open gate valve	0.3

EXAMPLE 8.1

A 35 mm Table X copper pipe passes water at 75 °C at a pressure drop rate of 200 Pa/m. Calculate the maximum water flow rate that the pipe can carry.

$$M = -\pi\rho\left(\frac{dp_1 \times d^5}{2\rho}\right)^{0.5} \times \log_{10}\left[\frac{k_s}{3.7d} + \frac{1.255\mu}{\rho} \times \left(\frac{2\rho}{dp_1 \times d^3}\right)^{0.5}\right]$$

From Table 8.1, at 75 °C, $\rho = 974.85 \, kg/m^3$, $\mu = 0.378 \, mPa \, s$. From Table 8.2, for a 35 mm Table X copper pipe, $d = 32.63 \, mm$. From Table 8.3, for copper pipe $k_s = 0.0015 \, mm$.

$$\rho = 974.85 \, kg/m^3$$
$$\mu = 0.378 \, mPa \, s$$
$$= 0.378 \times 10^{-3} \, Pa \, s$$
$$d = 32.63 \, m$$
$$= 0.032 \, 63 \, m$$
$$dp_1 = 200 \, Pa/m$$
$$k_s = 0.0015 \, mm$$
$$= 0.0015 \times 10^{-3} \, m$$

$$M = \rho\pi\left(\frac{dp_1 \times d^5}{2\rho}\right)^{0.5} \times \log_{10}\left[\frac{k_s}{3.7d} + \frac{1.255\mu}{\rho} \times \left(\frac{2\rho}{dp_1 \times d^3}\right)^{0.5}\right]$$

Arrange as

$$M = -\rho\pi\left(\frac{dp_1 \times d^5}{2\rho}\right)^{0.5} \times L$$

where

$$L = \log_{10}\left[\frac{k_s}{3.7d} + \frac{1.255\mu}{\rho} \times \left(\frac{2\rho}{dp_1 \times d^3}\right)^{0.5}\right]$$

$$= \log_{10}\left[\frac{0.0015 \times 10^{-3}}{3.7 \times 0.03263} + \frac{1.255 \times 0.378 \times 10^{-3}}{974.85} \times \left(\frac{2 \times 974.85}{(200 \times 0.03263)^3}\right)^{0.5}\right]$$

$$= \log_{10}(1.242 \times 10^{-5} + 4.866 \times 10^{-7} \times 529.716)$$
$$= \log_{10}(2.7018 \times 10^{-4})$$
$$= -3.568$$

$$M = \rho\pi \left(\frac{dp_1 \times d^5}{2\rho}\right)^{0.5} \times L$$

$$M = -974.85 \times \pi \times \left(\frac{200 \times 0.032\,63^5}{2 \times 974.85}\right)^{0.5} \times - 3.568\,\text{kg/s}$$

$$= 0.673\,\text{kg/s}$$

This is the maximum carrying capacity of the pipe at the specified pressure drop rate of 200 Pa/m. The required water flow rate is compared with the maximum carrying capacity, $Q_2\,\text{l/s}$, to establish the difference, or error:

$$\text{flow error } E = \frac{Q_1 - Q_2}{Q_1} \times 100\%$$

The user enters successive values for the water pressure drop rate until the error is less than $\pm 1\%$. The final value of the water pressure drop rate, $dp_1\,\text{Pa/m}$, is that used to calculate the system pressure drop and the pump pressure increase. The error percentage should be as small as can be achieved. Increments of 5 Pa/m for the pressure drop rate should be the minimum.

The volume flow rate of water through the pipe system is taken as the maximum value. This is because the user has found the nearest pressure drop rate to satisfy the design criteria. This can be done by manual calculation and inspection of published pipe sizing tables. The water volume flow rate is

$$Q_2 = M\frac{\text{kg}}{\text{s}} \times \frac{10^3\,\text{litre}}{1\,\text{m}^3} \times \frac{\text{m}^3}{\rho\,\text{kg}}$$

$$= \frac{M \times 10^3}{\rho}\frac{1}{\text{s}}$$

and the water velocity is

$$Q_2 = \frac{\pi d^2}{4}\,\text{m}^2 \times v\frac{\text{m}}{\text{s}} \times \frac{10^3\,\text{litre}}{1\,\text{m}^3}$$

$$v = \frac{4Q_2}{\pi d^2 \times 10^3}\frac{\text{m}}{\text{s}}$$

The equivalent length of the pipe is

$$l_e = \frac{0.81M^2}{dp_1\rho d^4}\,\text{m}$$

(CIBSE, 1986c)

Each pipe fitting provides a hydraulic resistance to the flow of water. This resistance is characterized by the velocity pressure loss factor k, as shown in Table 8.5. The overall equivalent length of the index pipe circuit is the sum of the straight pipe lengths and the total equivalent length of all the pipe fittings:

$$EL = L + kl_e\,\text{m}$$

The pressure drop through the index circuit is found from the resistance of the equivalent length of the pipework and fittings, plus the hydraulic resistance of other known plant or valves. The manufacturer of heating and cooling coils, heat exchangers, flow control valves, balancing valves and boilers provides the known pressure drop to add to the overall figure:

$$\text{total } dp = EL\,\text{m} \times dp_1 \frac{\text{Pa}}{\text{m}} \times \frac{1\,\text{kPa}}{10^3\,\text{Pa}} + \text{plant kPa}$$

EXAMPLE 8.2

An air heater coil provides 45 kW of heating to a ducted ventilation system. The total length of flow plus return pipe is 30 m. The water flow and return temperatures are 82 °C and 71 °C. The water velocity is not to exceed 1 m/s. The specific heat capacity of water is 4.186 kJ/kg K. The density of water at the mean water temperature is 974.85 kg/m^3. The dynamic viscosity of water at the mean water temperature is 0.378 mPa s. Copper Table X BS 2871 pipe is to be used. The absolute surface roughness of copper is 0.0015 mm.

The designer estimates that the heat emission from the pipes will be 8% of the installed heat load. A cast iron boiler is the heat source, and the pipe system has 12 bends, 4 gate valves and 4 branch tees. The hydraulic resistance of the heating coil is 6 kPa. The two-port flow control valve has a hydraulic resistance of 10 kPa. Calculate the required water flow rate in kg/s and l/s, the estimated pipe diameter, the maximum carrying capacity of a selected pipe diameter, the flow error, the flow occurring in the pipe, the water velocity, the equivalent length of the circuit, and the frictional resistance pressure drop through the pipework system.

The required water flow is

$$
\begin{aligned}
Q_1 &= \frac{H(100 + E)}{SHC(T_f - T_r) \times 100} \frac{\text{kg}}{\text{s}} \\
&= \frac{45 \times (100 + 8)}{4.186 \times (82 - 71) \times 100} \frac{\text{kg}}{\text{s}} \\
&= 1.0555\,\text{kg/s}
\end{aligned}
$$

Also:

$$
\begin{aligned}
Q_1 &= \frac{45 \times (100 + 8) \times 10}{4.186 \times (82 - 71) \times 974.85} \frac{\text{litre}}{\text{s}} \\
&= 1.0827\,\text{litre/s}
\end{aligned}
$$

The estimated pipe diameter is

$$
\begin{aligned}
d &= \left(\frac{4Q_1}{\pi v \times 1000}\right)^{0.5} \times 1000\,\text{mm} \\
&= \left(\frac{4 \times 1.0827}{\pi \times 1 \times 1000}\right)^{0.5} \times 1000\,\text{mm} \\
&= 37\,\text{mm}
\end{aligned}
$$

The next pipe diameter larger than this is 42 mm with an internal diameter of 39.63 mm. For a pressure drop rate dp_1 of 180 Pa/m in a 42 mm nominal diameter copper Table X BS 2871 pipe, the maximum water-carrying capacity is found, as before:

$$M = -\rho\pi\left(\frac{dp_1 \times d^5}{2\rho}\right)^{0.5} \times L$$

where

$$L = \log_{10}\left[\frac{k_s}{3.7d} + \frac{1.255\,\mu}{\rho} \times \left(\frac{2\rho}{dp_1 \times d^3}\right)^{0.5}\right]$$

$$= \log_{10}\left[\frac{0.0015 \times 10^{-3}}{3.7 \times 0.039\,63} + \frac{1.255 \times 0.378 \times 10^{-3}}{974.85} \times \left(\frac{2 \times 974.85}{175 \times (0.039\,63)^3}\right)^{0.5}\right]$$

$$= -3.6653$$

$$M = -\rho\pi\left(\frac{dp_1 \times d^5}{2\rho}\right)^{0.5} \times L$$

$$= -974.85\pi \times \left(\frac{175 \times (0.039\,63)^5}{2 \times 974.85}\right)^{0.5} \times -3.6653\,\text{kg/s}$$

$$= 1.0515\,\text{kg/s}$$

$$\text{flow error } E = \frac{Q_1 - M}{Q_1} \times 100\%$$

$$= \frac{1.0555 - 1.0515}{1.0555} \times 100\%$$

$$= 0.4\%$$

$$Q_2 = \frac{M \times 10^3}{\rho}\,\frac{\text{litre}}{\text{s}}$$

$$= \frac{1.0515 \times 10^3}{974.85}\,\frac{\text{litre}}{\text{s}}$$

$$= 1.0786\,\text{litre/s}$$

$$v = \frac{4Q_2}{\pi d^2 \times 10^3}\,\frac{\text{m}}{\text{s}}$$

$$= \frac{4 \times 1.0786}{\pi \times (0.039\,63)^2 \times 10^3}\,\frac{\text{m}}{\text{s}}$$

$$= 0.874\,\text{m/s}$$

The equivalent length of the pipe is

$$l_e = \frac{0.81M^2}{dp_1\rho d^4}\,\text{m}$$

$$= \frac{0.81 \times 1.0515^2}{175 \times 974.85 \times (0.039\,63)^4}\,\text{m}$$

$$= 2.13\,\text{m}$$

The pipe fittings velocity pressure loss factors have a total of

$$\Sigma k = 2.5 + 12 \times 1 + 4 \times 0.3 + 4 \times 1.5$$

$$= 21.7$$

$$EL = L + kl_e \text{ m}$$
$$= 30 + 21.7 \times 2.13 \text{ m}$$
$$= 76.2 \text{ m}$$

$$\text{total } dp = EL \text{ m} \times dp_1 \frac{\text{Pa}}{\text{m}} \times \frac{1 \text{ kPa}}{10^3 \text{ Pa}} + \text{plant kPa}$$
$$= 76.2 \text{ m} \times 175 \frac{\text{Pa}}{\text{m}} \times \frac{1 \text{ kPa}}{10^3 \text{ Pa}} + 6 + 10 \text{ Pa}$$
$$= 29.339 \text{ kPa}$$

PIPE HEAT EMISSION

The heat lost from heating hot water pipes, in watts per metre run of pipe per degree Kelvin difference between the mean water and ambient air temperatures, having 25 mm thick fibreglass rigid insulation, is given in Table 8.6 (CIBSE, 1986c).

Table 8.6   Heat emission from pipes insulated with 25 mm fibreglass

Nominal diameter (mm)	Heat emission (W/m K)
15	0.19
22	0.23
28	0.25
35	0.29
42	0.32
54	0.37
67	0.44
80	0.5
100	0.61
125	0.75
150	0.88

PIPELINE COST

The installed cost of a two-pipe system is evaluated from the cost of the pipe, thermal insulation, pipe fittings, supporting brackets, labour, overheads and profit. Costs are to be used for comparison, so that alternative routes, diameters and lengths may be tested quickly by the designer. After having found the relative cost of the accepted pipework design, the final costing can be completed. Such final costing includes fixed cost items that are not strongly influenced by the pipe layout. Pipe lengths and diameters that are accumulated from this worksheet can be used for the final costing. Typical pipe cost data is available either from within the user's company or from regularly published price books (Spon, 1992).

To provide a reasonable comparison of the installed cost of different pipe diameters, the following data is used. The data is for copper pipe to BS 2871, Table X, and an example is shown for a 35 mm diameter pipe. The user can construct similar tables of costs for other pipe materials, and use other sources of cost data as required.

The measured installed cost of 35 mm copper pipe is £12.92 per metre length. A 15% discount is normally applied and then an addition of 24.3% for waste, overheads and profit. The typical installation of a 5 m length of pipe in a heating system would have pipe fittings of a tee, two elbows, two pipe brackets and 25 mm thick thermal insulation. The tee costs £12.89, discount 35%, waste, overheads and profit 18.39%. An elbow costs £7.74, discount 35%, waste overheads and profit 18.39%. A single-pipe ring bracket costs £6.69, discount 25%, waste, overheads and profit 18.39%. Preformed glass fibre 30 mm thick thermal insulation for concealed service areas costs £4.73 per metre length, 0% discount, waste, overheads and profit 18.39%. The total installed cost of a 5 m length of 35 mm diameter copper Table X pipe is given by

$$
\begin{aligned}
\text{cost} = {} & (5 \times \text{£}12.92, -15\%, +24.3\%) + (\text{£}12.89, -35\%, +18.39\%) \\
& (2 \times \text{£}7.74, -35\%, +18.39\%) + (2 \times \text{£}6.69, -25\%, +18.39\%) \\
& + (5 \times \text{£}4.73, -0\%, +18.39\%) \\
= {} & \text{£}(68.25 + 9.92 + 11.91 + 11.88 + 28.00) \\
= {} & \text{£}129.96
\end{aligned}
$$

The average cost of one metre of 35 mm diameter installed copper pipe is

$$
\text{cost} = \frac{\text{£}129.96}{5\,\text{m}}
$$
$$
= \text{£}26.00 \text{ per metre}
$$

Copper pipe diameters of 15–67 mm are jointed with capillary fittings. Pipes of 76–159 mm have weldable joints. Typical costs are listed in Table 8.7.

Table 8.7   Installed costs of copper Table X pipework

Diameter (mm)	Installed cost (£/m)
15	13.70
22	16.10
28	19.45
35	26.00
42	29.55
54	37.00
67	52.00
80	77.40
100	110.85
125	172.40
150	216.60

EXAMPLE 8.3

A two-pipe heating system has copper Table X pipework. The lengths are 150 m of 15 mm, 80 m of 20 mm, 60 m of 28 mm, 40 m of 35 mm and 10 m of 42 mm. The heat emitters provide an output of 65 kW to the building. The flow and return temperatures used are 82 °C and 70 °C. The ambient air temperature around the pipes is 18 °C. Calculate the pipe heat emission as a percentage of the installed heat load, and the relative cost of the pipe system.

$$\text{temperature difference} = \frac{82 + 70}{2} - 18\,\text{K}$$

$$= 58\,\text{K}$$

The solution is presented in Table 8.8.

Table 8.8 Solution to Example 8.3

Diameter	Pipe length,	Heat emission		Installed cost	
(mm)	L (m)	(W/m K)	(kW)	(£/m)	(£)
15	150	0.19	1.653	13.70	2055
22	80	0.23	1.067	16.10	1288
28	60	0.25	0.870	19.45	1167
35	40	0.29	0.673	26.00	1040
42	10	0.32	0.186	29.55	295
		Total	4.449		5845

$$\text{pipe emission} = \frac{4.449\,\text{kW}}{65\,\text{kW}} \times 100\%$$

$$= 6.8\%$$

total cost of the pipe system = £5845

DATA REQUIREMENT

The user will enter new data for:

1. Your name in cell B4.
2. Job title in cell B5.
3. Job reference number in cell B6.
4. Filename in cell B9.
5. Water flow temperature in cell C44.
6. Water return temperature in cell C45.
7. Water density in cell C47.
8. Water dynamic viscosity in cell C48.
9. Specific heat capacity of the water in cell C49.
10. Temperature of the ambient air surrounding the pipework system in cell C50. This is the temperature that acts as the sink for the pipe heat emission. This may be the temperature within a builder's work duct or a ceiling space.
11. Roughness of the pipe internal surface in cell C51.
12. First estimate of the heat emission from the pipework system in cell C52. This can be left at the original value of 5%.
13. Total heat output from the heating system, H kW, in cell C54. This is the sum of the heat emitter outputs.
14. Type of pipe, nominal and internal diameters of the pipe to be used in cell range A69 to C79. COPY this block of data from the data bank that commences at line 161; the cell range from A69 to C79 must contain the correct range of pipe diameters to be used. The data that is entered here is automatically copied to other parts of the worksheet. This information is critical to the running of the worksheet.

15. The number of the first section of pipe on the index circuit in cell A91. The numbers provided for the various sections of pipe can remain.

16. A description of the first section of pipe may be entered into cell B91. The description is limited to 12 characters. This cell can remain as it is provided or left blank.

17. Length of pipe in the first section of the system in cell C91. A two-pipe system is entered as the total length of the flow pipe plus the return pipe in the first section.

18. The user must calculate the total heat supply into the pipe system from the design data, system drawings and schematic layouts. This is the sum of the heat demands. The heat-carrying capacity of the first section of the pipe system is entered in cell D91. The first section of pipe will often be closest to the heat exchanger or boiler that provides heat into the pipe distribution. Do not make any additions for the heat emission from the pipework system when entering this number, as that is done elsewhere on the worksheet.

19. Enter the limiting water velocity in cell F91. The velocity limit of 1.0 m/s provided will normally be suitable. Where the pipework is installed in noisy industrial environments, underground, or within concrete service ducts, higher velocity limits of 3 m/s or 6 m/s may be used. The user can easily discover the effect of these increased velocities on the cost of the system and the pump pressure rise by experimentation.

20. A first estimate for the rate of water pressure drop through the pipe in cell G91. This figure will be changed in response to the output from a further cell. A number within the range from 20 Pa/m to 1000 Pa/m can normally be used. Some designers choose to limit the pressure drop rate to 400 Pa/m, 0.4 kPa/m. There is no limit upon the user by the worksheet.

21. Nominal diameter of the pipe that is selected for the section of pipe in cell I91.

22. Internal diameter of the selected pipe in cell J91.

23. Velocity pressure loss factor for the pipe fittings in the first section of pipe in cell P91. This is the total of all the pipe fitting pressure loss factors in the flow plus return pipework. Do not include factors for plant items that are to be entered directly as a pressure drop in column S.

24. The specific pressure drop through items of plant in the section of pipe is entered in cell S91. The manufacturer of boilers, heat exchangers, valves, terminal heating and cooling units and finned coils will provide the hydraulic resistance value.

25. The input data for further sections of the index circuit is entered in the cell range A92 to S118.

26. The input data for pipe circuits that branch from the index circuit is entered in the cell range A130 to S139.

27. The pipe and fitting data that is to be used to evaluate the typical installed cost of 1 m of pipe is entered into the cell range M156 to M160. Change these numbers only when it is known that they are to be significantly different in the system that is being designed.

28. Material percentage discount that is to be applied to the listed prices is entered into the cell range P156 to P160.

29. Percentage to be added to the listed material prices for wastage, overhead charges and profit allowance is entered in the cell range S156 to S160.

30. The installed costs of copper pipe, elbows, tees, ring brackets and thermal insulation are entered into the cell range L170 to P180; update these prices from the reference source or your own company price data as appropriate.

When the section of pipe on any of the lines from 91 to 139 contains a specified pipe to be calculated, the cell ranges J91 to J100, J109 to J118 and J130 to J139 will contain the internal diameter of the selected pipe size. When the line is not needed to calculate a pipe diameter, its pressure drop will be zero. In order to stop an ERR error message appearing in column J and causing a subsequent error in other cells, column J must contain a valid number that is greater than zero. Enter 1.0 in column J whenever there is not a pipe in that section or line.

Enter in cell C52 the first estimate of the heat emission from the pipe system. The designer makes an inspired guess of this figure from experience. The percentage that is entered is used to calculate the water flow rates needed to provide the heat output from the final emitters and that lost from the pipe system. After this initial estimate, and having found the sizes for all the pipes, the worksheet automatically calculates the actual heat emission from the sizes and lengths used. The calculated heat emission is shown as output data in cell E58. This information must now be used. Enter the correct value in cell C52. This new entry causes the worksheet to recalculate all the water flow rates. Return to cell G91 and look at the revised value of the flow error in cell L91. It is now necessary to edit the pressure drop rate cells to minimize the flow error again. Enter new pressure drop rates to obtain a satisfactory solution. This should be achieved by the second data entry iteration. If the pipe diameters are changed for any reason, such as to reduce the installed cost, new data will have to be entered to minimize the flow error.

The reason for using a dedicated worksheet for this design work is that it allows the designer to try different combinations of pipe layout, diameter and pressure loss rate. This is where the power of a spreadsheet is demonstrated. The user should use this powerful tool to obtain the least-cost pipe system by trying different combinations of pipe layout and size.

The flow error, shown in cell L91, should be the smallest negative percentage that the designer can create. This means that the water flow capacity of the pipe selected is just larger than the flow required by the heat emitter. A percentage that is less than -1.0% should be achieved with increments of pressure drop rate of 5 Pa/m without difficulty.

OUTPUT DATA

1. The current date is given in cell B7.
2. The current time is given in cell B8.
3. The mean water temperature is given in cell E46.
4. Heat emission from all the pipework in the system is given in cell E56. All the pipes are taken as being thermally insulated.
5. The heat emission of the pipework, expressed as a percentage of the total of the heat emitter output, is given in cell E58.
6. The installed cost of the pipework system is given in cell E60.
7. The water flow rate required in the first section of pipe in the index circuit is given in cell E91. This is normally the total system flow, Q_1 l/s, in the flow pipe from the boiler or heat exchanger.

8. The estimated diameter, d mm, of the pipe to carry the required system water flow in the first section of pipe is given in cell H91.

9. The maximum water flow carrying capacity, M kg/s, of the selected pipe diameter and pressure drop rate is given in cell K91.

10. The percentage error in the water flow between the required flow and the maximum carrying capacity is given in cell L91.

11. The water flow capacity, Q_2 l/s, of the selected pipe diameter is given in cell M91.

12. The velocity of water, v m/s, in the selected pipe diameter is given in cell N91.

13. The equivalent length, l_e m, is given in cell 091.

14. The total equivalent length of the straight pipe and the pipe fittings, EL m, is given in cell Q91.

15. The pressure drop through the first section of pipe, dp kPa, is given in cell R91.

16. The total pressure drop through the first section of pipe, including the plant, dp kPa, is given in cell T91.

17. The pipe section number is copied from column A in cell V91.

18. The description of the first section of pipe is copied from column A in cell W91.

19. The output data for the other sections of pipe in the index circuit is given in the cell range E92 to W118.

20. The total length of pipe in the index circuit is given in cell C120.

21. The total equivalent length of the index circuit is given in cell Q120.

22. The pressure drop through the pipework in the index circuit is given in cell R120.

23. The total of the plant pressure drops in the index circuit is given in cell S120.

24. The total pressure drop through all the pipework and the plant in the index circuit is given in cell T120.

25. The data for the branch pipes from the index circuit is given in the cell range E130 to W140.

26. The total system water flow rate is given in cell D144. This is usually the water flow rate required for the circulating pump.

27. The required pump pressure rise is given in cell D146.

28. The mean water temperature in the system is given in cell D148. This is part of the specification of the pump.

29. The total heat load supplied by the heating system is given in cell D150.

30. The total heat load of the heat emitters and the pipework system is given in cell D152.

31. The total length of each pipe diameter used in the whole system is given in the cell range AD102 to AN102.

32. The installed cost of the pipework system is given in the cell range AD170 to AD182.

33. The total length of each pipe diameter used in the system is given in the cell range AA150 to AA160.

34. The total installed cost of 1 m of pipe for each pipe size that has been selected for this system by the user is shown in the cell range Q170 to Q180.

35. The heat emission from the total length of each pipe size that has been selected by the user is shown in the cell range AD150 to AD160.

36. The total length of pipe in the system that is used to calculate the heat emission is given in cell AA162.

37. The total heat emission from the pipe system is given in cell AD162.

38. The total length of pipe in the system that is used to calculate the total cost is given in cell AA182.
39. The installed cost of each pipe diameter is shown in the cell range AD170 to AD180.
40. The total cost of the pipework system is given in cell AD182.

FORMULAE

Representative samples of the formulae are given here. Each formula can be read on the spreadsheet by moving the cursor to the cell. The equation is presented in the form in which it would normally be written and in the format that is used by the spreadsheet.

Cell B7

@TODAY

This function produces the serial number of the current day and time. The cell is formatted to display the date.

Cell B8

@TODAY

This function produces the serial number of the current day and time. The cell is formatted to display the time.

Cell E45

@ABS(C44−C45)

The absolute value of the difference between the water flow and return temperatures is calculated:

$$\text{temperature difference} = |(t_f - t_r)|\, K$$

The absolute value of temperature difference is taken so that a positive value for the calculated water flow rate is produced for heating and cooling applications.

Cell E46

(C44+C45)/2

The mean water temperature of the heating system is calculated:

$$MWT = \frac{t_f + t_r}{2}\,°C$$
$$E46 = \frac{C44 + C45}{2}\,°C$$

Cell E56

+AD162

The pipe heat emission is copied from cell AD162 into cell E56.

Cell E58

+AD162*100/C54

The pipe heat emission is expressed as a percentage of the total emission from the terminal heat output units.

Cell E60

+AD182

The total cost of the installed pipework system is copied from cell AD182 into cell E60.

Cell E91

+D91*1000*(100+$C$50)/100/$C$49/$E$45/$C$47

The water flow rate, Q_1 l/s, that is required by the heat emitter is calculated. The flow rate is increased by a percentage that is declared for the heat emission from the pipework. This addition is refined after the first complete calculation of the pipe sizes and water flow rates. The user enters the calculated percentage after all the pipe sizes have been entered, to ensure that the correct allowance is made for the heat emission from the pipe system.

$$Q_1 = \frac{H\,\text{kW} \times 1000}{(t_f - t_r)\text{K}} \times \frac{\text{kg K}}{SHC\,\text{kJ}} \times \frac{100 + P\%}{100} \times \frac{\text{m}^3}{\rho\,\text{kg}}\frac{1}{\text{s}}$$

$$E91 = \frac{D91 \times 1000 \times (100 + \$C\$50)}{\$E\$45 \times 100 \times \$C\$49 \times \$C\$47}\frac{\text{kg}}{\text{s}}$$

Cell H91

1000*@SQRT(E91*4/1000/F91/@PI)

An initial estimate of the pipe diameter that is needed is made from

$$d = 1000 \times \left(\frac{4Q_1}{\pi v \times 1000}\right)^{0.5}\text{mm}$$

$$H91 = 1000 \times \left(\frac{4 \times E91}{\pi \times F91 \times 1000}\right)^{0.5}\text{mm}$$

Cell K91

−$C$47*@PI*@SQRT(G91*(J91/1000)^5/2/$C$47)*@LOG($C$51/3.7/J91 +(1.255*$C$48/1000/$C$47)* @SQRT(2*$C$47/G91/(J91/1000)^3))

The maximum water-carrying capacity of the selected pipe diameter and pressure drop rate is calculated from

$$M = -\rho\pi \left(\frac{dp_1 \times d^5}{2\rho}\right)^{0.5} \times \log_{10}\left[\frac{k_s}{3.7d} + \frac{1.255\mu}{\rho} \times \left(\frac{2\rho}{dp_1 \times d^3}\right)^{0.5}\right]$$

Arrange as

$$M = -\rho\pi \left(\frac{dp_1 \times d^5}{2\rho}\right)^{0.5} \times L$$

where

$$L = \log_{10}\left[\frac{k_s}{3.7d} + \frac{1.255\mu}{\rho} \times \left(\frac{2\rho}{dp_1 \times d^3}\right)^{0.5}\right]$$

$$= \log_{10}\left[\frac{C49}{3.7 \times J91} + \frac{1.255 \times C48}{C47} \times \left(\frac{2 \times C47}{G91 \times (J91)^3}\right)^{0.5}\right]$$

$$K91 = -C47 \times \pi \times \left(\frac{G91 \times (J91)^5}{2 \times C47}\right)^{0.5} \times L \text{ kg/s}$$

Note that the worksheet formula evaluates the whole equation without using the two parts that are shown here; this was done to aid the explanation only.

Cell L91

(E91 −K91)*100/E91

The flow error is calculated. This is the error between the water flow that is required to satisfy the heat load with the allowance for pipe heat emission, and the maximum water flow capacity of the selected pipe diameter at the pressure drop rate chosen. The user enters successive values for pressure drop rate until the error is less than ±1%.

$$\text{error} = \frac{Q_1 - Q_2}{Q_1} \times 100\%$$

where Q_1 = design water flow rate (l/s), and Q_2 = maximum water flow rate (l/s).

$$L91 = \frac{E91 - K91}{E91} \times 100\%$$

EXAMPLE 8.4

Calculate the flow error for a pipe that has a required water flow rate of 1.75 l/s and a maximum carrying capacity of 1.80 l/s at the design pressure loss rate.

$$\text{error} = \frac{Q_1 - Q_2}{Q_1} \times 100\%$$

$$= \frac{1.75 - 1.8}{1.75} \times 100\%$$

$$= -2.9\%$$

An error of greater than −1% should be reduced by further iteration of the pressure drop rate. This will ensure that the closest attainable match is made between the design flow rate and the actual flow rate that is selected for the pipe system. Often, a change of 5 Pa/m in the pressure drop rate will be sufficient, and the user should continue with entries for dp_1 Pa/m until the closest result is achieved. It will only take a few seconds of work by the user to achieve the best result.

Cell M91

+K91*1000/$C$47

The maximum water flow rate is converted into l/s.

$$Q_2 = M \frac{kg}{s} \times \frac{m^3}{\rho\, kg} \times \frac{1000\, litres}{1\, m^3} \frac{litres}{s}$$

$$M91 = \frac{K91 \times 1000}{\$C\$47} \frac{litres}{s}$$

Cell N91

+M91*4/1000/@PI/(J91/1000)^2

Water velocity is found from

$$v = \frac{4}{\pi} \times \frac{1}{d^2\, mm^2} \times \frac{10^6\, mm}{1\, m^2} \times Q_2 \frac{litres}{s} \times \frac{1\, m^3}{1000\, litres}$$

$$N91 = \frac{4 \times M91 \times 10^6}{\pi \times (J91)^2 \times 10^3} \frac{m}{s}$$

Cell O91

0.81*K91^2/G91/$C$47/(J91/1000)^4

The equivalent length value for the pipe is

$$L_e = \frac{0.81 M^2 \times (10^3)^4}{dp_1 \rho d^4}\ m$$

where L_e = length of pipe equivalent to one velocity head (m); M = mass flow rate of water (kg/s); dp_1 = frictional pressure drop rate (Pa/m); d = internal diameter of pipe (mm); and ρ = density of water (kg/m^3).

$$O91 = \frac{0.81 \times (K91)^2 \times (10^3)^4}{G91 \times \$C\$47 \times (J91)^4}\ m$$

EXAMPLE 8.5

Hot water flows at a rate of 2.73 kg/s through a medium-grade steel pipe at a pressure loss rate of 82.5 Pa/m. The water density is 974.3 kg/m^3. The pipe has a nominal size of 65 mm and an internal diameter of 68.65 mm. Calculate the equivalent length figure that will be used to find the frictional resistance of the pipe and the water velocity.

$$L_e = \frac{0.81 M^2 \times (10^3)^4}{dp_1 \rho d^4} \text{ m}$$

$$= \frac{0.81(2.73)^2 \times (10^3)^4}{82.5 \times 974.3 \times (68.65)^4} \text{ m}$$

$$= 3.38 \text{ m}$$

$$Q_2 = M \frac{\text{kg}}{\text{s}} \times \frac{\text{m}^3}{\rho \text{ kg}} \times \frac{1000 \text{ litres}}{1 \text{ m}^3}$$

$$= \frac{2.73 \times 1000}{974.3} \frac{\text{litres}}{\text{s}}$$

$$= 2.802 \text{ litres/s}$$

$$v = \frac{4Q_2 \times 10^6}{\pi d^2 \times 10^3} \frac{\text{m}}{\text{s}}$$

$$= \frac{4 \times 2.802 \times 10^6}{\pi \times 68.65^2 \times 10^3} \text{ m/s}$$

$$= 0.76 \text{ m/s}$$

This can be verified from the reference (CIBSE, 1986c).

Cell Q91

+C91+O91*P91

The total equivalent length of the pipe section is

$$EL = L + L_e k \text{ m}$$

where EL = total equivalent length of the index pipe section (m); L = measured length of the index pipe section (m); L_e = equivalent length of one velocity head for the section (m); and k = sum of the velocity pressure loss factors in the pipe section (dimensionless).

The total equivalent length of a section of pipework is the sum of the measured lengths of straight pipe and the equivalent length of straight pipe that is caused by the pipe fittings. The resistance to flow of the pipe fittings (the bends, tees, changes in diameter, boiler, radiator and valves) is converted into an equivalent length of straight pipe.

$$Q91 = C91 + O91 \times P91 \text{ m}$$

EXAMPLE 8.6

A section of the index pipe route in a heating system has a measured length of straight flow and return pipes of 15 m. The equivalent length of one velocity head for the pipe is 3.38 m. The sum of the velocity head pressure loss factors for the pipe section is 6.5. Calculate the frictional resistance of the pipe section.

$$EL = L + L_e k \text{ m}$$

$$= 15 + 3.38 \times 6.5 \text{ m}$$

$$= 36.97 \text{ m}$$

Thus this section of pipe has an equivalent length of 36.97 m of straight pipe. This equivalent length is multiplied by the pressure drop rate for the section to find the pipework resistance. The section may have additional hydraulic resistance from a heat changer or valve that has a known pressure drop.

Cell R91

+Q91*G91/1000

The pressure drop through the pipe is

$$dp = EL\,\text{m} \times dp_1 \frac{\text{Pa}}{\text{m}} \times \frac{1\,\text{kPa}}{10^3\,Pa}$$
$$R91 = Q91\,\text{m} \times G91\frac{\text{Pa}}{\text{m}} \times \frac{1\,\text{kPa}}{10^3\,Pa}$$
$$= Q91 \times G91\frac{1}{10^3}\,\text{kPa}$$

EXAMPLE 8.7

A section of a two-pipe heating hot water system has an equivalent length of 36.97 m of straight pipe and a design pressure drop rate of 82.5 Pa/m. Calculate the hydraulic resistance of the pipe section.

$$dp = EL\,\text{m} \times dp_1 \frac{\text{Pa}}{\text{m}} \times \frac{1\,\text{kPa}}{10^3\,Pa}$$
$$= 36.97\,\text{m} \times 82.5\frac{\text{Pa}}{\text{m}} \times \frac{1\,\text{kPa}}{10^3\,\text{Pa}}$$
$$= 3.05\,\text{kPa}$$

Cell T91

+R91+S91

The pressure drop through the section of pipe includes the straight pipe, pipe fittings and plant resistances such as boiler, valves and the index heat emitter.

$$dp = dp_1 + dp_2\,\text{kPa}$$
$$T91 = R91 + S91\,\text{kPa}$$

EXAMPLE 8.8

The final section of a two-pipe heating hot water system has a pipework hydraulic resistance of 3.05 kPa. This pressure drop includes the straight pipe and the pipe fittings. The pipe section connects to a heating coil in a ductwork system. The hydraulic resistance of the heating coil is 25 kPa at the design water flow rate. A two-port modulating valve controls the water flow rate through the coil. The valve has an authority of 0.5, meaning that it has a pressure drop of 50% of the resistance through the coil. Calculate the pressure drop that must be provided by the pump to overcome the hydraulic resistance of the pipe section and terminal equipment.

$$\text{section pressure drop} = 3.05\,\text{kPa}$$
$$\text{coil pressure drop} = 25\,\text{kPa}$$
$$\text{modulating valve pressure drop} = 0.5 \times 25\,\text{kPa}$$
$$= 12.5\,\text{kPa}$$
$$dp = dp_1 + dp_2 \text{ kPa}$$
$$dp_1 = 3.05\,\text{kPa}$$
$$dp_2 = 25 + 12.5\,\text{kPa}$$
$$= 37.5\,\text{kPa}$$
$$\text{Total } dp = 3.05 + 37.5\,\text{kPa}$$
$$= 40.55\,\text{kPa}$$

Cell V91

+A91

The pipe number is copied from cell A91 into cell V91 for information.

Cell W91

+B91

The section description is copied from cell B91 into cell W91 for information.

Cells E92 to W139

The formulae in line 91 are repeated for all the other sections of the pipe system.

Cell C120

@SUM (C91..C100)+@SUM(C109..C118)

The total length of the index circuit is found by summation of the section lengths.

Cell Q120

@SUM (Q91..Q100)+@SUM(Q109..Q118)

The total equivalent length of the index circuit is found from

$$EL = EL_{1-10} + EL_{11-20}\,\text{m}$$

Cell R120

@SUM(R91..R100)+@SUM(R109..R118)

The total pressure drop through the pipework in the index circuit is found from

$$dp = dp_{1-10} + dp_{11-20}\,\text{kPa}$$

Cell S120

@SUM(S91..S100)+@SUM(S109..S118)

The total of the plant pressure drops on the index circuit is found from

$$\text{plant } dp = dp_{1-10} + dp_{11-20} \text{ kPa}$$

Cell T120

@SUM(T91..T100)+@SUM(T109..T118)

The total pressure drop through the index circuit is the sum of the pressure drops for the pipework, fittings and plant:

$$\text{Total } dp = dp_{1-10} + dp_{11-20} \text{ kPa}$$

Cell D144

+E91

The total water flow rate through the pump is that passing through the first section of pipe. It is copied from cell E91 into cell D144.

Cell D146

+T120

The pressure rise at the circulating pump is equal to the drop of pressure in the index circuit from cell T120.

Cell D148

+E46

The mean water temperature is copied from cell E46 into cell D148.

Cell D150

+C54

The total heat load on the pipework system is copied from cell C54 into cell D150.

Cell D152

+C54+E56

The total heat load carried by the system is equal to the sum of the heat load provided at the terminal heat emitters plus the pipe heat emission:

$$\text{heat load} = \text{emitter kW} + \text{pipe emission kW}$$
$$D152 = C54 + E56 \text{ kW}$$

Cell AD88

+A69

The pipe diameter in cell A69 is copied from cell A69 into cell AD88. Cell A69 contains the smallest diameter of the range of pipe material that has been selected by the user.

There is no need for the user to look at the cells to the right of column X and below line 81. The cell range from Y81 to AO141 contain the comparison formulae to find the total length of each pipe size from the range that has been selected by the user in cells A69 to A79.

Note that the user must copy the correct range of pipe nominal and actual internal diameters into the cell range A69 to A79 for the type of pipe material to be used. These cells are used by the worksheet to find the total length of each pipe size used in the system.

Cells AE88 to AN88

The selected pipe diameters are copied into these cells as for cell AD88.

Cell Y91

+A91

The number of the first section of pipe is copied from cell A91 into cell Y91.

Cell range Y92 to Y100

The number of each subsequent section of pipe is copied into the cells as for cell Y91.

Cell Z91

+I91

The pipe diameter in the first section of the pipework system is copied from cell I91 into cell Z91.

Cells Z92 to Z100

The pipe diameters in the following sections of the pipework system are copied as for cell Z91.

Cell AA91

+C91

The length of the pipe in the first section of the pipework system is copied from cell C91 into cell AA91.

Cells AA92 to AA100

The pipe lengths in the following sections of the pipework system are copied as for cell AA91.

Cells AC91 to AC101

The selected pipe diameters are copied into these cells as for cell AD88.

Cell AD91

@IF($Z91=$AC91,$AA91,0)

This formula checks whether the pipe diameter in the first section of pipe is equal to the smallest diameter of the specified range of pipe sizes. If they are equal, the length of that pipe is copied into cell AD91 from where it was entered onto the worksheet by the user, in cell C91. If the test fails, the pipe size for that section has not been found, and zero length is put into cell AD91 against the smallest diameter in the range.

Cells AD92 to AD101

The same test is made for each pipe section to find the lengths of the smallest diameter in the first 10 sections of the index pipe system, as for cell AD91.

Cell AD102

@SUM(AD91..AD101)

The total length of the smallest pipe diameter in the first 10 sections of the index pipework is found by addition.

Cells AE91 to AN102

The index pipe lengths are found for each pipe diameter for the whole range of pipe sizes. This is carried out as for the smallest pipe size in the cell range AD91 to AD102.

Cells Y109 to AN120

This cell range finds the lengths of each pipe size in the second block of the index pipework system as for the first 10 sections.

Cells Y130 to AN141

This cell range finds the lengths of each pipe size in the branch pipes. The branch pipes connect into the index circuit but do not contribute to the pump pressure rise. The formulae are similar to those for the first block of 10 pipes of the index circuit in cells Y88 to AN102.

Cell Y150

+A69

The smallest pipe diameter is copied from cell A69 into cell Y150.

Cells Y151 to Y160

The pipe diameters are copied from the user's input data as for cell Y150.

Cell AA150

+AD$102+AD$120+AD$141

The total length of the smallest pipe diameter is added from the subtotals of each of the blocks of pipe sections:

$$L_{15} = L_1 + L_2 + L_3 \text{ m}$$

Cells AA151 to AA160

The total length of each pipe diameter is found as for cell AA150.

Cell AD150

+AA150*AC150*@ABS($E$46–$C$50)/1000

The heat emission to or from the smallest diameter pipework is found from

$$H = L \text{ m} \times E \frac{\text{W}}{\text{m K}} \times |(MWT - t_a)| \text{ K} \times \frac{1}{1000} \text{ kW}$$

where E = heat emission from pipe (W/m K); MWT = mean water temperature (°C); and t_a = ambient temperature (°C). The absolute temperature difference is used to calculate the heat emission to avoid a negative value for chilled water systems.

Cells AD151 to AD160

The heat emission to or from the subsequent pipe diameters is found as for the smallest size.

Cell AD162

@SUM(AA150..AA160)

The total heat emission to or from the whole of the pipework system is found.

Cell Y170

+A69

The smallest pipe diameter is copied from cell A69 into cell Y170.

Cells Y171 to Y180

The pipe diameters are copied from the user's input data as for cell Y170.

Cell AA170

+AD$102+AD$120+AD$141

The total length of the smallest pipe diameter is added from the subtotals of each of the blocks of pipe sections:

$$L_{15} = L_1 + L_2 + L_3 \, \text{m}$$

Cells AA171 to AA180

The total length of each pipe diameter is found as for cell AA170.

Cell AD170

+AA170*AC170

The installed cost of the smallest diameter pipework is found from

$$C = L\,\text{m} \times C_1 \, \frac{£}{\text{m}}$$

where C_1 = total installed cost per metre of pipework (£/m).

Cells AD171 to AD180

The total installed cost of the subsequent pipe diameters is found as for the smallest size.

Cell AA182

@SUM(AA170..AA180)

The total length of the installed pipework is found for information.

Cell AD182

@SUM(AD170..AD180)

The total installed cost of the whole of the pipework system is found.

Cell E60

+AD182

The total installed cost of the pipework system is copied from cell AD182. This information can be used to try different combinations of pipe sizes and route to assess the overall cost of alternative designs.

Cells K170 to K180

+A170

The range of pipe sizes that has been selected is copied into the cell range K170 to K180. This is to evaluate the installed cost for each pipe size.

Cell Q170

($M$156*L170*(1−$P$156/100)*(1+$S$156/100)+$M$158*N170*(1−$P$158/100)*(1+$S$158/100)+$M$157*M170*(1−$P$157/100)*(1−$S$157/100)+$M$159*O170*($P$159/100)*(1−$S$159/100)+$M$156*P170*(1-$P$160/100)*(1−$S$160/100))/$M$156

The total installed cost of a 1 m length of pipe is calculated from the elemental costs. These costs are for the typical installation in a two-pipe heating or chilled water system. A 5 m length of installed pipe is expected to have two elbows, one tee, two ring brackets and 25 mm thick fibreglass thermal insulation. The pipe length, number of pipe fittings, material price, discount and addition for waste, overhead charges and profit (Spon, 1992) can all be changed by the user. New data can be entered as indicated in the input data description. The cost of 1 m of installed pipe is found from

$$\text{cost} = \frac{\text{sum } \pounds(\text{pipe} + \text{elbows} + \text{tee} + \text{brackets} + \text{insulation})}{\text{pipe length m}}$$

The cost of each item is found from

$$\text{cost} = \text{number} \times \pounds\text{cost} \times \left(1 - \frac{\text{discount \%}}{100}\right) \times \left(1 + \frac{WOHP\%}{100}\right)$$

where $WOHP$ = material waste, overhead charge and profit (%).

Cells Y170 to Y180

+A69

The pipe diameters are copied into the cells range Y170 to Y180.

Cells AA170 to AA180

+AD$102+AD$120+AD$141

The total length of each pipe size is found in the cell range AA170 to AA180.

Cells AC170 to AC180

+Q170

The cost of 1 m of installed pipe is copied into the cell range AC170 to AC180.

Cells AD170 to AD180

+AA170*AC170

The total installed cost of each pipe size is calculated in the cell range AD170 to AD180.

$$\text{cost} = \text{length m} \times C\frac{\pounds}{m}$$

$$\pounds(\text{AD170}) = \text{AA170 m} \times \text{AC170} \frac{\pounds}{m}$$

Cell AD182

@SUM(AD170..AD180)

The total cost of the installed pipe system is calculated from the sum of the individual size costs.

QUESTIONS

All the numerical questions are to be evaluated on the worksheet.

1. State the factors that are taken into consideration when designing a two-pipe water circulation system. Include the design factors that are used within this chapter and the additional matters that relate to the space available for the pipe system in the building.

2. Explain what is meant by the index circuit.
3. Explain why the branch pipes from the index circuit do not contribute towards the pump pressure rise.
4. State how the heat emission from the hot water pipework and the total installed cost of the system are used by the design engineer.
5. The two-pipe heating system that is shown in Fig. 8.2 is to be installed in an office building. The heat emitters are fan coil units. The boiler flow and return water temperatures are 82 °C and 71 °C.

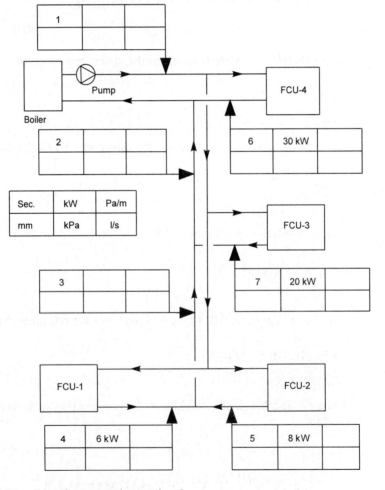

Fig. 8.2   Two-pipe fan coil heating system in question 5.

The ambient air temperature is 10 °C. The total pipe lengths, flow plus return pipework, and number of pipeline fittings are given in Table 8.9. The boiler is cast iron. Each heat emitter has a hydraulic resistance of 6 kPa. Copper pipe to BS 2871 Table X is to be used. The water velocity is not to exceed 1.3 m/s, and the pressure drop rate is not to exceed 800 Pa/m.

Table 8.9 Pipe data for the heating system in question 5

Section	1	2	3	4	5	6	7
Length, L (m)	20	10	15	12	8	6	4
Bends	6	0	0	4	4	4	4
Gate valves	3	0	0	2	2	2	2

Enter the data onto a copy of the original worksheet. Save this working copy under a new file name, A:PIPEQ5.WKS. Make sure that 1.0 is entered into cell ranges J95 to J100, J109 to J118 and J133 to J139. This ensures that a valid, but not significant, number is used for these unused pipe sections. Find the pipe diameters for the index circuit and for the three branch circuits.

Sketch the pipe schematic diagram, and enter the solutions into a data box for each section. Find the total system water flow rate, pump pressure rise and total system heat load.

6. A two-pipe heating system that serves a system of five panel radiators is shown in Fig 8.3. The boiler flow and return water temperatures are 82 °C and 71 °C. The ambient air temperature is 10 °C. The total pipe lengths and number of pipeline fittings are given in Table 8.10. The water velocity is not to exceed 1.2 m/s. The boiler is cast iron. Each radiator has two angle valves. Copper pipe to BS 2871 Table X is to be used. Find the pipe diameters for the index circuit and for the branch circuits. Sketch the pipe schematic diagram, and enter the solutions into a data box for each section. Find the total system water flow rate, pump pressure rise and total system heat load.

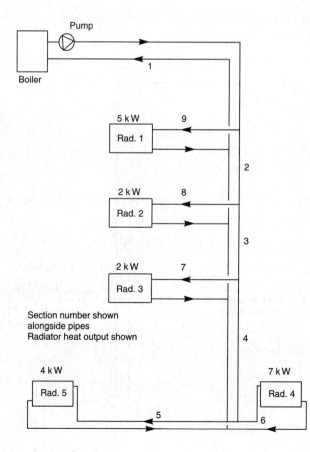

Fig. 8.3 Panel radiator system in question 6.

Table 8.10 Pipe data for the panel radiator heating system in question 6

Section	1	2	3	4	5	6	7	8	9
Length, L (m)	24	10	8	10	14	6	6	6	6
Bends	6	4	2	2	6	4	4	4	4
Gate valves	3	0	0	0	0	0	0	0	0

7. A two-pipe heating system that serves a group of three buildings is shown in Fig. 8.4. The boiler flow and return water temperatures are 80 °C and 70 °C. The ambient air temperature in the below-ground pipe service duct is 10 °C. The pipe lengths and number of pipeline fittings are given in Table 8.11. The water velocity is not to exceed 1.5 m/s. A fire tube horizontal steel boiler is used. The water flow rate is measured at the entry to each building for heat metering. There are three gate valves at the entry to each building, as shown. The index circuit in each building has a total of 200 m of pipework, four gate valves and a fan coil convector that has a hydraulic resistance of 10 kPa. The water flow meter has a hydraulic resistance of 15 kPa. Black medium-grade mild steel pipework to BS 1387 with screwed fittings is to be used. Find the pipe diameters for the index circuit and for the branch circuits. Sketch the pipe schematic diagram and enter the solutions into a data box for each section. Find the total system water flow rate, pump pressure rise and total system heat load and cost.

Table 8.11 Pipe data for the group heating system in question 7

Section	1	2	3	4	5
Length, L (m)	100	40	300	120	120
Bends	8	4	12	10	10
Gate valves	3	0	7	7	7
Plant kPa	0	0	10	10	10

8. A two-pipe heating system that serves four heater coils in a ducted heating and ventilation system is shown in Fig. 8.5. The boiler flow and return water temperatures are 80 °C and 70 °C. The ambient air temperature around the insulated pipe system is 12 °C. The pipe lengths and number of pipeline fittings are given in Table 8.12. The water velocity is not to exceed 1.0 m/s because of noise considerations. A cast iron boiler is used. There are three gate valves on each heater coil circuit. Heater coils HC-1, HC-2 and HC-3 have a hydraulic

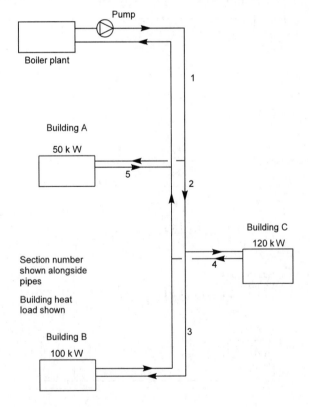

Fig. 8.4   Heating system for the group of buildings in question 7.

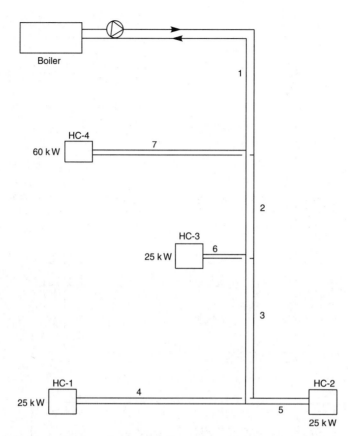

Fig. 8.5 Air heating coil pipework system in question 8.

resistance of 10 kPa. Heater coil HC-4 has a hydraulic resistance of 30 kPa. Black medium-grade mild steel pipework to BS 1387 with screwed fittings is to be used.

Make an assumption as to which will be the index pipe circuit, and find the pipe diameters. Inspect the pressure drops in the system, and decide which is the index circuit. If necessary, enter the correct data in the index circuit cells and for the branch circuits, to establish the correct data presentation. Sketch the pipe schematic diagram, and enter the solutions into a data box for each section. Find the total system water flow rate, pump pressure rise and total system heat load and cost.

9. A water-chilling refrigeration machine supplies the three cooling coils shown in Fig. 8.6 through a two-pipe system. Water leaves the chiller at 6 °C and returns at 11 °C. Copper pipes to BS 2871 Table X are used, and the pipe system data is given in Table 8.13. The pipes are insulated, and run through roof spaces where the air temperature is 25 °C. The water velocity is not to exceed 1 m/s. There are three gate valves on each cooling coil circuit. Cooling coils CC-2 and CC-3 have a hydraulic resistance of 10 kPa while coil CC-1 has a hydraulic resistance of 15 kPa. Sketch the pipe schematic diagram, and enter the solutions into a data box for each section. Find the total system water flow rate, pump pressure rise and total system heat load and cost.

Table 8.12 Pipe data for the heating coil system in question 8

Section	1	2	3	4	5	6	7
Length, L (m)	30	10	15	20	10	10	35
Bends	8	4	0	6	6	6	8
Gate valves	3	0	0	2	2	2	2
Plant kPa	0	0	0	10	10	10	30

Table 8.13 Pipe data for the chilled water cooling coil system in question 9

Section	1	2	3	4	5
Length, L (m)	10	20	15	6	6
Bends	4	4	6	4	4
Gate valves	3	0	2	2	2
Plant kPa	20	0	15	10	10

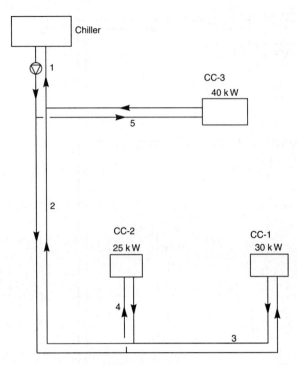

Fig. 8.6   Chilled water two-pipe system to air cooling coils in question 9.

SPREADSHEET **F**ACT 001 :

A SPREADSHEET IS AN
ELECTRONIC SHEET OF PAPER.

QUESTION : HOW BIG IS IT **?**

  (a) ONE SCREEN FULL   TICK ---
  (b) 100 SCREENS   ...
  (c) HAS NO LIMIT   ---
  (d) FOOTBALL OVAL   ---
  (e) PACIFIC OCEAN   ---
  (f) 20 BUSES   ---
  (g) DUNNO, PASS ME A STUBBIE ---

ANSWER : (a) IS CORRECT, BUT,
256 COLUMNS by 8192 ROWS
IS ABOUT THE PLAN AREA
OF 20 BUSES, OR 10500 SCREENS.

Spreadsheet fact 001.

9 Lighting

INTRODUCTION

The general artificial lighting of interiors is usually to provide a uniform illumination with fluorescent lamps. Desk level is frequently the illuminated working plane. This worksheet provides the lumen design method for general lighting of a single rectangular space. The design for additional areas and rooms is accomplished with repeated copies of this file. The user will save each room file under a unique name. The effect of localized lighting and glare is not calculated. Sufficient reference data is provided on the worksheet for the examples and questions and general design cases. The user will enter the required design illuminance and current photometric data for professional applications.

LEARNING OBJECTIVES

Study and use of this chapter will enable the user to:;

1. know the principles of the artificial lighting of interiors with the lumen method;
2. know the photometric data to be used for design;
3. use the utilization factor for installed lamp, luminaire and room combinations;
4. know the use of light loss factor;
5. calculate the number of luminaires and lamps for a lighting system;
6. know how to use luminaire space-to-height ratio;
7. design the layout of the luminaires;
8. calculate the installed electrical power for the lighting system;
9. know the electrical power load per square meter of floor area.

LUMEN DESIGN METHOD

The lumen design method for the artificial lighting of the interiors of rooms provides a uniform distribution of illuminance over the working plane. The working plane is frequently the same area as the floor, but at a raised height, typically at desk level. The intention of the method is to locate luminaires at a spacing between the rows and height above the working plane that will not create shadows or poorly illuminated areas (Chadderton, 1995, Ch. 11). Design values for illuminance vary from 50 to 2000 lx (lm/m^2), depending upon the detail to be viewed and the energy economy required (CIBSE, 1986a; Pritchard, 1985, Ch. 7).

The daylighting of interiors is accompanied by winter heat losses through the areas of glazing and the admission of solar heat gains during the warmer weather. Taken together with the consumption of electrical energy to provide the artificial illumination, it is clear that lighting is an energy-intensive service. The engineering designer aims to optimize the architectural, heating, air conditioning and electrical power components of the overall scheme in order to minimize the use of energy in the provision of an acceptable standard of service.

A selection of reference data is provided in the worksheet. The user will update the reference information and extend the range data in the worksheet in accordance with design office requirements. Lamp photometric data is on the reference screen in the same arrangement as is used in the calculation area of the worksheet. The user can copy the relevant data directly from the reference area to the calculation area with one COPY RANGE command. The data on the worksheet is for

1. illuminance levels in lux;
2. luminance factors for the room surfaces;
3. utilization factors for a typical combination of luminaire and room configuration (Philips, 1986);
4. fluorescent lamp data (GE, 1992).

The geometric configuration of the room is represented by the room index, *RI*:

$$RI = \frac{L_1 W_1}{H_3(L_1 + W_1)}$$

where *RI* = room index (dimensionless); L_1 = room length (m); W_1 = room width (m); H_3 = height from working plane to luminaire = $H_1 - H_2$ (m); and H_1 = height of the room (m), H_2 = height of working plane above the floor (m).

The surface finishes of the ceiling, walls and floor reflect the available light to create an overall illuminance. A luminance factor is attributed to each surface (BS

4800:1972). Typical values are 70% for a white ceiling, 50% for coloured walls, and 10% for flooring. The user can enter any value that is appropriate for the design.

Utilization factor is the ratio of the luminous flux received at the working surface to the installed flux. This ratio should be as high as possible for energy-efficient lighting. It is a combination of the lamp type, luminaire construction, room surface luminance factors and room dimensions. The designer normally refers to published utilization factors from manufacturers' data sheets. Sample values are provided in the worksheet. The data provided is for use within this chapter. The user will acquire current information for professional design.

The number of luminaires that are needed to provide the design illuminance are found from

$$\text{no. of luminaires} = \frac{\text{design lux} \times \text{working plane area } A_1 \text{ m}^2}{\text{luminaire lumens} \times UF \times LLF}$$

where A_1 = area of the working plane (m²); UF = utilization factor (dimensionless); LLF = overall light loss factor (dimensionless); and MF = maintenance factor (dimensionless).

$$LLF = \text{lamp lumen } MF\% \times \frac{\text{luminaire } MF\%}{100} \times \frac{\text{room surface } MF\%}{100}$$

Lamp lumen maintenance factor represents the performance of the lamp at 2000 h of use; this is typically 92% of the initial lumens (Pritchard, 1985). Frequent cleaning of the room surfaces and luminaires, at three-or six-monthly intervals, will ensure that the room surface and luminaire maintenance factors within clean environments have an average value of about 95%. Each design of luminaire has a recommended maximum spacing-to-height ratio. This defines the maximum allowable spacing between the rows of luminaires to provide a shadow-free distribution of light on the working plane:

$$SHR = \frac{S_1}{H_3}$$

where SHR = maximum spacing-to-height ratio (dimensionless); S_1 = space between rows of luminaires (m); and S_2 = space between luminaire perimeter row and wall (m).

When the number of luminaires has been calculated, the designer decides the distribution of the luminaires on a reflected ceiling plan. The lighting system is integrated with the position of the working planes, the air conditioning supply air diffusers and return air grilles, smoke and fire detectors, fire sprinkler heads, emergency and exit lighting, equipment being suspended from the ceiling, architectural features. The number of luminaires in a row and the number is determined by iteration of the options on the reflected ceiling p options are entered into the worksheet until a satisfactory solutio The worksheet shows the illuminance that will be provided by row spacing and the electrical power consumption by the illuminance produced is found from

$$\text{illuminance} = \frac{\text{no. of luminaires} \times \text{luminaire}}{A_1}$$

The electrical power consumption of the lighting system is

$$\text{power} = \text{luminaires} \times \text{lamps per luminaire} \times (W_1 + W_2)\,\text{W}$$

where $W_1 = $ lamp power (W) and $W_2 = $ lamp control gear power (W).

The lighting system electrical power consumption per square metre of room floor area is

$$\text{power} = \frac{\text{luminaires} \times \text{lamps per luminaire} \times (W_1 + W_2)}{A_1\,\text{m}^2}\,\text{W}$$

EXAMPLE 9.1

A uniform illuminance of 400 lx is to be provided in the first floor general office shown in Fig. 9.1. The working plane is to be taken as the same as the floor area. The lighting system will be permanently switched on from 8.00 a.m. to 6.00 p.m., 5 days per week for 52 weeks per year. The building has a complete internal cleaning twice per year. The ceiling has white acoustic tiles, luminance 70%, the walls are painted cream, luminance 50%, and the light grey carpet has a luminance of 10%. The maintenance factors are to be taken as 95% for the room surfaces and 95% for the luminaires. Fluorescent triphosphor Polylux 840 lamps are to be selected. The lamp lumen maintenance factor is 92% at 2000 h of use. Each lamp consumes 36 W plus 20 W by the control gear, is 1200 mm long, has an initial luminous flux of 3450 lm, a colour rendering index of 80 and a corrected colour temperature of 4000 K. The luminaires have two lamps each, and are 1300 mm long open-bottom reflectors that are surface mounted beneath the ceiling tiles. The maximum spacing-to-height ratio for the luminaires is 1.5. The manufacturer's utilization factor for the combination of lamp, luminaire and room data is 62%.

$$H_1 = 3\,\text{m}$$
$$H_2 = 0.8\,\text{m}$$
$$L_1 = 10\,\text{m}$$
$$W_1 = 10\,\text{m}$$
$$H_3 = H_1 - H_2\,\text{m}$$
$$= 3 - 0.8\,\text{m}$$
$$= 2.2\,\text{m}$$
$$RI = \frac{L_1 W_1}{H_3(L_1 + W_1)}$$
$$= \frac{10 \times 10}{2.2 \times (10 + 10)}$$
$$= 2.273$$
$$A_1 = 10\,\text{m} \times 10\,\text{m}$$
$$= 100\,\text{m}^2$$

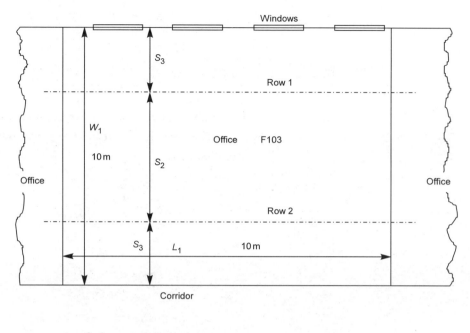

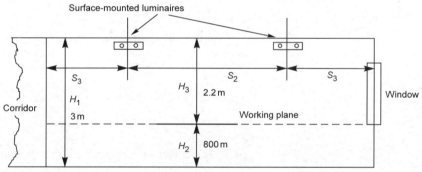

Fig 9.1 Office in Example 9.1

$$LLF = \text{lamp lumen } MF\% \times \frac{\text{luminaire } MF\%}{100} \times \frac{\text{room surface } MF\%}{100}$$

$$= 92\% \times \frac{95\%}{100} \times \frac{95\%}{100}$$

$$= 83.03\%$$

$$\text{luminaire lumens} = \frac{2 \text{ lamps}}{\text{luminaire}} \times \frac{3450 \text{ lm}}{\text{lamp}}$$

$$= 6900 \text{ lm}$$

$$\text{no. of luminaires} = \frac{\text{design lux} \times \text{working plane area } A_1 \text{ m}^2}{\text{luminaire lumens} \times UF \times LLF}$$

$$= \frac{400\,\text{lx} \times 100\,\text{m}^2}{6900\,\text{lm}} \times \frac{100}{62} \times \frac{100}{83.03}$$

$$= 11.26\,\text{luminaires}$$

12 luminaires are needed to provide 400 lx.

$$SHR = 1.5$$

$$S_1 = \text{maximum space between rows of luminaires (m)}$$

$$S_2 = \text{space between luminaire perimeter row and wall (m)}$$

$$= 0.5 S_1\,\text{m}$$

$$SHR = \frac{S_1}{H_3}$$

$$S_1 = SHR \times H_3\,\text{m}$$

$$= 1.5 \times 2.2\,\text{m}$$

$$= 3.3\,\text{m}$$

$$S_2 = 0.5 \times 3.3\,\text{m}$$

$$= 1.65\,\text{m}$$

$$\text{luminaire length} = 1300\,\text{m}$$

$$= 1.3\,\text{m}$$

$$\text{maximum luminaires in a row} = \frac{10\,\text{m}}{1.3\,\text{m}}$$

$$= 7\,\text{luminaires}$$

$$\text{required rows of luminaires} = 12\,\text{luminaires} \times \frac{1\,\text{row}}{7\,\text{luminaires}}$$

$$= 2\,\text{rows}$$

$$\text{possible number of luminaires} = 2\,\text{rows} \times \frac{7\,\text{luminaires}}{1\,\text{row}}$$

$$= 14\,\text{luminaires}$$

The spacing of two rows will be

$$S_1 = \frac{10\,\text{m}}{2\,\text{rows}}$$

$$= 5\,\text{m}$$

The proposed spacing is greater than the maximum allowable distance, 3.3 m, for the correct distribution of light. Try three rows of four luminaires per row, to provide the required 12 luminaires:

$$S_1 = \frac{10\,\text{m}}{3\,\text{rows}}$$

$$= 3.333\,\text{m}$$

$$S_2 = 0.5 \times 3.333\,\text{m}$$

$$= 1.667\,\text{m}$$

The row spacing is marginally wider than the maximum, and the luminaires are spread longitudinally along each row. The designer may be able to decide that this is an acceptable solution. The average illuminance upon the working plane is

$$\text{illuminance} = \frac{\text{no. of luminaires} \times \text{luminaire lumens} \times UF \times LLF}{A_1 \, \text{m}^2}$$

$$= \frac{12 \times 6900 \, \text{lm}}{100 \, \text{m}^2} \times \frac{62}{100} \times \frac{83.03}{100}$$

$$= 426 \, \text{lx}$$

The illuminance provided exceeds the design value by

$$\text{excess illuminance} = \frac{426 - 400}{400} \times 100\%$$

$$= 6.5\%$$

This is certainly acceptable. The designer may allow up to, say, 25% excess illuminance, 500 lx in this case. The lighting system electrical power consumption is

$$\text{power} = \text{luminaires} \times \text{lamps per luminaire} \times (W_1 + W_2) \, \text{W}$$

where W_1 = lamp power (W), and W_2 = lamp control gear power (W).

$$W_1 = 36 \, \text{W}$$

$$W_2 = 20 \, \text{W}$$

$$\text{power} = 12 \times 2 \times (36 + 20) \, \text{W}$$

$$= 1344 \, \text{W}$$

$$\text{power} = \frac{1344 \, \text{W}}{100 \, \text{m}^2}$$

$$= 13.44 \, \text{W/m}^2$$

DATA REQUIREMENT

The user will enter new data for:

1. Your name in cell B4.
2. Job title in cell B5.
3. Job reference in cell B6 and C6.
4. Filename in cell B10.
5. Room name in cell D25.
6. Room number in cell D26.
7. Room location in cell D27.
8. Room use in cell D28.
9. Hours of lighting use per day in cell D29.
10. Days of lighting use per week in cell D30.
11. Weeks of lighting use per year in cell D31.
12. Number of weeks between lamp and luminaire cleaning in cell D33.
13. Luminance of the ceiling in cell D34.
14. Luminance of the walls in cell D35.
15. Luminance of the floor in cell D36.
16. Maintenance factor for the room surfaces in cell D37.

17. Illuminance required for the working plane, in lumens, in cell D38.
18. Room length in cell D48.
19. Room width in cell D49.
20. Room height in cell D50.
21. Height of the working plane above the floor in cell D52. This is normally the desk height.
22. Length of the working plane in cell D56. This may not be the same as the room length. If the lighting is to provide the design illuminance over the whole floor area at the required height of the working plane, the two lengths are the same.
23. Width of the working plane in cell D57.
24. The type of lamp in cell D65. The range of cells from D65 to D75 can be copied from the reference data cells as required. Enter fluorescent, discharge, incandescent or halogen.
25. Lamp lumen maintenance factor in cell D66. Enter the manufacturer's factor when available. A typical maintenance factor is 92% for a discharge fluorescent lamp after 2000 h use (Pritchard, 1985).
26. Manufacturer's lamp code in cell D67.
27. Type of coating on the interior surface of the lamp tube in cell D68. This is for information only.
28. Trade name of the lamp in cell D69.
29. Electrical power consumption of the lamp in cell D70.
30. Lamp length in cell D71.
31. Initial luminous flux from the lamp in cell D72.
32. Colour rendering index for the lamp in cell D73. This is for information only.
33. Corrected colour temperature of the lamp in cell D74. This is for information only.
34. Electrical power consumption of the electrical control equipment for the lamp in cell D75. Discharge lamps have a ballast resistance which consumes power, in series with the lamp. Enter the manufacturer's data.
35. Type of luminaire in cell D84. This is for information only. Enter a brief description, such as OBR for open-bottom reflector or PD for prismatic diffuser, as required.
36. Number of lamps in the luminaire in cell D85.
37. Maximum space-to-height ratio for the luminaire in cell D88. This is used to calculate the maximum spacing of the rows of luminaires to provide an even illuminance on the working plane. If the rows are too far apart, less well-lit areas are produced between the rows. These lower-illuminance strips may be the walkways or on filing cabinets.
38. Luminaire length in cell D89. This may be taken as the lamp length plus, say, 100 mm.
39. Luminaire maintenance factor in cell D90. A 13- or 26-week cleaning cycle in a clean office environment would have a factor of 95%.
40. Utilization factor in cell D95. This is obtained from the reference data cells or the manufacturer's data. Each combination of lamp, number of lamps, luminaire design and room configuration has a different utilization factor. Enter the correct value for the design.
41. Enter the number of luminaires that have been selected in cell D105.

42. Enter the number of luminaires to be used in each row in cell D112.
43. Enter the number of rows of luminaires to be used in cell D113. Various combinations of lamp and luminaire type and number can be tried.

OUTPUT DATA

1. The current date is given in cell B7.
2. The current time is given in cell B8.
3. The hours of lighting use are given in cell F32.
4. The lighting hours between cleaning are given in cell F33.
5. The room name is given in cell F45.
6. The room number is given in cell F46.
7. The room location is given in cell F47.
8. The height of the luminaire from the working plane is given in cell F54.
9. The room index is given in cell F55.
10. The area of the working plane is given in cell F58.
11. The area of the floor is given in cell F59.
12. The lamp luminous flux is given in cell F86.
13. The luminaire luminous flux is given in cell F87.
14. The ceiling luminance is given in cell F91.
15. The wall luminance is given in cell F92.
16. The floor luminance is given in cell F93.
17. The room index is given in cell F94.
18. The overall light loss factor for the luminaire is given in cell F96.
19. The area of the working plane is given in cell F97.
20. The illuminance required for the working plane is given in cell F98.
21. The theoretical number of luminaires needed is given in cell F104. The calculated number is displayed without being rounded.
22. The illuminance provided by the selected number of luminaires is given in cell F106.
23. The maximum space-to-height ratio for the luminaires is given in cell F107.
24. The height of the luminaire above the working plane is given in cell F108.
25. The maximum allowable spacing between the rows of luminaires is given in cell F109.
26. The number of luminaires in a row is given in cell F112.
27. The number of rows of luminaires is given in cell F113.
28. The calculated number of luminaires is given in cell F114.
29. The spacing between the rows that has been produced by this design is given in cell F115.
30. The spacing of the perimeter rows from the walls is half of the spacing between the rows of luminaires. This is shown in cell F116.
31. The spacing produced by the design is checked against the maximum allowable spacing of the rows of luminaires. If the spacing is more than that allowed, 'too wide' is shown in cell F117. If the spacing is equal to or less than that allowed, 'correct' is shown in cell F117.
32. The illuminance produced in the room by the design is checked against the design illuminance that was specified. If the illuminance is higher than 25% above that required, 'too high' is shown in cell F118. If the illuminance is not

too high, 'acceptable' is shown in cell F118. The term 'acceptable' here assumes that the designer has not chosen to use fewer luminaires than necessary to achieve the design illuminance. Beware that any illuminance below the 25% excess will be called acceptable.

33. The installed electrical power consumption of the lighting system, including lamps and control gear, in watts, is shown in cell F119.
34. The installed electrical power consumption of the lighting system, including lamps and control gear, in watts per square metre of the floor area, is shown in cell F120.

FORMULAE

Representative samples of the formulae are given here. Each formula can be read by moving the cursor to the cell. The equation is presented in the form in which it would normally be written and in the format that is used by the spreadsheet.

Cell B7

@TODAY

This function produces the serial number of the current day and time. The cell is formatted to display the date.

Cell B8

@TODAY

This function produces the serial number of the current day and time. The cell is formatted to display the time.

Cell F32

+D29*D30*D31

The anticipated operation hours for the artificial lighting are

$$T_1 = \frac{10\,\text{h}}{\text{day}} \times \frac{5\,\text{day}}{\text{week}} \times \frac{52\,\text{week}}{\text{year}}$$
$$= 2600\,\text{h/yr}$$
$$F32 = D29 \times D30 \times D31\,\text{h/yr}$$

Cell F33

+D29*D30*D33

The lighting hours between cleaning are

$$T_2 = \frac{10\,\text{h}}{\text{day}} \times \frac{5\,\text{day}}{\text{week}} \times \frac{26\,\text{week}}{\text{year}}$$
$$= 1300\,\text{h/yr}$$
$$F33 = D29 \times D30 \times D33\,\text{h/yr}$$

Cell F45

+D25

The room name is copied from cell D25 into cell F45.

Cell F46

+D26

The room number is copied from cell D26 into cell F46.

Cell F47

+D27

The room location is copied from cell D27 into cell F47.

Cell F54

+D50−D52

The height from the working plane to the installation height of the luminaires is

$$H_3 = \text{room height } H_1 - \text{working plane height } H_2 \text{ m}$$
$$F54 = D50 - D52 \text{ m}$$

Cell F55

+D48*D49/F54/(D48+D49)

The dimensionless room index is a measure of the overall size and shape of the room:

$$RI = \frac{L_1 \text{ m} \times W_1 \text{ m}}{H_3 \text{ m} \times (L_1 + W_1) \text{ m}}$$
$$F55 = \frac{D48 \times D49}{F54 \times (D48 + D49)}$$

Cell F58

+D56*D57

The surface area of the working plane is

$$A_1 = L_2 \text{ m} \times W_2 \text{ m}$$
$$F58 = D56 \times D57 \text{ m}^2$$

The working plane area will be the same as the floor area when a uniform illuminance is to be provided for the whole of the room.

Cell F59

+D48*D49

The floor area of the room is

$$A_2 = L_1 \text{ m} \times W_1 \text{ m}$$
$$F59 = D48 \times D49 \text{ m}^2$$

Cell F86

+D72

The initial luminous flux from the specified lamp is copied from cell D72 into cell F86.

Cell F87

+F86*D85

The initial luminous flux emitted from the luminaire is

$$\text{luminaire lm} = \text{lamp lm} \times \text{no. of lamps in luminaire}$$
$$F87 = F86 \times D85 \text{ lm}$$

Cell F91

+D34

The ceiling luminance is copied from cell D34 into cell F91.

Cell F92

+D35

The wall luminance is copied from cell D35 into cell F92.

Cell F93

+D36

The floor luminance is copied from cell D36 into cell F93.

Cell F94

+F55

The room index is copied from cell F55 into cell F94.

Cell F96

+D66*D90*D37/10^4

Overall light loss factor for the lighting installation is

$$LLF = \text{lamp lm } MF\% \times \frac{\text{luminaire } MF\%}{100} \times \frac{\text{room surface } MF\%}{100}$$
$$F96 = \frac{D66 \times D90 \times D37}{10^4}\%$$

Cell F97

+F58

The working plane area is copied from cell F58 into cell F97.

Cell F98

+D38

The design illuminance is copied from cell D38 into cell F98.

Cell F104

+F98*F97*10^4/F87/D95/F96

The number of luminaires that are needed to provide the design illuminance is

$$\text{no. of luminaires} = \frac{\text{design lx} \times A_1\,\text{m}^2 \times 10^4}{\text{luminaire lx} \times UF \times LLF}$$
$$F104 = \frac{F98 \times F97 \times 10^4}{F87 \times D95 \times F96}$$

Cell F106

+D105*F87*D95*F96/10^4/F97

The illuminance provided by the selected number of luminaires is

$$\text{illuminance} = \frac{\text{no. of luminaires} \times \text{luminaire lm} \times UF \times LLF}{A_1\,\text{m}^2}$$
$$F106 = \frac{D105 \times F87\,\text{lm}}{F97\,\text{m}^2} \times \frac{D95}{100} \times \frac{F96}{100}$$

Cell F107

+D88

The maximum spacing-to-height ratio for the rows of luminaires is copied from cell D88 into cell F107.

Cell F108

+F54

The height from the working plane to the luminaires is copied from cell F54 into cell F108.

Cell F109

+F107*F108

The maximum spacing between the rows of luminaires is

$$S_1 = SHR \times H_3\,\text{m}$$
$$F109 = F107 \times F108\,\text{m}$$

Cell F112

@INT(D48*1000/D89)

The number of luminaires that can be installed in a row that is parallel to the room length is

$$\text{no. of luminaires} = \frac{L_1 \text{ m}}{L_3 \text{ m}} \times \frac{1000 \text{ m}}{1 \text{ m}}$$

$$F112 = D48 \times \frac{1000}{D89}$$

The integer of the result is taken, because only the whole number is relevant.

Cell F113

+D105/F112

The number of rows of luminaires that are required is

$$\text{no. of rows} = \text{total no. luminaires} \times \frac{1 \text{ row}}{\text{no.luminaires}}$$

$$F113 = \frac{D105}{F112}$$

Cell F114

@INT(F112*D113)

The number of luminaires that are required is

$$\text{no. of luminaires} = \text{no. luminaires in row} \times \text{no. rows selected}$$

$$F114 = F112 \times D113$$

The integer of the result is taken, because only the whole number is relevant.

Cell F115

+D49/D113

The row spacing produced is

$$S_2 = \frac{W_1 \text{ m}}{\text{no. rows}}$$

$$F115 = \frac{D49}{D113} \text{ m}$$

Cell F116

+F115/2

The spacing between the perimeter row and the long wall is

$$S_3 = 0.5 S_2 \text{ m}$$

$$F116 = \frac{F115}{2} \text{ m}$$

Cell F117

@IF(F115>F109,"Too wide","Correct")

The row spacing that has been produced must not exceed the maximum luminaire spacing. If S_2 is greater than S_1, 'Too wide' is shown in cell F117. If S_2 is equal to, or less than, S_1, 'Correct' is shown in cell F117.

Cell F118

@IF(F106>=1.25*D38,"Too high","Acceptable")

The illuminance that will be generated by the proposed lighting system is compared with the design illuminance. The result of selecting the next whole number of luminaires above the theoretical quantity means that there will be more illuminance than required. The designer can choose whether to accept the higher illuminance. This cell provides a guide to that decision. When the illuminance is equal to or less than 125% of the design value, 'Acceptable' is shown in cell F118. Edit the 1.25 value as required.

Cell F119

+D85*F114*D70+D75*D85*F114

The installed electrical power consumed by the lighting system is

$$\text{power} = \text{no. lamps per luminaire} \times \text{no. luminaires} \times \text{lamp power}$$
$$+ \text{no. lamps per luminaire} \times \text{no. luminaires}$$
$$\times \text{control gear power W}$$
$$\text{F119} = \text{D85} \times \text{F114} \times \text{D70} + \text{D75} \times \text{D85} \times \text{F114 W}$$

Cell F120

+F119/F59

The electrical power loading from the lighting system per square metre of floor area is

$$\text{load} = \frac{\text{installed W}}{A_2 \, \text{m}^2}$$
$$\text{F120} = \frac{\text{F119 W}^2}{\text{F59 m}^2}$$

Questions

1. A uniform illuminance of 450 lx is to be provided in the office module shown in Fig. 9.2. The working plane is to be taken as the same as the floor area. The lighting system will be permanently switched on from 8.30 a.m. to 7.00 p.m., 5 days per week, for 50 weeks per year. The building has a complete internal cleaning twice per year. The ceiling has white acoustic tiles (luminance 80%), the walls are painted cream (luminance 60%), and the light grey carpet has a luminance of 10%. The maintenance factors are to be taken as 95% for the room surfaces and 95% for the luminaires. Fluorescent warm white triphosphor Polylux 840

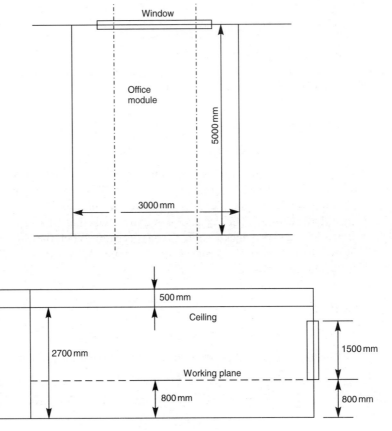

Fig 9.2 Office module in question 1.

lamp number 5 is to be selected. The lamp lumen maintenance factor is 92% at 2000 h of use. Each luminaire has one lamp, and is a 1300 mm long prismatic diffuser that is recessed into the ceiling space. The maximum spacing-to-height ratio for the luminaires is 1.43. The manufacturer's utilization factor for the combination of lamp, luminaire and room data is 45%.

2. A uniform illuminance of 400 lx is to be provided in the ground-floor open-plan office shown in Fig. 9.3. The working plane is to be taken as the same as the floor area, and is at a height of 800 mm. The lighting system will be permanently switched on from 8.00 a.m. to 6.00 p.m., 5 days per week, for 52 weeks per year. The building has a complete internal cleaning twice per year. The ceiling has white acoustic tiles (luminance 70%), the walls are painted cream (luminance 50%), and the light grey carpet has a luminance of 10%. The maintenance factors are to be taken as 95% for the room surfaces and 95% for the luminaires. The lamp

control gear has a power consumption of 20 W. The maximum spacing-to-height ratio for the luminaires is 1.9. Use the data provided on the worksheet to propose a suitable design for the lighting system.

3. An illuminance of 1000 lx is to be provided on an inspection table that is 5 m long, 2 m wide, and at a working plane height of 1 m above the floor. The floor has dimensions of 10 m × 5 m, and the room height is 3.5 m. The lighting system will be permanently switched on from 8.00 a.m. to 6.00 p.m., 5 days per week, for 52 weeks per year. The building has a complete internal cleaning twice per year. The ceiling has white acoustic tiles (luminance 70%), the walls are painted cream (luminance 50%), and the light grey carpet has a luminance of 10%. The maintenance factors are to be taken as 95% for the room surfaces and 95% for the luminaires. The maximum spacing-to-height ratio for the luminaires is 1.25. Use the data provided on the work-

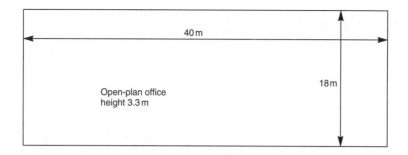

Fig 9.3 Open-plan office in question 2.

sheet to propose a lighting design for the inspection table.

4. A uniform illuminance of 100 lx is to be provided in a corridor that is 20 m long, 2 m wide and 3 m high in a hospital. The lighting system will be permanently switched on. The building has a complete internal cleaning every 3 months. The ceiling has white acoustic tiles (luminance 70%), the walls are painted cream (luminance 50%), and the light grey carpet has a luminance of 10%. The maintenance factors are to be taken as 95% for the room surfaces and 95% for the luminaires. The maximum spacing-to-height ratio for the surface-mounted luminaires is 2.1. Use the data provided on the worksheet to propose a lighting design for the corridor.

That is not the way to open a data file!

10 Electrical cable sizing

INTRODUCTION

This worksheet finds the size of cable that is required for single-phase, three-phase and extra low voltage single-phase electrical power and lighting circuits. The user enters the power load, power factor and the measured length of the conductor. The worksheet evaluates the phase current, the maximum allowable length of the conductor, the voltage drop, and whether each cable size will meet the resistance criteria. Cable data from any source can be entered and used. Sufficient reference data is provided in the worksheet for the examples and questions and general design cases. The user will enter the required reference data for professional applications.

LEARNING OBJECTIVES

Study and use of this chapter will enable the user to:

1. know the data needed to calculate the size of an electrical cable;
2. know the voltage drop limits for sizing cable;
3. know the current limits for cables;
4. know the cable sizes used;
5. calculate the phase current for single-and three-phase circuits;
6. calculate the maximum allowable length of cable to meet the voltage drop limitation;
7. calculate the voltage drop in a cable;
8. calculate the percentage voltage drop in a cable.

Key terms concepts

CABLE DESIGN METHOD

The current in a cable is found from the electrical power consumption of the load that is supplied, the voltage that is applied to the circuit, and the power factor of the load. This design current is compared with the current capacity of cables to find a suitable cross-sectional area. The voltage drop along the length of the proposed cable is calculated from the known resistance. The voltage drop in a cable from the supply point to the final load is limited, by regulation, to 4% of the applied voltage. The cable support method and how the cables are bunched together are taken into account when using the published cable data (IEE, 1991). Table 10.1 shows values of current rating and voltage drop for a selection of cables. A larger number of cables are listed on the worksheet for design applications. The simple approach (Jenkins, 1991) to the selection of cables is used, with the inclusion of the power factor (Chadderton, 1995, Ch. 13). Note that mV/A m is the resistance of 1 m of cable in milliohms (mΩ).

Table 10.1   Electrical cable capacities for unenclosed copper cables, twin-sheathed in PVC, clipped to the surface of the building

Nominal cross-sectional area of conductor (mm^2)	*Maximum current rating, I_b* (A)	*Voltage drop* (mV/A m)
1	15	44
1.5	19.5	29
2.5	27	18
4	36	11
6	46	7.3
10	63	4.4
16	85	2.8

(IEE, 1991)

The phase current in a three-phase 415 V circuit is

$$\text{current} = \frac{\text{power W}}{415\,\text{V} \times PF \times \sqrt{3}}\,\text{A}$$

The current in a single-phase 240 V circuit is

$$\text{current} = \frac{\text{power W}}{240\,\text{V} \times PF}\,\text{A}$$

The current in a single-phase 12 V circuit is

$$\text{current} = \frac{\text{power W}}{12\,\text{V} \times PF}\,\text{A}$$

where power = output electrical power (W), and *PF* = power factor (dimensionless).

The maximum allowable reduction in voltage in a circuit is 4% of the incoming voltage (IEE, 1991). The voltage drops are:

1. 415 V three phase, $V = \dfrac{4}{100} \times 415\,\text{V}$

   $= 16.6\,\text{V}$

2. 240 V single phase, $V = \dfrac{4}{100} \times 240\,\text{V}$

   $= 9.6\,\text{V}$

3. 12 V single phase, $V = \dfrac{4}{100} \times 12\,\text{V}$

   $= 0.48\,\text{V}$

The maximum length of cable that can be used so that the voltage drop limit is not exceeded is

$$L_1 = \frac{\text{maximum voltage drop allowed V} \times 10^3\,\text{m}}{\text{load current A} \times \text{voltage drop mV/A m}}$$

When the length of the conductor is known, from measurement on the building drawings or along the cable route, the voltage drop produced in that circuit is

$$\text{voltage drop} = L_1\,\text{m} \times I_b\,\text{A} \times \text{voltage drop} \frac{\text{mV}}{\text{A m}} \times \frac{1}{10^3}\,\text{V}$$

where I_b = design phase current (A).

EXAMPLE 10.1

A three-phase 415 V fan motor has an electrical power rating of 7.5 kW. The motor power factor is 0.75. The conductor from the main distribution board has a measured length of 60 m. The cable is installed in cable trays with other circuits. Use the cable data in Table 10.1 to find a suitable cable size.

The solution to this example is shown in the original copy of the file A:CABLE.WKS. The cable data in the worksheet is from the 16th edition of the reference regulation.

Allowable voltage drop is

$$415\,\text{V three phase,} \quad V = \frac{4}{100} \times 415\,\text{V}$$

$$= 16.6\,\text{V}$$

Phase current is

$$I_b = \frac{\text{power W}}{415\,\text{V} \times PF \times \sqrt{3}}\,\text{A}$$

$$= \frac{7.5 \times 10^3\,\text{W}}{415\,\text{V} \times 0.75 \times \sqrt{3}}\,\text{A}$$

$$= 13.9\,\text{A per phase}$$

Try a 1 mm^2 cable.

allowable voltage drop rate $= 44$ mV/A m
maximum current rating $= 15$ A
maximum cable length is

$$L_1 = \frac{\text{maximum voltage drop allowed V} \times 10^3}{I_b \text{ A} \times \text{voltage drop mV/A m}} \text{ m}$$

$$= \frac{16.6 \text{ V} \times 10^3}{13.9 \text{ A} \times 44 \text{ mV/A m}} \text{ m}$$

$$= 27 \text{ m}$$

The resistance of this cable is too high, because the allowable voltage drop will allow only 27 m of cable, whereas 60 m is required. If this cable was used, the voltage drop would be

$$\text{volt drop} = L_1 \text{ m} \times I_b \, A \times \text{voltage drop} \frac{\text{mV}}{\text{A m}} \times \frac{1}{10^3} \text{ V}$$

$$= 60 \text{ m} \times 13.9 \text{ A} \times 44 \frac{\text{mV}}{\text{A m}} \times \frac{1}{10^3} \text{ V}$$

$$= 36.7 \text{ V}$$

This would exceed the allowable drop of 16.6 V. Try the 2.5 mm^2 cable; allowable drop is 18 mV/A m.

$$L_1 = \frac{\text{maximum voltage drop allowed V} \times 10^3}{I_b \text{ A} \times \text{voltage drop mv/A m}} \text{ m}$$

$$= \frac{16.6 \text{ V} \times 10^3}{13.9 \text{ A} \times 18 \text{ mV/A m}} \text{ m}$$

$$= 66.3 \text{ m}$$

$$\text{voltage drop} = L_1 \text{ m} \times I_b \, A \times \text{voltage drop} \frac{\text{mV}}{\text{A m}} \times \frac{1}{10^3} \text{ V}$$

$$= 60 \text{ m} \times 13.9 \text{ A} \times 18 \frac{\text{mV}}{\text{A m}} \times \frac{1}{10^3} \text{ V}$$

$$= 15 \text{ V}$$

The 2.5 mm^2 cable has a maximum current capacity of 27 A and it produces a voltage drop that is within the 4% allowed. The percentage volt drop is

$$\text{voltage drop} = \frac{15}{415} \times 100\%$$

$$= 3.6\%$$

The proposed design complies with the allowable voltage drop and the current capacity.

DATA REQUIREMENT

The user will enter new data for:

1. Your name in cell B4.
2. Job title in cell B5.

3. Job reference in cells B6 and C6.
4. Filename in cell B9.
5. The electrical power load in cell C27.
6. Power factor in cell C29.
7. Conductor length to be installed in cell C31.
8. Circuit voltage in cell C33.
9. Cable number in column A. A consecutive series of numbers are in the worksheet, and these can remain unaltered.
10. A description of the cable installation in cells C35, D35, E35, F35 and G35. Enter a description such as: grouped cable, clipped direct. Enter the correct number of characters in each cell.
11. The normal range of cable cross-sectional areas is provided in column B from cell B53. New ranges of cable sizes can be entered as required.
12. The maximum current rating of each cable size is provided in column C from cell C53. New data can be entered in these cells as required.
13. The voltage drop in mV per A per m length of cable is provided in column D from cell D53. New data can be entered in these cells as required.

OUTPUT DATA

1. The current date is given in cell B7.
2. The current time is given in cell B8.
3. The voltage in a three-phase supply is given in cell D43.
4. The maximum allowable percentage voltage drop in a circuit is given in cell D44.
5. The maximum allowable voltage drop in a circuit is given in cell D45.
6. The power load carried by the cable is given in cell E53.
7. The power factor of the connected load is given in cell F53.
8. The phase current is given in cell G53.
9. The maximum length of the conductor is given in cell H53.
10. The measured length of the conductor is given in cell I53.
11. The voltage drop in the conductor is given in cell J53.
12. The voltage drop expressed as a percentage of the voltage that is applied to the circuit is given in cell K53.
13. The acceptability criteria for the voltage drop (either 'Allowed' or 'Too high') are shown in cell L53.
14. The current limit acceptability criteria for the current carried by the cable are given in cell M53. Either 'Allowed' or 'Too high' is shown.
15. The cell range E54 to M79 contains the equivalent output data to that in line 53.
16. The cell range A83 to M100 contains the equivalent output data to that on line 53 for 240 V single-phase circuits.
17. The cell range A103 to M117 contains the equivalent output data to that on line 53 for 12 V extra low voltage single-phase circuits.

FORMULAE

Representative samples of the formulae are given here. Each formula can be read by moving the cursor to the cell. The equation is presented in the form in which it would normally be written and in the format that is used by the spreadsheet.

Cell B7

@TODAY

This function produces the serial number of the current day and time. The cell is formatted to display the date.

Cell B8

@TODAY

This function produces the serial number of the current day and time. The cell is formatted to display the time.

Cell D45

+D43*D44/100

The maximum voltage drop that is allowable in a cable is found from the allowable percentage drop from

$$\text{voltage} = 415\,\text{V} \times \frac{4}{100}\,\text{V}$$

$$= 16.6\,\text{V}$$

$$\text{D45} = \text{D43} \times \frac{\text{D44}}{100}\,\text{V}$$

Cell range C46 to G46

+$C$35

The description of the cable installation is copied from cell C35 into C46; this is repeated along the range.

Cell E53

+$C$27

The circuit power load is copied from cell C27 into cell E53.

Cell F53

+$C$29

The power factor of the load is copied from cell C29 into cell F53.

Cell G53

+E53/$D$43/F53/@SQRT(3)

The phase current is

$$I_b = \frac{\text{power W}}{\text{volts} \times PF \times \sqrt{3}} \text{ A}$$

$$G53 = \frac{\text{E53 W}}{\text{D43 V} \times \text{F53} \times \sqrt{3}} \text{ A}$$

Cell H53

+$D$45*1000/G53/D53

The maximum allowable length of the conductor in a 415 V three-phase circuit is

$$L_1 = \frac{\text{maximum voltage drop allowed V} \times 10^3}{I_b \text{ A} \times \text{voltage drop mV/A m}} \text{ m}$$

$$= \frac{16.6 \text{ V} \times 10^3}{I_b \text{ A} \times \text{voltage drop mV/A m}} \text{ m}$$

$$H53 = \frac{\text{D45 V} \times 10^3}{\text{G53 A} \times \text{D53 mV/A m}} \text{ m}$$

Cell I53

+$C$31

The measured length of conductor is copied from cell C31 into cell I53.

Cell J53

+I53*G53*D53/1000

The actual voltage drop in the cable is

$$\text{voltage drop} = L_2 \text{ m} \times I_b \text{ A} \times \frac{\text{mV}}{\text{A m}} \times \frac{1}{1000} \text{ V}$$

$$J53 = I53 \text{ m} \times G53 \text{ A} \times \frac{\text{D53 mV}}{\text{A m}} \times \frac{1}{1000} \text{ V}$$

Cell K53

+J53*100/$D$43

The percentage voltage drop that is produced is

$$\text{voltage drop} = \text{voltage drop} \times \frac{100}{\text{voltage}} \%$$

$$\text{voltage drop} = J53 \text{ V} \times \frac{100}{\text{D43}} \%$$

Cell L53

@IF (K53<=$D$44,"Allowed","Too high")

This tests whether the voltage drop that is produced in the cable exceeds the maximum allowable voltage drop. If the actual voltage drop in cell K53 is less than or equal to the limit in cell D44, the cable volt drop is 'Allowed'. If the voltage drop exceeds the limit, the next message, 'Too high', is shown in cell L53.

Cell M53

@IF (G53<=C53,"Allowed","Too high")

This tests whether the current in the cable exceeds the maximum allowable current. If the circuit current in cell G53 is less than or equal to the limit in cell C53, the cable current is 'Allowed'. If the current exceeds the cable limit, the next message, 'Too high', is shown in cell M53.

Cell range E54 to M117

The formulae from line 53 are repeated in this cell range for the additional cables and single phase circuits.

EXAMPLE 10.2

A single-phase 240 V exhaust fan motor has an electrical power rating of 250 W. The motor power factor is 0.7. The conductor from the main distribution board has a measured length of 25 m. The cable is installed in cable trays with other circuits. Use the data in Table 10.1 to find a suitable cable size.

Allowable voltage drop is

$$240 \text{ V phase}, V = \frac{4}{100} \times 240 \text{ V}$$
$$= 9.6 \text{ V}$$

Phase current is

$$I_b = \frac{250 \text{ W}}{240 \text{ V} \times 0.7} \text{ A}$$
$$= 1.49 \text{ A}$$

The 1 mm^2 cable has an allowable drop of 44 mV/A m:

$$L_1 = \frac{9.6 \text{ V} \times 10^3}{1.49 \text{ A} \times 44 \text{ mV/A m}} \text{ m} = 146 \text{ m}$$

$$\text{voltage drop} = 25 \text{ m} \times 1.49 \text{ A} \times 44 \frac{\text{mV}}{\text{A m}} \times \frac{1}{10^3} \text{ A}$$
$$= 1.64 \text{ V}$$

The 1 mm^2 cable has a maximum current capacity of 15 A, and it produces a voltage drop that is within the 4% allowed. The percentage voltage drop is

$$\text{voltage drop} = \frac{1.64}{240} \times 100\%$$

$$= 0.7\%$$

The proposed design complies with the allowable voltage drop and the current capacity.

EXAMPLE 10.3

A single-phase 12 V lamp has an electrical power rating of 50 W. The lamp power factor is 1.0. The conductor from the main distribution board has a measured length of 5 m. The cable is clipped onto the building surfaces alongside other cables. Use the data in Table 10.1 to find a suitable cable size to use.

Allowable voltage drop is

$$12\,\text{V single phase,}\quad V = \frac{4}{100} \times 12\,\text{V}$$

$$= 0.48\,\text{V}$$

Phase current is

$$I_b = \frac{50\,\text{W}}{12\,\text{V} \times 1.0}\,\text{A}$$

$$= 4.2\,\text{A}$$

The 2.5 mm² cable has an allowable drop of 18 mV/A m:

$$L_1 = \frac{0.48\,\text{V} \times 10^3}{4.2\,\text{A} \times 18\,\text{mV/A m}}\,\text{m}$$

$$= 6.35\,\text{m}$$

$$\text{voltage drop} = 5\,\text{m} \times 4.2\,\text{A} \times 18\,\frac{\text{mV}}{\text{A m}} \times \frac{1}{10^3}\,\text{V}$$

$$= 0.38\,\text{V}$$

The 2.5 m² cable has a maximum current capacity of 27 A, and it produces a voltage drop that is within the 4% allowed. The percentage voltage drop is

$$\text{voltage drop} = \frac{0.38}{12} \times 100\%$$

$$= 3.15\%$$

The proposed design complies with the allowable voltage drop and the current capacity.

Questions

Use the data on the worksheet to answer the questions.

1. A three-phase 415 V air-handling unit fan motor has an input electrical power rating of 9.6 kW. The motor power factor is 0.85. The conductor from the main distribution board has a measured length of 37 m. The cable is installed in cable trays with other circuits. Find a suitable cable size.

2. A single-phase 240 V fan coil unit motor has an input electrical power rating of 375 W. The motor power factor is 0.65. The conductor from the main distribution board has a measured length of 18 m. The cable is installed in cable trays with other circuits. Find a suitable cable size.

3. A single-phase 12 V lamp has an input electrical power rating of 35 W. The lamp power factor is 1.0. The conductor from the main distribution board has a measured length of 7.5 m. The cable is clipped onto the building surfaces alongside other cables. Find a suitable cable size.

When I leave here...

References

Bird, J. O. and May, A. J. C. (1991) *Mathematical Formulae*, Longman Scientific and Technical, Harlow.

BSI (1972) BS 4800: 1972 *Paint colours for building purposes*, British Standards Institution, London.

Chadderton, D. V. (1995) *Building Services Engineering*, 2nd edn, E & F N Spon, London.

Chadderton, D. V. (1997) *Air Conditioning, A Practical Introduction*, 2nd edn, E & F N Spon, London.

CIBSE (1986a) *CIBSE Guide A*, Chartered Institution of Building Services Engineers, London.

CIBSE (1986b) *CIBSE Guide B*, Chartered Institution of Building Services Engineers, London.

CIBSE (1986c) *CIBSE Guide C*, Chartered Institution of Building Services Engineers, London.

Fantech (1993) *Fans By Fantech, A Manual of Selection Data and Recommended Practice*, 1st edn, Fantech Pty Ltd, 13–19 Dunlop Road, Mulgrave, Victoria 3170, Australia.

GE (1992) *GE Lighting Buyers Guide*, GE, Mitcham, Survey.

IEE (1991) *IEE Regulations for Electrical Installations*, 16th edn, Institution of Electrical Engineers, London.

Jenkins, B. D. (1991) *Electrical Installation Calculations for Compliance with the 16th Edition of the IEE Wiring Regulations*, Blackwell Scientific Publications, Oxford.

Philips (1986) *Philips Lighting Manual*, 4th edn

Pritchard (1985) *Lighting*, 3rd edn, Longman Group Ltd, Harlow, Essex

Rogers, G.F.C. and Mayhew, Y.R. (1987) *Thermodynamic and Transport Properties of Fluids*, SI Units, 3rd edn, Basil Blackwell, Oxford.

Spon (1992) *Spon's Mechanical and Electrical Services Price Book*, E & F N Spon, London.

FURTHER READING

ASHRAE (1989) *ASHRAE Fundamentals Handbook*, American Society of Heating, Refrigeration and Air Conditioning Engineers, Atlanta, GA.

AIRAH (1982) *Application Manual*, Australian Institute of Refrigeration Air Conditioning and Heating Inc.

Building Services, The CIBSE Journal, Chartered Institution of Building Services Engineers, monthly.

Eastop, T. D. and McConkey, A. (1986) *Applied Thermodynamics for Engineering Technologists*, 4th edn, Longman, Harlow, pp. 550–624

Rayner, J. (1971) *Basic Engineering Thermodynamics*, Longman, Harlow.

Answers

CHAPTER 2

9. $U = 2.066\,\text{W/m}^2\,\text{K}$, $C = 4.097\,\text{W/m}^2\,\text{K}$, $W = 171.2\,\text{kg/m}^2$
10. $U = 1.08\,\text{W/m}^2\,\text{K}$, $C = 1.548\,\text{W/m}^2\,\text{K}$, $W = 464.9\,\text{kg/m}^2$
11. $U = 1.134\,\text{W/m}^2\,\text{K}$, $C = 1.467\,\text{W/m}^2\,\text{K}$, $W = 411.7\,\text{kg/m}^2$
12. $U = 0.273\,\text{W/m}^2\,\text{K}$, $C = 0.282\,\text{W/m}^2\,\text{K}$, $W = 70\,\text{kg/m}^2$
13. $U = 0.306\,\text{W/m}^2\,\text{K}$, $C = 0.325\,\text{W/m}^2\,\text{K}$, $W = 389.4\,\text{kg/m}^2$
14. $U = 0.21\,\text{W/m}^2\,\text{K}$, $C = 0.217\,\text{W/m}^2\,\text{K}$, $W = 65.3\,\text{kg/m}^2$
15. $S = 0.9862$, $\tanh(S) = 0.7557$
16. $S = 0.992$, $U = 0.197\,\text{W/m}^2\,\text{K}$
17. $S = 0.997$, $U = 0.083\,\text{W/m}^2\,\text{K}$
18. $S = 0.964$, $U = 0.528\,\text{W/m}^2\,\text{K}$
19. $U_e = 0.581\,\text{W/m}^2\,\text{K}$, $R_g = 0.721\,\text{m}^2\,\text{K/W}$, $U = 0.45\,\text{W/m}^2\,\text{K}$
20. $U_e = 0.429\,\text{W/m}^2\,\text{K}$, $R_g = 4.907\,\text{m}^2\,\text{K/W}$, $U = 0.145\,\text{W/m}^2\,\text{K}$
21. $U_e = 0.114\,\text{W/m}^2\,\text{K}$, $R_g = 3.89\,\text{m}^2\,\text{K/W}$, $U = 0.089\,\text{W/m}^2\,\text{K}$
22. $U = 4.67\,\text{W/m}^2\,\text{K}$, $W = 14.7\,\text{kg/m}^2$
23. $U = 0.363\,\text{W/m}^2\,\text{K}$, $C = 0.384\,\text{W/m}^2\,\text{K}$, $W = 357.9\,\text{kg/m}^2$
24. $U = 0.289\,\text{W/m}^2\,\text{K}$, $C = 0.303\,\text{W/m}^2\,\text{K}$, $W = 136.3\,\text{kg/m}^2$
25. $U = 0.289\,\text{W/m}^2\,\text{K}$, $C = 0.307\,\text{W/m}^2\,\text{K}$, $W = 87.3\,\text{kg/m}^2$
26. $U = 0.538\,\text{W/m}^2\,\text{K}$, $C = 0.586\,\text{W/m}^2\,\text{K}$, $W = 363.1\,\text{kg/m}^2$
27. $U = 0.489\,\text{W/m}^2\,\text{K}$, $C = 0.566\,\text{W/m}^2\,\text{K}$, $W = 45.7\,\text{kg/m}^2$
28. $U = 0.874\,\text{W/m}^2\,\text{K}$, $C = 1.158\,\text{W/m}^2\,\text{K}$, $W = 422.4\,\text{kg/m}^2$
29. $U = 1.202\,\text{W/m}^2\,\text{K}$, $C = 1.582\,\text{W/m}^2\,\text{K}$, $W = 277.4\,\text{kg/m}^2$
30. $U = 0.186\,\text{W/m}^2\,\text{K}$, $C = 0.192\,\text{W/m}^2\,\text{K}$, $W = 167.2\,\text{kg/m}^2$
31. $U = 0.285\,\text{W/m}^2\,\text{K}$, $C = 0.3\,\text{W/m}^2\,\text{K}$, $W = 422.2\,\text{kg/m}^2$
32. $U = 0.251\,\text{W/m}^2\,\text{K}$, $C = 0.263\,\text{W/m}^2\,\text{K}$, $W = 95.2\,\text{kg/m}^2$
33. $U_f = 5.88\,\text{W/m}^2\,\text{K}$, $U_g = 2.929\,\text{W/m}^2\,\text{K}$, $U = 3.313\,\text{W/m}^2\,\text{K}$, $W = 30.5\,\text{kg/m}^2$
34. $U_f = 1.29\,\text{W/m}^2\,\text{K}$, $U_g = 1.897\,\text{W/m}^2\,\text{K}$, $U = 1.81\,\text{W/m}^2\,\text{K}$, $W = 58.5\,\text{kg/m}^2$

CHAPTER 3

9. $Q_{ps} = 20.221\,\text{kW}$, $Q_p = 27.767\,\text{kW}$ total heat, $N = 3.8$ air changes/h, duct diameter $d = 696\,\text{mm}$.
13. $Q_{ps} = 6.036\,\text{kW}$, $Q_p = 6.881\,\text{kW}$ sensible heat, $N = 4.3$ air changes/h, duct diameter $d = 429\,\text{mm}$.
14. $Q_{ps} = 100.825\,\text{kW}$, $Q_p = 107.883\,\text{kW}$ sensible heat, $N = 1.5$ air changes/h, duct diameter $d = 1448\,\text{mm}$.

CHAPTER 4

1. Carbon C 11.459, hydrogen H_2 34.335, sulphur S 4.292, carbon monoxide CO 2.446, methane CH_4 17.167, octane C_8H_{18} 15.064
2. 15.59 kg air/kg oil
3. 13.219 kg air/kg benzene
4. 13.47 kg air/kg oil; 15.62% CO_2, 0.07% SO_2, 84.31% N_2
5. (a) 1.2% H_2, 1.3% CH_4, 25.4% CO, 1.2% C_2H_6, 5.3% O_2, 65.7% N_2
   (b) Stoichiometric air 1.22 kg air/kg gas; actual air-to-fuel ratio 1.59 to 1; DFG 18.78% CO_2, 4.69% O_2, 76.52% N_2
6. (a) $C_{10}H_{22} + 15.5O_2 = 10CO_2 + 11H_2O$
   (b) 14.991 kg air/kg fuel in cell D65
   (c) 17.2%
7. 28.69 kg air/kg gas; 2.26% CO_2, 6.19% O_2, 21.67% H_2O, 69.88 N_2
8. (a) 14.39% CH_4, 70.77% CO, 5.13% H_2, 1.01% O_2, 8.69% N_2 by mass
   (b) Stoichiometric air 5.92 kg air/kg gas; actual air-to-fuel ratio 11.84 to 1; 8.53% CO_2, 10.73% O_2, 80.74% N_2
9. 14.35 kg air/kg oil; 45.34%
10. Air-to-fuel ratio 17.59 kg air/kg petrol; 11.12% CO_2, 10.85% H_2O, 3.31% O_2, 74.72% N_2
11. Air-to-coal ratio 15.92 kg air/kg coal; 12.05% CO_2, 7.16% O_2, 80.79% N_2; GCV 34.2 MJ/kg, NCV 32.8 MJ/kg
12. Air-to-coal ratio 13.48 kg air/kg coal; 15.27% CO_2, 4.28% O_2, 80.46% N_2
13. Stoichiometric air 11.49 kg air/kg coal; actual air 18.38 kg air/kg coal; 11.82% CO_2, 7.99% O_2, 0.05% SO_2, 80.14% N_2
14. Stoichiometric air 12.07 kg air/kg coal; actual air 17.5 kg air/kg coal; 9.4% CO_2, 6.19 O_2, 9.39% H_2O, 75.02% N_2
15. 11.97%
16. Stoichiometric air 11.48 kg air/kg coal; actual air 14.93 kg air/kg coal; 14.6% CO_2, 4.93% O_2, 0.02% SO_2, 80.44% N_2
17. 4.8%
18. Stoichiometric air 21.13 kg air/kg fuel; actual air 25.35 kg air/kg fuel; WFG analysis 2.08% CO_2, 26.83% H_2O, 3.02% O_2, 68.07% N_2; GCV 84.5 MJ/kg, NCV 72.5 MJ/kg
19. The fuel is 52.2% carbon, 13% hydrogen and 34.8% oxygen by mass. Stoichiometric air 8.97 kg air/kg fuel; actual air 11.21 kg air/kg fuel; DFG analysis 11.83% CO_2, 4.45% O_2, 83.72% N_2, GCV 26.8 MJ/kg
20. 80.55% N_2 in the dfg, actual air supply 14.94 kg air/kg fuel; 23% excess air; 10.73% O_2 for 100% excess air

CHAPTER 5

4. 0.3318 W/m^3 K
6. $U_1 = 2.45$ W/m^2 K, $U_2 = 1.67$ W/m^2 K, $Q_p = 449$ W
7. $U_1 = 0.93$ W/m^2 K, $U_2 = 0.22$ W/m^2 K, $Q_p = 178$ W
8. $U_1 = 1.63$ W/m^2 K, upward $U_2 = 0.453$ W/m^2 K, downward $U_2 = -0.627$ W/m^2 K, lounge loss $Q_p = 221.6$ W, bedroom loss $Q_p = -221.6$ W. The bedroom loss is negative, meaning a gain, and this equals the heat loss from the

lounge. It is necessary to utilize all the decimal places available to obtain an accurate result for Q_p in this example. Only one extra decimal place has been displayed.

9. $U_1 = 2.42\,\text{W/m}^2\,\text{K}$, $U_2 = 1.53\,\text{W/m}^2\,\text{K}$, $Q_p = 586\,\text{W}$
10. $15.7°\text{C}$
11. $4.39\,\text{W/m}^2\,\text{K}$, single glazing, single sheet of corrugated steel
12. $3.6\,\text{W/m}^2\,\text{K}$
13. $14.9°\text{C}$
14. $t_{ei} = 19.1°\text{C}$, $t_{ai} = 20.7°\text{C}$, $t_r = 19.3°\text{C}$. The mean radiant and environmental temperatures are similar, so the space has a source of radiant heat emission. The air and resultant temperatures are practically the same, so there is a considerable amount of convective heat output. It is likely that the heat emitter is a panel or column radiator.
15. $t_{ei} = 15.1°\text{C}$, $t_{ai} = 18.5°\text{C}$, $t_r = 13.5°\text{C}$
16. $t_{ei} = 14.7°\text{C}$, $t_{ai} = 12.2°\text{C}$, $t_r = 15.8°\text{C}$
17. $\Sigma(AU) = 309.93\,\text{W/K}$, $Q = 0.25\,\text{m}^3/\text{s}$, $F_1 = 0.97$, $F_2 = 1.09$, $t_{ei} = 20.3°\text{C}$, $t_{ai} = 23°\text{C}$, $t_s = 18.9°\text{C}$, $t_r = 19°\text{C}$, $Q_p = 15\,038\,\text{W}$, $25.1\,\text{W/m}^3$, $75.2\,\text{W/m}^2$
18. $K = 70\%$, $E = 30\%$, living room 3281 W, kitchen 1641 W, hall 411 W, bedroom 1 1030 W, bedroom 2 690 W, 5 fictitious rooms, total 10.758 kW
19. Total heat loss $= 90.606\,\text{kW}$, total $Q_p = 111.667\,\text{kW}$, $33.6\,\text{W/m}^2$, $11.2\,\text{W/m}^3$
20. $4°\text{C}$
21. (a) $0°\text{C}$
    (b) Roof space is always at the same temperature as the outdoor air.
    (c) $12°\text{C}$.
    (d) The uninsulated roof space offers no protection against freezing temperatures. The exterior roof surfaces need to be insulated to reduce the risk of water services freezing.

CHAPTER 6

4. At 2600 rev/min the fan and system operating state is 900 l/s and a fan total pressure rise of 899 Pa. The fan delivers insufficient air flow for the design. At 960 rev/min the fan and system operate at 351 l/s and 121 Pa.
5. (a) At 2900 rev/min the fan and system operating state is 2300 l/s and a fan total pressure rise of 2000 Pa. The fan delivers excessive air flow for the design. This may cause incorrect supply air temperatures, increased room air change rates, some draughts and excessive noise from turbulence.
   (b) At 1300 rev/min the fan and system operate at 1009 l/s and a fan total pressure of 500 Pa. This is insufficient air flow.
   (c) A fan speed of 2800 rev/min produces an air flow of 2270 l/s in the ductwork system with a fan total pressure rise of 1900 Pa. This satisfies the commissioning criteria. Fan static pressure is 1850 Pa, fan velocity pressure is 100 Pa, air velocity leaving the fan is 13 m/s, motor power consumed is 14 kVA, and on a 415 V supply the current is 34 A.
6. (a) At 2900 rev/min, fan and system operating point is 2450 l/s and a fan total pressure rise of 1500 Pa. The fan passes excessive air flow for the design. Motor power absorbed is 7673 VA.

(b) Extract air flow is to be 2200 l/s. At fan speed of 2700 rev/min, $Q = 2235$ l/s, $FTP = 1300$ Pa, $FSP = 1250$ Pa, $FVP = 50$ Pa, discharge air velocity = 9.4 m/s, input power = 7 kVA, current = 17 A.

(c) At the commissioning air flow rate of 2200 l/s, the fan develops an FTP of 1300 Pa while the ductwork system absorbs 1000 Pa. The excess 300 Pa may need to be removed with a damper during commissioning if it is significant.

7. (a) $Q = 2530$ l/s, $FTP = 3970$ Pa, $FSP = 3876$ Pa, $FVP = 94$ Pa, power = 21 772 V A, current = 52.5 A.

(b) $Q = 2310$ l/s, fan speed = 2000 rev/min, $FTP = 3450$ Pa, $FSP = 3366$ Pa, $FVP = 84$ Pa, discharge air velocity = 11.8 m/s, input power = 13 230 V A, current = 31.9 A

(c) Excess 450 Pa.

8. (a) $Q = 3400$ l/s, $FTP = 1950$ Pa, $FSP = 1600$ Pa, $FVP = 400$ Pa, power = 2020 V A, current = 5 A.

(b) $Q = 3360$ l/s, fan speed = 2300 rev/min, $FTP = 1840$ Pa, $FSP = 1493$ Pa, $FVP = 347$ Pa, discharge air velocity = 24 m/s, input power = 18 090 V A, current = 46 A.

(c) Excess 340 Pa.

9. (a) $Q = 1130$ l/s, $FTP = 800$ Pa, $FSP = 780$ Pa, $FVP = 20$ Pa, power = 3.8 kVA, current = 16 A.

(b) $Q = 1275$ l/s, fan speed = 1700 rev/min, $FTP = 1050$ Pa, $FSP = 1017$ Pa, $FVP = 33$ Pa, discharge air velocity = 7.4 m/s, input power = 5.35 kVA, current = 23 A.

(c) excess 300 Pa.

10. (a) $Q = 4600$ l/s, $FTP = 205$ Pa, $FSP = 144$ Pa, $FVP = 61$ Pa, power = 3300 V A, current = 14 A.

(b) +31%

CHAPTER 7

5. 1.252 kg/m^3
6. 59.2 °C d.b.
7. −119 Pa
8. 11.8 m/s
9. 295 l/s, 754 l/s, 4.25 m^3/s, 6.79 m^3/s
10. 1.225 kg/m^3, 19 Pa, 139 Pa
11. 1.132 kg/m^3, 5.27 m/s, −44 Pa

12.

Node	P_t(Pa)	P_v(Pa)	P_s(Pa)
1	120	18	102
2	111	18	93
3	108	37	71
4	96	37	59

13.

Node	P_t(Pa)	P_v(Pa)	P_s(Pa)
1	350	35	315
2	346	35	311
3	340	147	193
4	302	147	155

14.

Node	P_t(Pa)	P_v(Pa)	P_s(Pa)
1	200	55	145
2	135	55	80
3	119	11	108
4	109	11	98

Static regain $P_{s3} - P_{s2} = 28\,\text{Pa}$

15.

Node	P_t(Pa)	P_v(Pa)	P_s(Pa)
1	−25	80	−105
2	−48	80	−128
3	−93	3	−96
4	−93	3	−96

Static regain $P_{s3} - P_{s2} = 32\,\text{Pa}$

16.

Node	P_t(Pa)	P_v(Pa)	P_s(Pa)
1	250	24	226
2	244	24	220
3	241	62	179
4	188	62	126
5	169	16	152
6	165	16	149

Static regain $P_{s3} - P_{s2} = -41\,\text{Pa}$

17.

Node	P_t(Pa)	P_v(Pa)	P_s(Pa)
1	75	12	63
2	72	12	60
3	70	12	59
4	67	12	56
5	62	15	47
6	59	15	44

Static regain $P_{s3} - P_{s2} = -1\,\text{Pa}$

18. $v_1 = 8.7\,\text{m/s}$, $p_{v1} = 45\,\text{Pa}$, $p_{s1} = -275\,\text{Pa}$, $p_{t1} = -230\,\text{Pa}$, $v_2 = 13.8\,\text{m/s}$, $p_{v2} = 113\,\text{Pa}$, $p_{s2} = 217\,\text{Pa}$, $p_{t2} = 330\,\text{Pa}$, $FTP = 560\,\text{Pa}$, $FVP = 113\,\text{Pa}$, $FSP = 447\,\text{Pa}$

19. There is more than one correct solution to this question. Section 1–11 1300 mm × 1300 mm, 11–24 950 mm × 900 mm, 24–52 600 mm × 550 mm, 24–62 750 mm × 700 mm, duct at node 53 900 mm × 300 mm, node 54 1100 mm × 1000 mm. Fan total pressure 533 Pa at 4 m^3/s and 20 °C d.b.

20. There is more than one correct solution to this question. Section 3–4 600 mm × 600 mm, 5–6 950 mm × 600 mm, 7–8 950 mm × 750 mm, 9–10 1300 mm × 750 mm, 12–39 1400 mm × 1400 mm, section 11 800 mm × 750 mm, 1–3 950 mm × 900 mm, 50–5 950 mm × 550 mm, 60–7 650 mm × 550 mm, 70–9 1200 mm × 550 mm, fan total pressure 229 Pa at an air flow of 5.5 m^3/s and 20 °C d.b.

21. Section 1–11 1000 mm × 950 mm, 11–24 700 mm × 700 mm, 24–27 600 mm × 550 mm, 50–52 450 mm × 450 mm, 53–54 600 mm × 550 mm, 27–30 450 mm × 450 mm, 60–62 500 mm × 300 mm, 63–64 500 mm × 450 mm, 33–82 400 mm × 400 mm, 70–72 350 mm × 300 mm, 73–74 600 mm × 300 mm, 33–82 400 mm × 400 mm, 83–84 600 mm × 400 mm, fan total pressure 318 Pa, fan air flow 2.3 m^3/s at 20 °C.

22. Section 2–5 1250 mm × 1200 mm, 6–11 600 mm × 600 mm, 12–15 800 mm × 950 mm, fan total pressure 215 Pa, fan air flow 3.5 m^3/s at 20 °C

23. Section 2–11 1000 mm × 1000 mm, nominal fan section 550 mm × 500 mm, 12–13 700 mm × 600 mm, diffuser 14 1000 mm × 1000 mm, grille 15–16 1050 mm × 950 mm, fan total pressure 265 Pa, fan air flow 2.4 m^3/s at 30 °C

24. Supply duct section 2–6 750 mm × 700 mm, fan section 6–11 350 mm × 400 mm, duct 12–13 500 mm × 500 mm, diffuser section 15 800 mm × 650 mm, opening 16–17 200 mm × 250 mm, supply fan total pressure 283 Pa delivering 1.2 m^3/s at 14 °C d.b. Room static air pressure is +10 Pa. Exhaust duct section 20 700 mm × 650 mm, 21–23 500 mm × 450 mm, fan section 24 350 mm × 350 mm, 25–26 500 mm × 450 mm, 27 700 mm × 650 mm, exhaust system pressure drop 68 Pa, exhaust fan total pressure 58 Pa at air flow of 1.08 m^3/s and 24 °C d.b.

25. Supply duct section 1–2 400 mm × 350 mm, 2–3 300 mm × 250 mm, 4–5 700 mm × 650 mm, 6–9 950 mm × 900 mm, fan section 11 500 mm × 450 mm, duct 12–13 700 mm × 600 mm, diffuser section 14 950 mm × 900 mm, supply fan total pressure 291 Pa delivering 2 m^3/s at 35 °C d.b. Exhaust duct section 20 950 mm × 900 mm, 21–22 650 mm × 650 mm, fan section 23 500 mm × 450 mm, 24–25 700 mm × 600 mm, 26–27 250 mm × 250 mm, 50–51 650 mm × 550 mm, exhaust fan total pressure 161 Pa at air flow of 2 m^3/s and 24 °C d.b.

26. Supply duct section 1–2 600 mm × 450 mm, 2–3 400 mm × 350 mm, 3–4 850 mm × 750 mm, 4–11 1200 mm × 1100 mm, fan section 11 600 mm × 550 mm, 12–24 800 mm × 800 mm, 24–27 750 mm × 700 mm, 27–30 700 mm × 600 mm, 30–33 600 mm × 550 mm, 33–36 500 mm × 450 mm, 36–38 350 mm × 300 mm, diffuser section 38–40 500 mm × 450 mm, all branch ducts are the same sizes as section 36–40, supply fan total pressure 412 Pa delivering 3 m^3/s at 14 °C d.b. Exhaust duct section 100–101 550 mm × 400 mm, 101–102 350 mm × 300 mm, 102–103 550 mm × 400, 103–104 600 mm × 550 mm, 104–111 850 mm × 750 mm, fan section 111 600 mm × 550 mm, 111–124 850 mm × 750 mm, 124–140 400 mm × 350 mm, exhaust louvre 600 mm × 450 mm, recirculation duct 124–150 750 mm × 700 mm, exhaust fan total pressure 103 Pa at air flow of 3 m^3/s and 24 °C d.b.

CHAPTER 8

5. The pipe heat emitter loads, pipe diameters and pressure drop rate used are shown in Fig. 8.7. Total system water flow is 1.456 l/s, pump pressure rise is 39.578 kPa, the total heat load including the pipe heat emission is 65.315 kW, total installed pipe cost is £1,646.10. When comparing the results with published CIBSE 1986 C4 pipe data tables, use the water flow mass rates M kg/s from the worksheet.

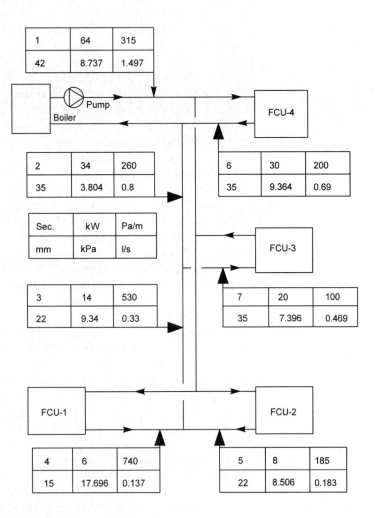

Fig. 8.7   Solution to question 5.

6. The pipe heat emitter loads, pipe diameters and pressure drop rates used are shown in Fig. 8.8. Total system water flow is 0.385 l/s, pump pressure rise is 50.154 kPa, the total heat load including the pipe heat emission is 17.249 kW, total installed pipe cost is £1,334.09.

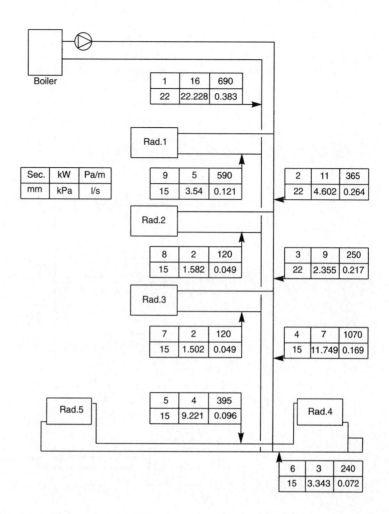

Fig. 8.8 Solution to the panel radiator system in question 6.

7. The pipe heat emitter loads, pipe diameters and pressure drop rate used are shown in Fig. 8.9. Total system water flow is 7.032 l/s, pump pressure rise is 152.019 kPa, the total heat load including the pipe heat emission is 286.913 kW, total installed pipe cost is £29,500.70.

8. The index circuit comprises sections 1 and 7. The pipe heat emitter loads, pipe diameters and pressure drop rate used are shown in Fig. 8.10. Total system water flow is 3.38 l/s, pump pressure rise is 41.04 kPa, the total heat load including the pipe heat emission is 137.914 kW, total installed pipe cost is £4,709.01.

9. The index circuit comprises sections 1, 2 and 3. The pipe heat emitter loads, pipe diameters and pressure drop rate used are shown in Fig. 8.11. Total system water flow is 4.56 l/s, pump pressure rise is 42.337 kPa, the total heat

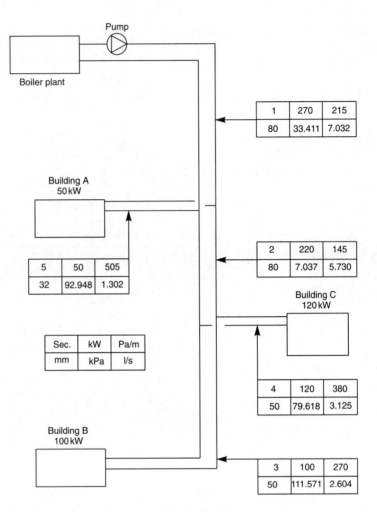

Fig. 8.9 Solution to the heating system in question 7.

load including the pipe heat emission is 95.406 kW, total installed pipe cost is £3,103.33.

CHAPTER 9

1. 2 rows of 3 luminaires at 90° to the perimeter wall; $S_2 = 1.5$ m; electric power load 22.4 W/m²
2. $RI = 5$; $UF = 65\%$; 4 rows of 15 luminaires; $S_2 = 4.5$ m; lamp type 3; 424 lx produced; 10.8 kW electrical power; 15 W/m²
3. $RI = 1.3$; $UF = 49\%$; 2 rows of 5 luminaires; 2 north light lamps per luminaire; 1000 W installed electrical power
4. $RI = 0.6$; $UF = 33\%$; 5 luminaires in one row; 118 lx installed; 280 W electrical power

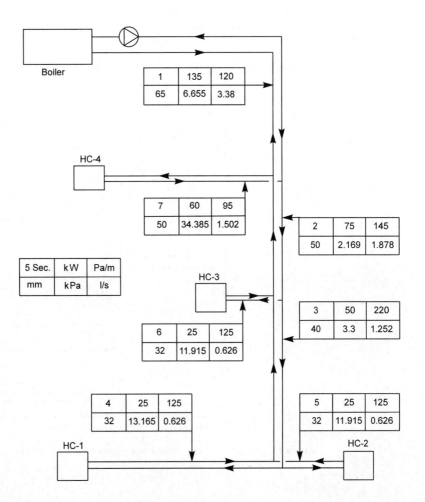

Fig. 8.10 Solution to the air-heating coil pipework system in question 8.

CHAPTER 10

1. $2.5\,\text{mm}^2$, $20.6\,\text{A}$, $53.7\,\text{m}$, $11.4\,\text{V}$, 2.8%
2. $1\,\text{mm}^2$, $2.4\,\text{A}$, $95.1\,\text{m}$, $1.8\,\text{V}$, 0.8%
3. $2.5\,\text{mm}^3$, $2.9\,\text{A}$, $9.7\,\text{m}$, $0.4\,\text{V}$, 3.1%

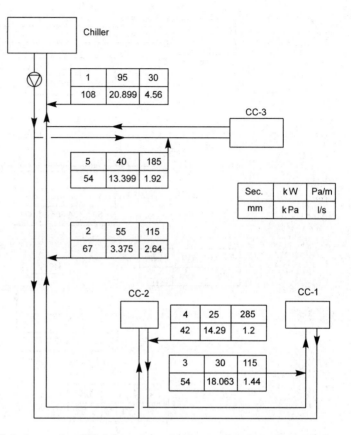

Fig. 8.11   Solution to the chilled water two-pipe system for air-cooling coils in question 9.

Index

You are taking...